Ventilation for Control of the Work Environment

William A. Burgess
Harvard University
School of Public Health
Boston, Massachusetts

Michael J. Ellenbecker
University of Lowell
Department of Work Environment
Lowell, Massachusetts

Robert D. Treitman
Interleaf, Inc.
Cambridge, Massachusetts

WILEY

A WILEY-INTERSCIENCE PUBLICATION

John Wiley & Sons

NEW YORK·CHICHESTER·BRISBANE·TORONTO·SINGAPORE

Library of Congress Cataloging in Publication Data:
Burgess, William A. 1924-
    Ventilation for control of the work environment/by William A.
    Burgess, Michael J. Ellenbecker, Robert D. Treitman.
p. cm.
    "A Wiley-Interscience publication."
    Includes bibliographies and index.
    ISBN 0-471-89219-X
    1. Factories--Heating and ventilation. I. Ellenbecker, Michael
J. II. Treitman, Robert D. III. Title.
TH7684.F2B86   1989
697.9′2--dc19

Printed in the United States of America

10 9 8 7 6 5 4 3 2 1

Preface

We estimate that as many as 50,000 new industrial exhaust ventilation systems are installed annually in the United States alone. If we assume that the average capacity is 1000 cubic feet per minute (cfm) with installation costs of $5 per cfm and yearly operating expenses of $1 per cfm, it is apparent that this represents a considerable investment. However, many of these systems are improperly designed, not installed according to the plans, or poorly maintained, leading to tremendous loss as a result of inefficiency.

In an attempt to improve this situation, we wrote this book to provide information to those responsible for the specification, design, installation, and maintenance of industrial ventilation systems. This audience includes plant engineers, industrial hygienists, and students of industrial ventilation. We hope to provide a theoretical background to the plant engineer who, while intimately familiar with the equipment and hardware, may not have an appreciation of the principles of ventilation. For the industrial hygienist, we have attempted to bring the concepts and principles of ventilation into an industrial setting. Finally, for the student, we have written a text book that demonstrates both the theoretical and practical aspects of the subject.

This book is not a reference manual. We wrote it to be used with the 20th edition of the ACGIH *Industrial Ventilation Manual* and its self-study companion by D. Jeff Burton. Throughout the text, we refer the reader to specific designs, charts, and tables in the *Ventilation Manual*. We did not want to repeat the information provided there, but rather concentrated on explaining the rationale *behind* the material and providing a framework within which to use the *Ventilation Manual*. We have successfully used this approach in the instruction of hundreds of students. If the ACGIH manual is not currently on the reader's bookshelf, we strongly recommend its purchase. Successful use of this book requires it. The selection of a system of units to be used in this book presented difficulties. The use of Système Internationale d'Unites (SI) units is almost universal in scientific and engineering disciplines outside of the United States. In this country, unfortunately, practicing ventilation engineers and contractors still use the English Engineering System. In addition the ACGIH Ventilation Manual uses English Units throughout. Given the choice between prodding the field into the modern era and making this book easy to use, we selected the latter course and employed English Engineering Units.

We have not attempted to write a fluid mechanics textbook, nor did we try to prepare a "troubleshooting" guide to solving ventilation problems. This is also not a compendium of design solutions for a myriad of specific situations, nor is it a catalog of charts, nomograms, and tables. The first two chapters of the book provide background material, with the introductory chapter presenting an overview of industrial ventilation and its historical and technical relationship with other methods of environmental control. The principles of fluid mechanics, required information for proper application of design techniques, are presented in Chapter 2.

We present methods and instrumentation for measuring airflow in Chapter 3. Although this may seem out of place, we have found not only that it dovetails well with the principles of airflow presented in Chapter 2 but that it improves the reader's ability to grasp concepts presented in subsequent chapters. The topic of general ventilation (also called dilution ventilation) is covered in Chapter 4.

Hood design, selection, and performance are addressed in Chapters 5, 6, and 13, respectively. We have given a great deal of attention to the subject of hoods because they are a critical part of any local exhaust system. Special attention is given to chemical laboratory hoods and associated ventilation systems (Chapter 7) because of the great number of these in use. Chapters 8 and 9 cover the design of local exhaust ventilation systems, building on the principles provided in Chapter 2. Examples of single- and multiple-hood systems designs are presented.

Two other components of the local exhaust system are the fan or blower and the air cleaning device. Information on fan design and performance is found in Chapter 10, and a summary presentation on air-cleaning devices is covered in Chapter 11, although the reader should be fully aware that this is only an overview of this subject.

One aspect of an industrial ventilation system often overlooked is the provision of air to replace the air that is mechanically exhausted, either locally or generally. The rationale for designing and installing replacement-air systems is provided in Chapter 12, with specific guidelines and examples. Replacement air requires conditioning to maintain a comfortable and healthy thermal environment. In Chapter 14 we present information on physiological response to heat and humidity as well as techniques for evaluation of the thermal environment. The technology and equipment for conditioning the air supplied to the workplace are discussed in Chapter 15.

Finally, Chapter 16 presents the topic of reentry of exhausted air back into the building. The problem of air from exhaust stacks reentering the building either through windows or via mechanical air intakes presents unique challenges to ventilation engineers as well as building architects. We thank Martin Horowitz for preparation of this chapter.

We acknowledge those who provided many helpful criticisms of early drafts of this book: Dr. Stephen Rudnick, Dr. P. Barry Ryan, and Louis DiBerardinis. We are grateful to Robert Gempel for the series of design problems in Chapters 8 and 9. We also acknowledge Ms. Ellen Bryant and Ms. Cheryl McGoverny for the many hours of assistance in the preparation of the manuscript. Special thanks to the numerous students who have helped to hone this material into what we hope is an instructive text.

WILLIAM A. BURGESS

MICHAEL J. ELLENBECKER

ROBERT D. TREITMAN

Boston, Massachusetts
March 1989

Contents

List of Units

atm	atmosphere
Btu/h	British thermal units per hour
°C	degree Celsius
cfm, ft^3/min	cubic feet per minute
clo	thermal insulation of clothing, 1 clo = 0.88 ft^2-h-°F/BTU
cm	centimeter
d	day
°F	degree Fahrenheit
fpm, ft/min	feet per minute
ft	foot
ft^2	square foot
ft/sec	feet per second
ft-lb/min	foot-pounds per minute
g	gram
g/min	grams per minute
gal	gallon
h	hour
hp	horsepower
in.	inch
in. Hg	inch of mercury
in. H_2O	inch of water
in. H_2O/fpm	inches of water per foot per minute
in. H_2O/fpm-lb/ft^2	inches of water per foot per minute per pound per square foot
°K	Kelvin
kcal/h	kilocalories per hour
L	liter
lb/ft^2	pound per square foot
lbf	pound-force
lbm	pound-mass
lpm	liters per minute
m	meter
m^2	square meter
m/s	meters per second
met	unit of metabolism: 1 met = 50 kcal/m^2/h
mg	milligram
mg/m^3	milligrams per cubic meter
min	minute
ml	milliliter
mm Hg	millimeters of mercury
mph	miles per hour
mppcf	million particles per cubic foot

Pa	pascal
ppm	parts per million
psf	pounds per square foot
psi	pounds per square inch
psia	psi-absolute
psig	psi–gage
°R	degree Rankine
rpm	revolutions per minute
s	second
W	watt
y	year
μm	micrometer
μg/m^3	micrograms per cubic meter

1

Ventilation for Control

The provision of a safe and healthful working environment entails three primary components: (a) awareness of potential hazards (*recognition*); (b) assessment of these hazards (*evaluation*); and (c) abatement of these hazards (*control*). Although a great deal of attention has been given to awareness and assessment, including toxicological research, epidemiological studies, standards setting, and environmental monitoring, the single most crucial element in the program—the reduction or elimination of the problem—has curiously been ignored. In a survey of industrial hygiene literature, 40% of the papers published in one journal over a $3\frac{1}{2}$-year period addressed monitoring, 24% addressed physical effects and epidemiology, 8% covered personal protection, and less than 8% were devoted to environmental control (Hammond, 1980). As Hammond (1980) states:

> One would hope in 10–20 yr time to be able to look back and find the monitoring and the environmental control rankings reversed, with the ongoing and necessary epidemiology holding its central position. This would place monitoring nearer its correct position as a back up to good environmental control.

To place the use of ventilation in perspective with other mechanisms for environmental control, an introduction to those mechanisms follows.

1.1 CONTROLS

The means used to control exposures to harmful materials or conditions in the occupational environment can be categorized as engineering or administrative.

In nearly all cases, the most effective approach is the combination of controls into an integrated package. The elimination of an offending agent from the workplace, accompanied, if necessary, by its replacement with a safer material, should be considered first in any effort to control the environment. The substitution of less hazardous materials has become quite common in industry, as knowledge of dangers from certain materials becomes available. For example, hydrocarbon-solvent-based paints are being replaced by water-based paints. The use of asbestos is being severely curtailed. Purchasers of some organic solvents are specifying that these solvents contain only trace amounts of benzene contamination. The principle of elimination and substitution applies to equipment and processes as well as materials. Newer machinery is often designed to minimize dust generation and release, for example. Processes can be modified to incorporate contaminant-reduction techniques. The introduction of a raw material in pellet form is less likely to generate dust than the same material presented as a powder.

If it is not feasible to eliminate the contaminant from the workplace, the next step is to isolate it from the workers who frequent the area. Distance and physical barriers, preferably around the process, but possibly around the workers, can provide protection. In either case, this method of control is usually accompanied by a ventilation system. When the process is isolated, the emphasis is on exhausting the contaminated air from the process. In contrast, when the worker is isolated, the emphasis should be on supplying clean air to the worker's station.

The use of ventilation is ubiquitous in the modern workplace. Virtually every industrial and commercial facility contains some form of ventilation system for environmental control. The intent may be comfort (temperature, humidity, odors), safety (flammable vapors), or health (toxic particles, gases and vapors, airborne contagions). The last resort for preventing exposures to toxic chemicals is personal protection, in the form of respirators and protective garments. Respiratory protection is used when all other controls are inadequate or when the possible failure of those controls would produce a hazardous situation.

Administrative controls such as worker rotation through hazardous areas can also be implemented. In the nuclear power industry, exposure to radiation is limited on a three-month as well as an annual basis. Any worker achieving the maximum permissible exposure before the end of the pertinent period is transferred to a low-radiation work area for the balance of time. In hot environments, workers should be allowed to rest in a cool area on a frequent basis throughout the work shift to allow time for the body to recover from the thermal stress. Other administrative controls include biological monitoring, worker education, and equipment maintenance. In all cases, administrative controls should be combined with attempts to reduce the hazard through engineering controls.

When choosing the most appropriate approaches to controlling the workplace environment, the nature of the hazard is of paramount importance. For example, is it an airborne material, such as a particle, gas, or vapor? Or is it a physical agent, such as heat, noise, or radiation? A source of ionizing radiation is often an ideal candidate for isolation and enclosure, whereas a toxic particulate may be easier to control with ventilation. As we show in subsequent chapters, the design of any effective ventilation system must incorporate knowledge of contaminant form. In addition to this, one must consider how the contaminant is produced and what properties it may have as it is released. Is it emitted from a hot process in a heated plume? Is it emitted on a steady basis or are sudden "puffs" of material anticipated? Is it released from a known location? Does it have any imparted velocity? Is the emission rate known or predictable?

1.2   VENTILATION FOR CONTROL

A review of the literature reveals that there was a great deal of interest in the theoretical and engineering aspects of industrial ventilation in the late 1930s and the early 1940s. The pace of the activity increased with the onset of World War II, with many of the articles covering industrial processes with direct defense applications, such as shipyard welding, rubber life-raft manufacturing, and synthetic-rubber production. It was during this time that pioneering industrial hygienists and engineers such as Phillip Drinker, Theodore Hatch, Allen Brandt, Constantin Yaglou, Leslie Silverman, W. C. L. Hemeon and J. M. DallaValle were all very active. Much of the information was eventually incorporated into the first edition of the *Industrial Ventilation Manual*, published by the American Conference of Governmental Industrial Hygienists (ACGIH) in 1951.

The New York State Department of Labor's Division of Industrial Hygiene and Safety Standards has published a *Monthly Review* beginning in the 1920s. Articles appearing in the late 1940s and early 1950s presented several theoretical and practical guides for the use of ventilation for contaminant control. In the mid-1950s, the Division of Occupational Health of the Michigan Department of Health began publication of *Michigan's Occupational Health*, a newsletter by industrial hygienists and other occupational health professionals that contained numerous articles on practical applications of industrial ventilation. The Michigan Department of Health contributions by James Barrett, Bernie Bloomfield, and Marvin Schuman were joined by those of George Hama, Knowlton Caplan, and Ken Robinson and others to provide core material for the ACGIH's manual as it evolved in the 1950s.

1.3   CURRENT APPLICATIONS

Industrial ventilation systems include a wide variety of applications, ranging from a simple exhaust fan in the ceiling to sophisticated high-velocity/low-volume exhaust systems for hand tools (Fig. 1.1). Ventilation systems are used in nearly every occupational setting, from coal mines and foundries to hospital operating rooms and high-technology clean rooms. Virtually every building has some form of ventilation system, whether it be solely for climate control or for

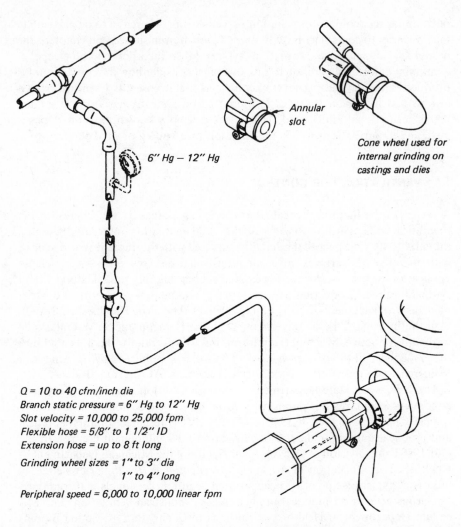

Annular slot

Cone wheel used for internal grinding on castings and dies

6" Hg − 12" Hg

Q = 10 to 40 cfm/inch dia
Branch static pressure = 6" Hg to 12" Hg
Slot velocity = 10,000 to 25,000 fpm
Flexible hose = 5/8" to 1 1/2" ID
Extension hose = up to 8 ft long

Grinding wheel sizes = 1" to 3" dia
⠀⠀⠀⠀⠀⠀⠀⠀⠀⠀⠀⠀⠀⠀1" to 4" long

Peripheral speed = 6,000 to 10,000 linear fpm

Figure 1.1 Low-volume/high-velocity hood for an internal grinder. Note the large static pressure and extremely high slot velocity. (From USHEW/NIOSH Publication No. 76–102, "Recommended Industrial Ventilation Guidelines," 1976.)

contaminant removal. Even residential structures contain ventilation systems (e.g., forced-air heating or cooling, kitchen exhaust fans, or simple window fans).

The following applications demonstrate the various forms that an industrial ventilation system can take. Ventilation engineers categorize systems as being either general (dilution) or local exhaust ventilation systems. The most basic form of ventilation is *general ventilation*, consisting simply of an exhaust fan pulling air out of the workplace and exhausting it to the outdoors. A general ventilation system may include a replacement air system, replacement air distribution ducting, and in rare situations, air-cleaning equipment on the exhaust stream. As discussed in Chapter 4, general exhaust ventilation can be used if the contaminant(s) of interest is not highly toxic and if the rate of generation is predictable. It is not usually the system of first choice to the ventilation designer, but may be the most practical for a situation where there are many contaminant sources scattered throughout the workplace or where the sources are mobile (e.g., forklift trucks in a warehouse).

Local exhaust ventilation (LEV) implies an attempt to remove the contaminant at or near the point of release, thus minimizing the opportunity for the contaminant to enter the workplace air. The ability of a LEV system to accomplish this task is dependent on its proper design, construction, and operation. The nominal LEV system includes an exhaust hood, ducting, a fan, and an exhaust outlet. As with general exhaust ventilation, additional components, such as replacement air systems and air-cleaning devices, may (and should) be included. Local exhaust systems are used in a wide variety of settings, from research laboratory hoods to commercial kitchens to foundries. LEV systems can, and should, be used in the vast majority of situations in preference over general exhaust.

In addition to the nominal system described above, there are a number of special types of industrial ventilation systems, used for particular applications and types of equipment. A *low-volume/high-velocity system* involves the positioning of a small hood adjacent to, or surrounding, the point of contaminant generation. A relatively high capture velocity (10,000 to 15,000 fpm) is attained at a low air flow (60 to 150 cfm) by designing a small hood opening. These systems operate at much higher static pressures than traditional ventilation systems but

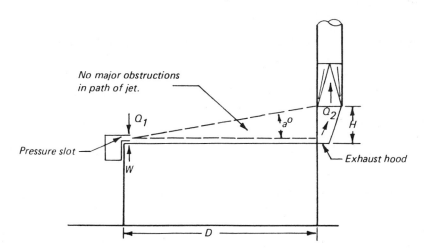

Figure 1.2   Push-pull hood for an open-surface tank. The pressure slot (left) directs a stream of air across the tank surface, pushing contaminants toward the exhaust hood (right). (From USHEW/NIOSH Publication No. 76–102, "Recommended Industrial Ventilation Guidelines," 1976.)

have the distinct advantage of minimizing the total exhaust flow, thus reducing the need for expensive replacement air.

Push-pull hoods are used on wide, open-surface tanks where exhaust slots on either side would be inadequate to draw air from the center of the tank. Instead, one side of the tank is fitted with a source of supply air while the other remains as an exhaust, as shown in Fig. 1.2. A jet of air from the supply side is blown across the tank surface and collected in the exhaust hood.

Clean rooms, increasingly common in the electronics and computer industry, represent a situation where the objective includes not only the removal of any generated material but also the exclusion of materials that may be present in the incoming air. The replacement air must be filtered and treated to prevent contamination of the product being manufactured.

1.4 CASE STUDIES

One of the more clear examples of the effectiveness of ventilation as a prime factor in the reduction of an industrial disease problem is the case of the Vermont granite workers exposed to silica in the first half of the twentieth century. Around the beginning of the century, pneumatic tools were introduced into the granite-cutting industry. These tools were capable of generating large amounts of airborne dust, much more than had been produced with the hand tools used previously. The net result of this new technology was a dramatic rise in the death rate attributable to tuberculosis* among granite cutters using these new tools, at a time when the national tuberculosis mortality rate was steadily declining (Fig. 1.3). The association of the mortality rate with dust level was quite dramatic (Fig. 1.4). The pneumatic tool users and cutters (group A) had the highest dust levels and the highest death rates. Lower concentrations and mortality rates were observed in group B (surface machine operators) and group C (those exposed to general plant dust). The lowest mortality rates were observed in group D, workers who were exposed to less-than-average dust concentrations, such as personnel associated with sandblasting, an operation that had always been done with local exhaust ventilation. These data led the state of Vermont to require workplace controls in the granite-cutting sheds to reduce the dust concentration to below 10 million particles per cubic foot (mppcf).† In the late 1930s, local exhaust ventilation was installed as the primary workplace control. The

---

* Silicosis, a fibrosis of the lung tissue caused by the inhalation of quartz dust, is often associated with tuberculosis because it favors the growth of the tubercle bacilli. Furthermore, advanced silicosis and the early stages of tuberculosis produce similar x-ray images, complicating the diagnosis. Silicosis, while disabling, is not usually fatal in itself. However, the accompanying tuberculous infection is.

† For typical granite dust, 10 mppcf, as defined with an impinger collection and light microscopy analysis, is approximately equal to 0.1 mg of quartz per cubic meter of respirable dust.

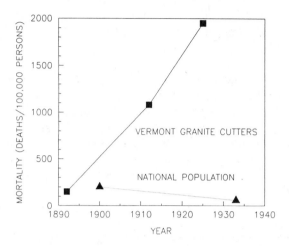

Figure 1.3 Annual data on mortality from tuberculosis for Vermont granite workers and the national population from 1890 to 1935, as power tools were being introduced to the industry. [Data from Albert E. Russell (1936), ''Silicosis and Other Dust Diseases, '' The Harold S. Boquist Second Annual Memorial Lecture, University of Minnesota.]

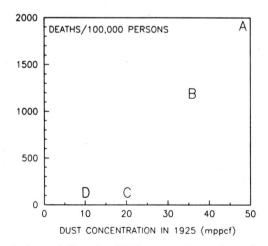

Figure 1.4 Tuberculosis mortality rates (1930) among Vermont granite workers shown in relation to dust concentration (1925) for each of four occupational groups. A, pneumatic tool users; B, surface machine operators; C, workers exposed to plant dust; D, workers at ventilated operations. [Data from Albert E. Russell (1936), ''Silicosis and Other Dust Diseases,'' The Harold S. Boquist Second Annual Memorial Lecture, University of Minnesota.]

immediate effect on the dust concentrations was a 10 to 80% reduction between 1937 and 1940 (Fig. 1.5). For all four occupational groups, the concentrations were reduced below 10 mppcf. Further decreases were noted in 1955 and 1972 as more processes became dust controlled. By 1967 it was recognized that silicosis had been virtually eradicated in the Vermont granite sheds.

Similar data from underground iron-ore mines in the Lake Superior area indicate the tremendous impact that a good ventilation system can have for reducing dust levels (Urban, 1950). A combination of ventilation and wet drilling produced a reduction from 50 mppcf to 5 mppcf between 1920 and 1945, as shown in Fig. 1.6. In 1934, wet drilling and a "moderate ventilation control" program were initiated. In 1939, increased demand for ore raised production rates. Simultaneously, ventilation was improved, with larger air volumes being

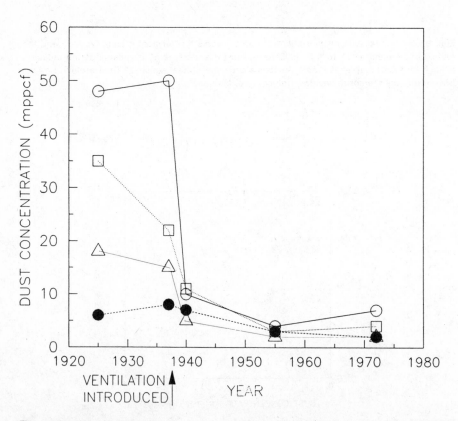

Figure 1.5 Dust concentrations for four occupational groups in Vermont granite mills. Open circles, group A; open squares, group B; open triangles group C; filled circles, group D. The introduction of exhaust ventilation in the late 1930s is accompanied by a dramatic decline in exposures for the heavily exposed groups. [Data are from G. P. Theriault, W. A. Burgess, L. J. DiBerardinis, and J. M. Peters (1974), "Dust Exposure in the Vermont Granite Sheds," *Arch. Environ. Health*, **28**:12.]

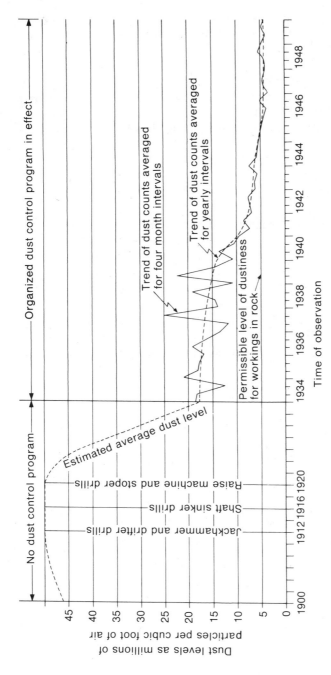

Figure 1.6 Results of a two-stage dust control program in the underground iron-ore mines of the Lake Superior region. The wet drilling and modest ventilation controls introduced in the early 1930s effected a noticeable improvement. The improved ventilation in the late 1930s spurred a more rapid decline in dust concentrations. [From E. C. J. Urban (1950), "The Control of Certain Health Hazards Encountered in Underground Metal Mines," *Am. Ind. Hyg. Assoc. Q.* **10**:201.]

moved through the mines. In Fig. 1.6, the effect of the improved ventilation is quite noticeable.

In shipyards, welding in poorly ventilated areas presented a serious health problem, particularly when the metal being welded was galvanized steel. In one report, tests were performed to determine the reduction in zinc and lead fumes (Rosenfeld, 1944). Without mechanical ventilation, the air was found to contain $4.4\,mg/m^3$ of lead fume and over $100\,mg/m^3$ of zinc fume. The introduction of an exhaust duct to within 20 in. of the arc dropped the air concentrations to $0.16\,mg/m^3$ of lead fume and $1.8\,mg/m^3$ of zinc. Bringing the exhaust duct within 6 in. of the arc showed a further reduction, to 0.044 and $0.92\,mg/m^3$ of lead and zinc fume, respectively.

1.5 SUMMARY

There is no doubt that exhaust ventilation is one of the most effective tools available to the industrial hygienist for controlling the workplace environment. Its effectiveness can be enhanced by combining it with other control methods. However, it must be designed, installed, and operated intelligently if its full potential is to be realized. Errors made in the design phase can be quite costly. Any system must be checked after installation to assure that it conforms to the design specifications and that the concentration of the contaminant of interest is reduced to the specified level. Finally, lack of maintenance on any system will produce a degradation in performance over time.

REFERENCES

Hammond, C. M. (1980), "Dust Control Concepts in Chemical Handling and Weighing," *Ann. Occup. Hyg.*, **23**:95–109.

Rosenfeld, Y. (1944), "Control of Welding Fumes—in Shipbuilding Operations," *Am. Ind. Hyg. Assoc. Q.*, **5**:103.

Urban, E. C. J. (1950), "The Control of Certain Health Hazards Encountered in Underground Metal Mines," *Am. Ind. Hyg. Assoc. Q.*, **10**:201.

2

Principles of Airflow

Under normal conditions of temperature and pressure, matter can exist in three states: solid, liquid, and gas. Liquids and gases are both fluids and are thus governed by many of the same laws of physics. The fundamental differences between solids and fluids are that the former have definite shapes and are able to sustain shear forces while at rest, whereas the latter assume the shape of the container in which they reside and can sustain shear forces only while in motion. Within the class of fluids, liquids and gases differ in their ability to form interfaces with other liquids or gases (gases do not form interfaces with other gases), in the way they fill a volume (gases disperse evenly throughout), and in their compressibility (liquids cannot be compressed significantly). Other properties, such as flow patterns, boundary layers, and shear forces in motion, are common to all fluids and depend on factors other than the physical state.

The study of fluids and their properties dates back to the time of Archimedes (285–212 B.C.), who developed the displacement principle, and to Roman hydraulic engineers, who managed to design and construct extensive water supply systems with only a rudimentary understanding of the flow–friction relationship. Daniel Bernoulli (1738) and Leonhard Euler (1755) developed the equations of motion and the energy relationships that are used today (and bear their names). In the nineteenth century D'Alembert, Navier, Stokes, Reynolds, and Prandtl advanced the science of fluid flow by exploring the laws of motion, the two flow regimes (turbulent and laminar), and boundary layer phenomena. The similarities between liquids and gases allow the use of equations and formulas developed by hydraulic engineers to be adapted to the study of airflow. Thus it should not be disconcerting to the ventilation designer to hear expressions such as "wetted perimeter." Most of the theory behind industrial ventilation is rooted in fluid mechanics, and many of the terms, definitions, and equations are borrowed from the research and practice of that discipline.

2.1   AIRFLOW

Air movement can be *restrained* or *unrestrained*, the former referring to air moving in a conduit (duct or pipe) and the latter involving air moving in a large space (room, building, or outdoors). The industrial ventilation designer will encounter both situations, although the properties pertaining to restrained flow are of greater importance. A thorough understanding of the basic laws governing fluid flow is valuable to the ventilation engineer so that the design equations are properly interpreted and implemented, even though the actual design of most systems can follow a straightforward approach. For the purposes of this chapter, the following assumptions will be made:

- In most ventilation systems, air can be considered to be incompressible; that is, its density will not vary appreciably from point to point within a system.
- Environmental parameters will deviate only slightly from a temperature of 68°F and a pressure of 1 atm. These values will be termed *standard conditions* in this chapter.
- Air is a homogeneous fluid; its composition is spatially and temporally consistent.

Exceptions to these assumptions will be noted. Knowledge of a fluid's density, pressure, and temperature is required to describe a resting fluid adequately. When considering fluid motion, additional properties, such as velocity, viscosity, and frictional energy losses, become important. Each of these parameters is discussed in the following sections.

2.2   DENSITY

One important property of a fluid is its *density*, a measure of the amount of mass in a unit volume. The density of dry air ρ_{air} at standard conditions is 0.075 lbm/ft^3.* It is helpful to express the density in the same units as those used in other ventilation measurements. The density of any gas, including air, is a

---

*The unit lbm, pound-mass, is a measure of mass, not weight. The corresponding weight measure is the pound-force lbf or pound lb. Under normal effects of gravity on earth, if an object has a mass of 1 lbm, it will weigh 1 lbf. The general relationship between these units derives from Newton's first law: force is equal to the product of mass and acceleration. When the force is weight, the acceleration is that due to gravity, g. Thus weight is equal to the product of mass and g, which, in the English units traditionally used in ventilation, is 32.17 ft/sec^2. Given the fact that under normal gravitational conditions, a 1-lbm object weighs 1 lbf, it can be seen that 1 lbf = (32.17 ft/sec^2) (1 lbm). This relationship will be used at several points in this chapter.

function of the temperature and pressure at which it exists. The formula relating these parameters is the *ideal gas law*:

$$\rho R_a T = P \tag{2.1}$$

where $\rho =$ density, lbm/ft^3,
$P =$ absolute pressure, lbf/ft^2
$T =$ absolute temperature, °R,
$R_a =$ gas constant for air, 53.3 ft-lbf/lbm-°R

Because pressure differences from point to point in a ventilation system are, with the exception of high-velocity/low-volume systems, rarely as large as 5% (on an absolute basis), the effects of pressure on the density of air are unimportant and can usually be ignored. Similarly, unless the air being handled has been heated or cooled significantly (a 50°F change from room temperature is required to effect a 10% change in density), density changes due to temperature can also be ignored. If the airstream temperature is outside the range 20 to 120°F, density corrections are appropriate.

2.3 CONTINUITY RELATION

A simple, yet often neglected or misunderstood principle of fluid flow in ducts requires that within the system, matter must be conserved. In other words, in a continuous circuit the same quantity of gas must flow throughout. Known as the *principle of continuity*, the *continuity relation*, or the *law of conservation of matter*, this can be represented mathematically as

$$\int \rho V_r \, dA = \text{constant} \tag{2.2}$$

where $\rho =$ density of fluid, lbm/ft^3
$V_r =$ velocity of fluid at an axial position r in the conduit, ft/min
$A =$ cross-sectional area where the velocity is measured, ft^2

This means that if the density and velocity of the airstream were measured at all points in a cross section of the conduit, the integral of the product ρV_r is constant, regardless of location. With the assumption that the density of the air is constant throughout the conduit and the definition of the average velocity at any cross section V in the conduit to be

$$V = \frac{1}{A} \int V_r \, dA \tag{2.3}$$

Equation 2.2 then reduces to

$$\rho V A = \text{constant throughout system} = M \tag{2.4}$$

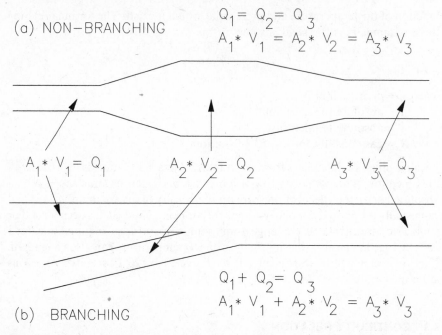

(a) NON-BRANCHING

$$Q_1 = Q_2 = Q_3$$
$$A_1 * V_1 = A_2 * V_2 = A_3 * V_3$$

$$A_1 * V_1 = Q_1$$

$$A_2 * V_2 = Q_2$$

$$A_3 * V_3 = Q_3$$

$$Q_1 + Q_2 = Q_3$$
$$A_1 * V_1 + A_2 * V_2 = A_3 * V_3$$

(b) BRANCHING

Figure 2.1 In a nonbranching system (a), the continuity relation for incompressible flow requires that the flow be constant at all points. When the duct diameter increases ($A_2 > A_1$), the velocity must decrease ($V_2 < V_1$). Conversely, when the diameter decreases ($A_3 < A_2$), the velocity must increase accordingly ($V_3 > V_2$). Since $Q_1 = Q_2 = Q_3$, then $A_1 V_1 = A_2 V_2 = A_3 V_3$. In a branching system (b), the continuity relation requires that the flow in a main (Q_3) be equivalent to the sum of the flow in the submains ($Q_1 + Q_2$). Similarly, the area–velocity products can be summed ($A_1 V_1 + A_2 V_2 = A_3 V_3$).

where M is the *mass flow*, in lbm/min. This equation can be rearranged to

$$\frac{M}{\rho} = VA = Q = \text{constant throughout system} \qquad (2.5)$$

or

$$V_1 A_1 = Q_1 = V_2 A_2 = Q_2 \qquad (2.6)$$

where Q is the *volumetric flow*, in ft^3/min, or cfm.* At all points in a nonbranching system, the flow rate Q, the product of the average velocity V across any cross section and the area A of the duct at that cross section, is the same (Fig. 2.1a). In a branching system, the flows in parallel submains and branches must be added to determine the flow in the resulting main duct (Fig. 2.1b).

---

*The subscripts are used to denote positions along the length of the conduit. It is conventional to use lower numbers to indicate upstream locations. Thus the fluid is assumed to be flowing from position 1 toward position 2.

2.4 PRESSURE

A second fundamental property of air is *pressure,* the force per unit area being exerted normal to that area. Pressure is also defined as the energy per unit volume. *Atmospheric pressure* results from the weight (force) of the air mass above us and is usually measured in reference to a total vacuum (zero pressure). When a pressure is measured in this manner, it is referred to as *absolute pressure.* Because the air pressures in a ventilation system differ only slightly from atmospheric, it can be useful to measure them relative to atmospheric pressure rather than with respect to zero. In these cases, pressure is termed *gage* (or *gauge*) *pressure.*

This is somewhat analogous to measuring temperature on a relative scale (Fahrenheit or Celsius) rather than an absolute scale (Rankine or Kelvin), with one important difference. In the case of pressure, the ambient reference is not constant. It varies as a result of elevation and meteorological conditions. As an example, when one measures pressure in automobile or bicycle tires, one is measuring gage pressure. This gage pressure is the difference between the internal tire environment and the ambient atmosphere. The barometric pressure is an absolute pressure. It is the difference between ambient and a total vacuum. In ventilation measurements, pressure differences between the atmospheric environment and the internal system environments are more important than the absolute magnitudes. Therefore, gage pressure is most commonly specified and reported.

2.4.1 Pressure Units

Four sets of units are commonly used to quantify pressure. If a vertical column of air measuring 1 square inch were to be weighed at sea level under standard conditions, one would find that there are 14.7 lbf of air in the volume extending from the ground through the atmosphere. Thus the air pressure at sea level is 14.7 pounds per square inch (psi). This absolute pressure is often referred to as *standard atmospheric pressure.* Often, a "g" or "a" is appended to the unit psi to differentiate between gage and absolute pressure. Under standard conditions, 14.7 psia = 0 psig. The psi unit is most often used when describing rather large pressure differences.

A second unit results from the common use of an instrument for measuring absolute pressure, the mercury (Hg) barometer. This device consists of an evacuated tube inverted in a pool of mercury (Fig. 2.2). Atmospheric pressure, exerting itself on the open pool, forces the mercury to rise in the column. At sea level, under "normal" weather conditions, the column would rise 29.92 in. above the surface of the open pool. The common unit of measurement of pressure is *inches of mercury,* abbreviated as "in. Hg", and standard atmospheric pressure is 29.92 in. Hg. The unit is actually a surrogate for pressure and is often referred to as a unit of "pressure head".

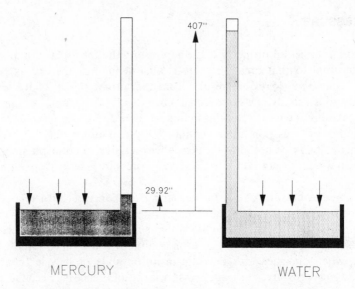

MERCURY   WATER

Figure 2.2 The absolute pressure of air can be measured by inverting an evacuated tube in a pool of liquid (shown here using both mercury and water). Due to the density differences between the liquids, the columns rise to different heights above the pool surfaces. At standard atmospheric pressure (1 atm), a column of mercury will rise 29.92 in. while a column of water will reach an elevation of 407 in.

A third unit commonly used in ventilation measurements is the *inch of water*, (in. H_2O). If the barometer described above were filled with water instead of mercury, the column would rise to a height of 407 in., due to the difference in density between mercury and water. Inches of mercury and inches of water are measurement-related units of pressure. Gage pressures are often expressed in these units because they can be measured directly using either unit with a fluid-filled manometer. In reality, these are not actually pressure units but are surrogates for the pressure exerted by a column of fluid 1 in. high. They are *head units*, as shown in a later section.

The unit of pressure in the International System of Units (SI) is the *pascal* (Pa). The pascal is equal to 1 newton per square meter. Conveniently, 1 atm is equal to 1.0×10^5 Pa. Thus 14.7 psi, 29.92 in. Hg, 407 in. H_2O, and 1.0×10^5 Pa are equivalent quantities for atmospheric pressure under standard conditions.

Pressures existing in a ventilation system are nearly always measured and expressed as inches of water rather than mercury, to facilitate accurate determin-ation of small pressure differences. On a manometer it is visually much easier to measure 1.4 in. H_2O than the equivalent 0.1 in. Hg. Some commercially available manometers use fluids less dense than water, yet are scaled to read directly in in. H_2O. These provide even better resolution of small pressure differences than do water-filled manometers.

2.4.2 Types of Pressure

Pressure, a measure of force per unit area, is also equivalent to energy per unit volume, since energy is the product of force and distance. In still air, pressure is omnidirectional and can be calculated from the ideal gas law, $P = \rho RT$. This hydrostatic pressure is associated with the *potential energy* of the fluid. However, because the ventilation engineer is concerned primarily with air in motion, other pressures must be considered. These pressures are associated with the *kinetic energy* of the moving fluid stream and are subject to different measurement techniques.

The *static pressure* (p_s) of a moving airstream represents the component associated with the potential energy. It is exerted equally in all directions, regardless of the direction or magnitude of flow. It is a function of the density and the temperature of the gas and exists in both moving and nonmoving bodies of air. The pressure measured by an automobile or bicycle tire gauge is static pressure. In a ventilation system, it is pressure that tends to either burst (when greater than ambient) or collapse (when less than ambient) the duct. The static pressure is nearly always measured relative to ambient conditions and is therefore a gage pressure.

Velocity pressure (p_v) is associated with the kinetic energy of the moving airstream and is exerted only in the direction of flow. While it, like the static pressure, is affected by the gas density and temperature, p_v is primarily a function of the airstream velocity. If the air is not moving, $p_v = 0$. The velocity pressure is measured by mathematically or mechanically subtracting the static pressure

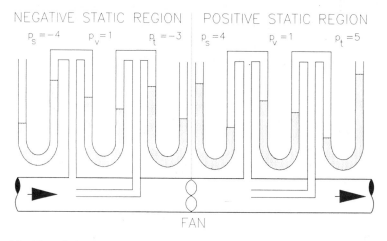

Figure 2.3   The relationships between total (p_t), velocity (p_v), and static (p_s) pressures are shown for both the negative- and positive-pressure situations. On the negative-pressure or suction side of the fan, the static and total pressures must both be negative, while the velocity pressure is always positive. On the positive-pressure side, the total and static pressures are positive with respect to ambient.

from the total pressure (defined below). Velocity pressure is an absolute pressure; it cannot have a negative value.

The *total pressure* (p_t) reflects the combined static and velocity components and is measured as the pressure exerted in the direction of flow. While the total and static pressures can be either negative or positive with respect to ambient conditions, the velocity pressure must always be positive. The total pressure must always be greater than the static pressure for a moving fluid and equal to the static pressure for a nonmoving fluid. In ventilation measurements, total and static pressures are nearly always measured relative to atmospheric while velocity pressure is usually measured by directly comparing total pressure and static pressure, as shown in Fig. 2.3. The standard procedure for determining these pressures is to measure p_s with a pressure tap perpendicular to the direction of airflow and p_t with an impact tube placed parallel to flow such that the moving air impinges directly on the opening. The two pressures, p_t and p_s, are measured with manometers and the velocity pressure is determined by subtracting the latter from the former. If one manometer is used to compare p_t and p_s simultaneously, the difference in the levels of the two legs of the manometer, displayed in in. H_2O, is the velocity pressure. The procedure is explained in detail in Chapter 3.

2.5   HEAD

Although we always use pressure in our discussions, the concept of *head* is encountered occasionally, so its relationship to pressure is introduced here. Fluid head was first used in describing liquid flow where gravity was used as the driving force. Total head h_t is a measure of the total energy of the fluid per unit weight:

$$h_t = \frac{p}{\rho g} + \frac{V^2}{2g} + z \qquad (2.7)$$

where p = static pressure
   ρ = fluid density
   V = average fluid velocity
   g = acceleration due to gravity
   z = elevation (above a reference point)

The individual terms in the equation, the pressure head $p/\rho g$, the velocity head $V^2/2g$, and the elevation head z, each has the unit of length, commonly feet or inches, of fluid. Given that head denotes energy per unit weight of fluid in motion and energy is measured in foot-pounds (ft-lbf), it can be seen that head would be in terms of ft-lbf/lbf, or simply, feet. The velocity head is the kinetic energy per unit weight, the pressure head is the potential energy (due to pressure differences) per unit weight, and the elevation head is the potential energy (due to gravity) per unit weight. For example, the kinetic energy E_k of a body is defined as

$$E_k = \tfrac{1}{2}mV^2 \tag{2.8}$$

where m is the mass and V is the velocity. Dividing both sides of this equation by the mass and then converting the mass m to weight by multiplying by the acceleration due to gravity g, we obtain

$$\frac{E_k}{mg} = \frac{V^2}{2g} = \text{velocity head} \tag{2.9}$$

Similarly, the static pressure p_s is the potential energy per unit volume v:

$$p_s = \frac{E_p}{v} \tag{2.10}$$

The density was defined as the mass per unit volume, hence

$$p_s = \frac{\rho E_p}{m} \tag{2.11}$$

The weight w is the product of the mass m and g:

$$p_s = \frac{g\rho E_p}{w} \tag{2.12}$$

or

$$\frac{p}{\rho g} = \frac{E_p}{w} \tag{2.13}$$

That is, the static pressure head is the potential energy per unit weight.

Equation 2.7 can be simplified by (1) assuming a constant elevation at reference (i.e., $z = 0$), and (2) multiplying both sides of the equation by the product of the fluid density ρ and acceleration due to gravity g, so that

$$\rho g h_t = p + \rho \frac{V^2}{2} \tag{2.14}$$

These terms are now pressure terms, with

$$\rho g h = \text{total pressure } (p_t)$$
$$p = \text{static pressure } (p_s)$$
$$\rho V^2/2 = \text{velocity pressure } (p_v)$$

Thus pressure is equivalent to the product of head, density, and the acceleration due to gravity.

2.6 ELEVATION

In simplifying Eq. 2.7, elevation differences in the system were ignored. The elevation term z is significant only when the density of the fluid stream differs from that of the surroundings. Expressed in pressure units with the elevation term included, Eq. 2.7 becomes

$$\rho g h_t = p + \rho \frac{V^2}{2} + \rho g z \qquad (2.15)$$

Under standard conditions, the density of air is 0.075 lbm/ft^3. Each foot of elevation difference at sea level represents a static pressure change of 0.014 in. H$_2$O. (This rate of change is not linear with elevation, due to the variation in density.) However, because this change occurs both inside and outside an air-handling system, any inside pressure measurements made relative to outside would not reflect it. For example, a 70-ft rise in elevation would at first seem to introduce a 1 in. H$_2$O pressure differential. However, there is an

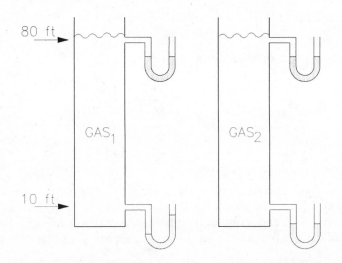

Figure 2.4 The importance of the elevation term is shown for fluids with different densities. Two 80-ft-high chambers have rigid bottoms and walls while the tops are flexible, impermeable membranes that act as weightless pistons. Gases with different densities are introduced into the chambers so that the pressure at the top is balanced with ambient. The gas introduced into the first chamber (GAS$_1$) has a density twice that of air, while the density of the gas in the second chamber (GAS$_2$) is the same as that of air. If the atmospheric pressure at a height of 80 ft is 406 in. H$_2$O, the pressure at 10 ft would be 407 in. H$_2$O. Inside the GAS$_2$ chamber, the absolute pressures would be the same as ambient: 407 in. H$_2$O at 10 ft and 406 in. H$_2$O, at 80 ft. Measured with respect to ambient, p_s at both positions is zero. Inside the GAS$_1$ chamber the absolute pressure at the 80-ft position is the same as ambient, but at the 10-ft location, it is higher, 408 in. H$_2$O. The relative, or gage, static pressure at 80 ft is 0; at 10 ft it is 1 in. H$_2$O.

accompanying change in atmospheric (ambient) pressure external to the system. Since the reference atmosphere also "loses" 1 in. H_2O pressure, any measurement of system p_s or p_t made relative to ambient will not be affected by the elevation.

If the fluid being moved in the system has a density different from that of the external environment, elevation changes may be important. As shown in Fig. 2.4, two 80-ft chambers are filled with gases of different densities. The chambers have rigid bottoms and walls while the tops are flexible, impermeable membranes so that the pressure at the top of each column is identical with the surroundings. The gases are introduced into the chambers so that the membrane is perfectly flat, assuring that the pressure at the top is balanced with ambient. The gas inside the first chamber (GAS_1) has a density twice that of air ($\rho_1 = 0.15$ lbm/ft^3), while the density of the gas in the second chamber (GAS_2) is the same as that of air (0.075 lbm/ft^3).

If the atmospheric pressure at a height of 80 ft was 406 in. H_2O, the pressure at 10 ft would be 407 in. H_2O. Inside the GAS_2 chamber, the absolute pressures would be the same as ambient: 407 in. H_2O at 10 ft and 406 in. H_2O at 80 ft. Because static pressure is measured with respect to ambient, p_s at both heights is zero. For this situation, where the system fluid density is the same as ambient fluid density, the elevation term can be ignored. In the other chamber, however, the elevation aspect becomes important. Although the absolute pressure at the 80-ft mark is the same as that of the second column, the absolute pressure at the 10-ft point of the column 1 is higher, 408 in. H_2O. The relative, or gage, static pressure at 80 ft is 0; at 10 ft it is 1 in. H_2O. In the case of fluids with densities different from those of air, the elevation-pressure term can be quite important and should not be ignored when there are system inlets and outlets that have significant vertical separation.

2.7   PRESSURE RELATIONSHIPS

As stated previously, the velocity pressure is a function of the velocity and fluid density:

$$p_v = \rho \frac{V^2}{2} \qquad (2.16)$$

Rearranging gives us

$$V^2 = \frac{2p_v}{\rho}$$

$$= \frac{2p_v}{0.075 \text{ lbm/ft}^3}$$

Since 1 lbf $=(32.17 \text{ ft/sec}^2)$lbm,

$$V^2 = p_v \left[\frac{(2)(32.17)(3600)}{0.075} \right] \frac{\text{ft}^4\text{-sec}^2}{\text{min}^2\text{-sec}^2\text{-lbf}}$$

Replacing the real pressure unit lbf/ft^2 with the commonly used head unit in. H$_2$O, where 1 in. H$_2$O is the head corresponding to a pressure of 5.2 lbf/ft^2*, we obtain

$$V^2 = p_v \left[\frac{(2)(32.17)(3600)(5.2)}{0.075} \right] \frac{\text{ft}^4}{\text{min}^2\text{-ft}^2\text{-in. H}_2\text{O}}$$

$$= 1.6 \times 10^6 \, p_v \frac{\text{ft}^2}{\text{min}^2\text{-in. H}_2\text{O}}$$

$$V = 4000 \sqrt{p_v} \tag{2.17}$$

where the velocity pressure is measured in in. H$_2$O and the velocity is expressed in feet per minute (fpm). This equation is very handy for determining velocity from velocity pressure measurements made at standard conditions.

The relationships between velocity, velocity pressure, and static pressure are derived from *Bernoulli's equation*, which states, in terms of head, that fluids flowing from point 1 to point 2 obey the formula

$$\frac{p_1}{g\rho_1} + \frac{V_1^2}{2g} = \frac{p_2}{g\rho_2} + \frac{V_2^2}{2g} + \text{losses}_{1-2} \tag{2.18}$$

where $p_1, p_2 =$ static pressures at points 1 and 2
$\quad \rho_1, \rho_2 =$ densities at points 1 and 2
$\quad V_1, V_2 =$ average velocities at points 1 and 2
$\quad g =$ acceleration due to gravity
$\text{losses}_{1-2} =$ head losses due to friction and turbulence between points 1 and 2

The head losses represent the conversion of potential energy to heat, which in a ventilation system serves no beneficial purpose. This form of the equation assumes that no work is being done on the system (i.e., no input of energy in the form of a fan or blower between points 1 and 2) as well as no changes in elevation. Assuming that the density is constant ($\rho_1 = \rho_2 = \rho$), then Eq. 2.18, now expressed in pressure units by multiplying all terms by ρg, is

$$p_1 + \rho \frac{V_1^2}{2} = p_2 + \rho \frac{V_2^2}{2} + \rho g \,(\text{losses}_{1-2}) \tag{2.19}$$

*Recalling that 14.7 psi $= 407$ in. H$_2$O, then
$\qquad$ 1 in. H$_2$O $= (14.7 \text{ lbf/in}^2)/407$
$\qquad\qquad\qquad = (0.0361 \text{ lbf/in}^2)(12 \text{ in.}/\text{ft})^2$
$\qquad\qquad\qquad = 5.2 \text{ lbf/ft}^2$

The first term on either side is the static pressure; the second is the velocity pressure; and the last, ρg (losses$_{1-2}$), represents the *change* in total pressure. Equation 2.19 can be rewritten as

$$p_{s,1} + p_{v,1} = p_{s,2} + p_{v,2} + \Delta p_t \qquad (2.20)$$

The relationships between total, static, and velocity pressure can be addressed by considering the simple system shown in Fig. 2.5. There is a reduction in the cross-sectional area of the duct on the intake side of the fan, where the static pressure would be negative (with respect to ambient). Assuming that the only losses in the system are due to friction and ignoring the losses associated with the shock in the abrupt reduction, the graphical representation of the pressures in the system is as shown.

While the static and total pressures drop through the system, the velocity pressure remains constant within each section. The increase in velocity pressure at the transition from the larger section to the smaller section results from the change in velocity, which is a function of the flow Q and the duct area A. Replacing V with Q/A in the velocity-to-velocity pressure relationship (Eq. 2.17)

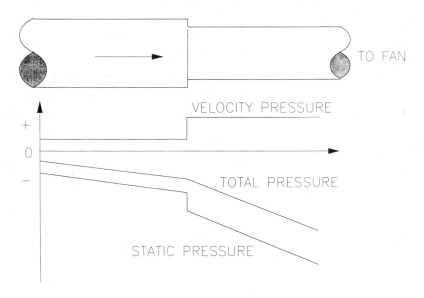

Figure 2.5 The relationships between the three pressures (static, velocity, and total) are shown for a converging system. As the cross-sectional area decreases, the velocity and velocity pressure increase, since $V = Q/A$ (the continuity relation) and $V = 4000\sqrt{p_v}$. The frictional losses through the duct are shown as a gradually sloping line when the velocity is low and a more severely sloping line when the velocity increases in the smaller duct. The slope of the line is proportional to p_v. The total pressure is the arithmetic sum of p_v and p_s. Therefore, the increase in p_v at the abrupt reduction is accompanied by a decrease in p_s so that there is no sudden change in p_t. (The losses associated with turbulent shock are not shown.)

yields

$$p_v = \left(\frac{Q}{4000A}\right)^2 \tag{2.21}$$

Since Q is constant throughout a nonbranching system (due to the conservation of mass), for a given flow, the change in velocity pressure is dependent solely on the change in the cross-sectional area between two points in that system. If the area is constant, then p_v is constant. If the area changes, p_v must also change, as

$$p_{v,2} = p_{v,1}\left(\frac{A_1}{A_2}\right)^2 \tag{2.22}$$

2.7.1 Reynolds Number

Before considering the causes of losses associated with flow through the system, an understanding of the two types of flow, laminar and turbulent, is helpful. *Laminar flow* (also known as streamline or *viscous flow*) refers to the parallel streamlines gas molecules are observed to follow at low velocities. There is no lateral or radial mixing of the fluid in laminar flow. As the velocity increases, the flow becomes more chaotic and considerable cross-current mixing occurs. At higher velocities, the streamlines disappear altogether and the flow becomes fully *turbulent*.

While the specific transition point between laminar and turbulent flow is difficult to define, it is known to be dependent on the density, viscosity, and velocity of the fluid as well as the geometry of the conduit within which the fluid is moving. These factors are combined in a term developed by Osborne Reynolds and known as the *Reynolds number* (Re), which is computed as

$$\text{Re} = VD\frac{\rho}{\mu} \tag{2.23}$$

where $V =$ average velocity of the fluid
$\rho =$ density of the fluid
$\mu =$ viscosity of the fluid
$D =$ diameter of the conduit

If a consistent set of units is used (e.g., $[D] =$ ft, $[V] =$ ft/sec, $[\rho] =$ lbm/ft^3, and $[\mu] =$ lbm/ft-sec), the Reynolds number is dimensionless. In general, if the Reynolds number for a given flow regime is less than 2000, the flow will probably be laminar. If Re is greater than 4000, the flow will be turbulent. When $2000 < \text{Re} < 4000$, the type of flow is less predictable, depending on other factors, such as obstructions or directional changes in flow. Because the relationship between static pressure losses and velocity depends on the type of flow encountered, the derivation of a general formula requires one to know whether normal ventilation systems operate in the laminar or turbulent flow regions. To gain a better appreciation of the relationship between velocity and Reynolds number for air, Eq. 2.23 can be rearranged to

$$VD = \text{Re} \frac{\mu}{\rho} \qquad (2.24)$$

The viscosity of air at standard conditions is 1.2×10^{-5} lbm/ft-sec and the density is 0.075 lbm/ft^3. Substituting these values into Eq. 2.24 yields

$$VD = \left[\frac{(1.2 \times 10^{-5} \text{ lbm/ft-sec}) (60 \text{ sec/min})}{0.075 \text{ lbm/ft}^3} \right] \text{Re}$$

$$= 9.6 \times 10^{-3} \text{ Re ft}^2/\text{min} \qquad (2.25)$$

For air at standard conditions, turbulent flow will exist when

$$4000 < \text{Re}$$

$$< \frac{VD}{9.6 \times 10^{-3} \text{ ft}^2/\text{min}}$$

or,

$$VD > 40 \text{ ft}^2/\text{min} \qquad (2.26)$$

Thus any flow with a velocity–diameter product greater than 40 ft^2/min would be turbulent. Because industrial ventilation systems typically operate at velocities greater than 1000 fpm and duct diameters are usually at least 0.25 ft, it is unlikely that laminar flow would be encountered. There may be certain, isolated cases (some air-cleaning devices operate at very low velocities) where laminar conditions must be considered. Nevertheless, as most industrial ventilation system display the turbulent regime, turbulent flow will be assumed in the determination of pressure losses.

2.8 LOSSES

As a fluid moves through a system, it will encounter resistance to flow. This resistance will result in decreases in static and total pressure throughout the system, which the air-moving device (fan) will be expected to overcome. These pressure drops, or losses, represent the conversion of potential energy into heat (first law of thermodynamics), which is nonproductive in a ventilation system. This resistance can arise from two general mechanisms: (1) friction associated with shearing stresses and turbulence within the duct, and (2) shock from sudden velocity (speed or direction) changes or flow separation at elbows, branches, and transitions.

2.8.1 Frictional Losses

Frictional pressure losses result from "rubbing" the fluid against the walls of the conduit as well as against itself. The magnitude of these losses depends primarily on the type of flow regime (laminar or turbulent) in the system. The system displayed in Fig. 2.6 shows the velocities at different radial positions across the duct. The lengths of the arrows indicate the relative velocities at that point. At the edge of the duct, next to the wall, there is an extremely thin boundary layer of stagnant air, the thickness of which depends on Re. The air velocity at the rigid boundary is zero. In the center of the duct, the air is moving at its maximum velocity. Between these two points, there is a velocity gradient that depends on the flow regime. In laminar flow, this gradient introduces shear stresses that produce frictional losses. In both laminar and turbulent flow, the gradient at the wall produces shear stresses.

If only turbulent flow is considered (the predominant flow in ventilation systems), the velocity profile is flat in the center of the duct and sharp near the boundary, at which point it drops rapidly to zero. This sharp gradient at the edge produces a large shear stress. This wall–fluid shear is more pronounced in turbulent flow, leading to a larger energy loss than in laminar flow. The chaotic flow also introduces shear stresses within the fluid as the slow- and fast-moving streams are continually being intermixed.

The pressure loss associated with frictional energy losses (conversion to heat) is given by the *Darcy–Weisbach relation*, in pressure terms:

$$\Delta p_f = f\frac{L}{D}\left(\rho\frac{V^2}{2}\right)$$

$$= f\frac{L}{D}p_v \tag{2.27}$$

where Δp_f = pressure loss due to friction
$\quad f$ = friction factor
$\quad L$ = length of conduit
$\quad D$ = diameter of conduit

Values of the friction factor depend on the Reynolds number and, in turbulent flow, the roughness of the conduit interior wall. In laminar flow,

$$f = \frac{64}{\mathrm{Re}} \tag{2.28}$$

Because Re is linear with velocity, the frictional pressure loss in *laminar* conditions is also linear with velocity.

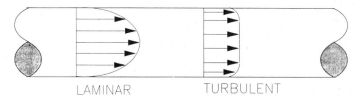

Figure 2.6 Flow-velocity profiles of laminar (Re < 2000) and turbulent (Re > 4000) flow regimes. The turbulent flow profile is nearly uniform across the radius, whereas the laminar profile is more parabolic.

For turbulent flow, the relationship is more complicated. Values of f are usually obtained from the Moody chart (Fig. 2.7), which provides friction factors as a function of Re and relative roughness of the conduit. As can be seen from this chart, at high Re and high relative roughness (a region known as *fully rough flow*), the friction factor becomes independent of Re. In the transition between fully rough flow and laminar flow, the curves for determining f are complex. Regardless of the origin of the friction *factor*, the friction *loss* is expressed as Eq. 2.27, with the loss proportional to the velocity pressure. In practice, this equation and the Moody chart are not directly used by the designer of ventilation systems. Instead, charts and nomograms have been developed for air

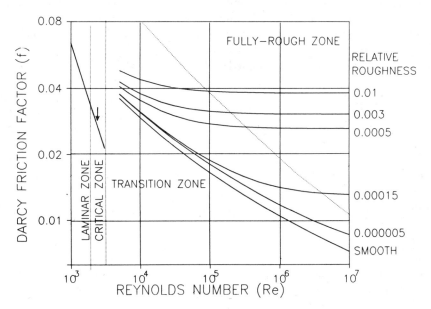

Figure 2.7 The Moody chart relates the friction factor to the Reynolds number and the relative roughness. For laminar flow conditions, the factor is dependent on Re and independent of relative roughness. In fully rough flow, the friction factor becomes independent of Re, depending only on relative roughness. Between these flow regimes, f depends on both parameters.

passing through the standard, galvanized ductwork commonly used. Correction factors can be used to compensate for duct materials with different roughnesses. Alternatively, the designer can use data supplied by the duct manufacturer. To obtain approximate values for the term (fp_v/D), which has the unit of in. H_2O per 100 ft, the designer is referred to the *Ventilation Manual*, Fig. 5–21*a* and *b*. The values provided by these charts have been adjusted from the theoretical equations based on empirical data. The formula on which they are based is provided below the chart. This term is used in the design technique known as the *equivalent foot method*, which is discussed in subsequent chapters.

A second approach to system design, known as the *velocity pressure method*, employs the term f/D, which represents the frictional pressure loss, in in. H_2O per velocity pressure per length of duct. When multiplied by the velocity pressure in the duct section and the duct length, this provides the expected pressure drop. Values for f/D are presented in Fig. 5–18*a* and *b* of the *Ventilation Manual*. Although more details regarding these two design methods are provided in subsequent chapters, the reader should remember that the pressure drop due to friction through straight ducts is a function of roughness, diameter, length, and velocity (or velocity pressure).

2.8.2 Shock Losses

A second type of energy loss results from turbulence or "shock" in the fluid stream, which causes a violent mixing of the fluid and subsequent eddy formation. These disturbances are usually associated with redirection of flow or sudden changes in the duct size that consequently cause drastic velocity changes. The well-established velocity profiles are distorted and pressure losses result. These shock losses are also associated with merging airstreams. While the shock losses are highly variable, depending on the geometry of the system where the turbulence occurs, they are consistent in their dependence on the velocity pressure.

A second cause of shock loss is the flow separation observed primarily in contractions and expansions. Separation occurs when the primary fluid stream no longer follows the wall and breaks away, leaving irregular eddies in the wake region (Fig. 2.8). Reconsidering the earlier example (Fig. 2.5), the losses due not only to friction but also those resulting from the shock caused by the sudden contraction are included (Fig. 2.9). As before, the velocity pressure changes only as a result of the decrease in area and subsequent increase in velocity. However, the total pressure decreases as a result of that portion of the loss in static pressure associated with the shock losses at the point of the velocity transition. For a sudden contraction such as this, the magnitude of the loss is proportional to the velocity pressure in the smaller section $p_{v,2}$ according to the formula

$$\Delta p_{\text{loss}} = K p_{v,2} \tag{2.29}$$

where K is primarily a function of the ratio of the duct areas (A_2/A_1) and taper

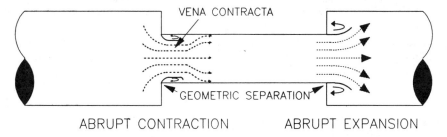

Figure 2.8 Flow separation is shown for contractions and expansions. There are energy (pressure) losses associated with this separation that are part of the shock losses for the fitting. A tapered fitting reduces the separation and thus the loss.

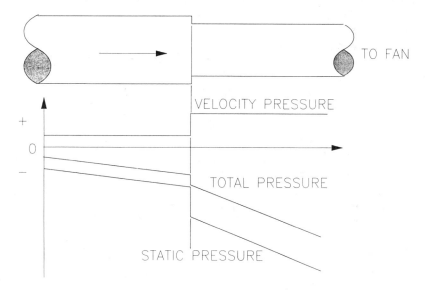

Figure 2.9 This is the same system as shown in Fig. 2.5, but the losses incurred due to "shock" at the transition are incorporated. The velocity pressure is not affected, but the static and total pressures both reflect the loss as a sudden drop. The extent of the loss is a function of the ratio of the areas and the velocity pressure.

angle according to Fig. 5–19 of the *Ventilation Manual*. In addition to the shock loss associated with the duct size change, there is also the static pressure reduction required to boost the velocity pressure from $p_{v,1}$ to $p_{v,2}$. The static pressure difference for a fitting can thus be described as

$$\Delta p_s = -(\Delta p_{\text{loss}} + \Delta p_v) \qquad (2.30)$$

where $\Delta p_s = p_{s,2} - p_{s,1}$ and $\Delta p_v = p_{v,2} - p_{v,1}$. The static pressure is reduced by the sum of the loss at the fitting and the velocity pressure change. The loss associated with the velocity pressure change is often referred to as an *inertial* loss. Shock losses occur at nearly every disturbance in the system, including hood entries, branch entries, reductions, expansions, elbows, and takeoffs. Most of the guidelines to reduce the magnitude of the transition static pressure loss involve reducing the severity of the disturbance.

2.9   LOSSES IN FITTINGS

The pressure loss encountered in straight ducts was shown to be dependent on the velocity pressure, the duct diameter and length, and the friction factor. Most of the losses encountered in other parts of a ventilation system (e.g., expansions, contractions, elbows, hood entries, and branch entries) are also proportional to the velocity pressure and are treated as such by expressing them as

$$\Delta p_{\text{loss}} = K \frac{\rho V^2}{2} = K p_v \qquad (2.31)$$

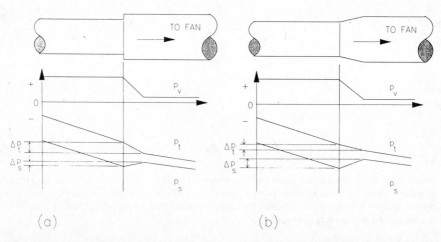

Figure 2.10 In an abrupt expansion (*a*), the velocity, and the resulting velocity pressure, decrease in accordance with the continuity equation, although some flow separation does occur. The magnitude of the decrease is in proportion to the ratio, *r*, of the areas. When the velocity pressure drops, some of the energy is regained as static pressure while the remainder is lost to turbulent shock. The quantity regained is shown as Δp_s while the frictional loss is Δp_t. The rate of static pressure (and total pressure) loss in the larger section is less than that in the smaller section. This is because the loss rate is proportional to the velocity pressure. When the velocity pressure decreases, the loss rate does as well. In the tapered expansion(*b*), the static pressure regain is larger because the gradual widening decreased the flow separation. [These pressure profiles apply to the case where the expansion is on the negative-pressure side of the air-moving device (AMD). If the expansion were on the outlet side, the total and static pressure profiles would be positive, not negative, but they would have the same shapes.]

The value of K depends primarily on the type and geometry of the fitting under consideration. Tables containing nominal values for most types of fittings can be found in the *Ventilation Manual*.

2.9.1   Expansions

The expansion of the cross-sectional area of a duct may be either abrupt or tapered (Fig. 2.10). The latter reduces the losses that result from flow separation. In either case, as the air emerges from the smaller duct section into the larger, the velocity is reduced. The impact of the faster-moving airstream on the slower one and the flow separation lead to shock losses. The Borda loss equation for expansions states that the total pressure loss depends on the square of the difference between the velocities in the two sections:

$$\Delta p_{\text{loss}} = \frac{\rho(V_1 - V_2)^2}{2} \tag{2.32}$$

where $\Delta p_{\text{loss}} =$ drop in total pressure
$\quad\quad V_1 =$ velocity in smaller (upstream) section
$\quad\quad V_2 =$ velocity in larger (downstream) section

The continuity relation requires that $A_1 V_1 = A_2 V_2$, so

$$\Delta p_{\text{loss}} = \frac{\rho}{2}\left(V_1 - \frac{A_1}{A_2}V_1\right)^2$$

$$= p_{v,1}\left(1 - \frac{A_1}{A_2}\right)^2$$

$$= p_{v,1}(1 - r)^2 \tag{2.33}$$

where r is the ratio of the areas (A_1/A_2). The change in velocity pressure, a portion of which will be "regained" as static pressure, is simply

$$\Delta p_v = p_{v,2} - p_{v,1} \tag{2.34}$$

Recalling Eq. 2.22, we have

$$\Delta p_v = r^2 p_{v,1} - p_{v,1}$$

$$= p_{v,1}(r^2 - 1) \tag{2.35}$$

In the ideal case, all of the velocity pressure loss would be recaptured as a static pressure regain, leaving the total pressure loss close to zero.

$$\Delta p_s = -(\Delta p_{\text{loss}} + \Delta p_v)$$

$$= -[p_{v,1}(1 - r)^2 + p_{v,1}(r^2 - 1)]$$

$$= p_{v,1}\, 2r(1 - r) \tag{2.36}$$

The efficiency of the expansion, defined as the ratio of the static pressure regain (a "negative" loss) and the velocity pressure loss, is a measure of the extent of regain:

$$\text{efficiency} = \frac{\Delta p_s}{\Delta p_v}$$

$$= \frac{p_{v,1} \, 2r(1-r)}{p_{v,1}(1-r^2)}$$

$$= \frac{2r(1-r)}{(1-r)(1+r)}$$

$$= \frac{2r}{1+r} \tag{2.37}$$

Efficiency therefore decreases as the ratio decreases, with a value of 100% at $r=1$ (no expansion) and a value of 0% at $r=0$ (a duct delivering air into a large room, for example).

Tapered expansions are used to reduce the extent of the shock by decreasing the flow separation and improve the efficiency of the static pressure regain. The pressure losses in tapered expansions are dependent on the angle of the expansion as well as the ratio of the initial and final cross-sectional areas. A table of values for the static regain in tapered expansions is provided in the *Ventilation Manual*, Fig. 5–19.

2.9.2   Contractions

A duct contraction, like an expansion, can be either sudden or gradual, with the latter being more efficient (Fig. 2.11). When the cross-sectional area decreases, the velocity (and the velocity pressure) must increase in accordance with the continuity relation. An accompanying decrease in static pressure occurs to accommodate the velocity pressure increase and the losses due to the shock losses. As with any fitting involving velocity changes, the general form of the equation describing the static pressure decrease is

$$\Delta p_s = -(\Delta p_{\text{loss}} + \Delta p_v) \tag{2.30}$$

Hood Entries. A hood entry (as opposed to a branch entry) is a special type of contraction, where the room is considered to be the larger section and the duct the smaller, as shown in Fig. 2.11a. The static pressure differential between the room and the duct is called the entry loss h_e and the total static decrease is given by

$$\Delta p_s = -(h_e + \Delta p_v) \tag{2.38}$$

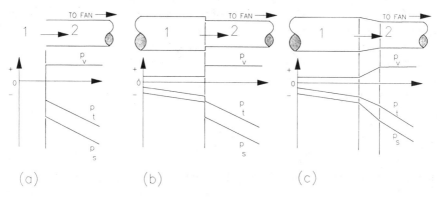

Figure 2.11 Three types of contraction are shown: an entry, an abrupt in-line contraction, and a tapered in-line contraction. At the entry (*a*), the three pressures (p_v, p_s, and p_t) are assumed to be zero. As the air enters the system, the static pressure must be sufficient to accelerate the air up to duct velocity as well as handle any energy loss associated with the entry. The latter loss is known as the hood entry loss (h_e), as shown. Similarly, for an abrupt in-line contraction (*b*), the static pressure must accommodate the increased velocity pressure and the transitional frictional losses. These frictional losses can be reduced by using a gradual, tapered fitting (*c*).

Because both room and duct static pressures are measured relative to ambient,

$$\Delta p_s = p_{s,2} - 0 = p_{s,2} \qquad (2.39)$$

Similarly, because the room air has no velocity,

$$\Delta p_v = p_{v,2} - 0 = p_{v,2} \qquad (2.40)$$

Equation 2.38 reduces to

$$p_{s,h} = -p_{s,2} = p_v + h_e \qquad (2.41)$$

where the hood static pressure is $p_{s,h}$ and the duct velocity pressure is simply $p_v = p_{v,2}$. Because the hood static pressure is traditionally expressed as a positive value even though it is a negative gage pressure, the definition of $p_{s,h} = -p_{s,2}$ accounts for this. As with all other losses in turbulent conditions, the hood entry loss h_e is treated as being directly proportional to the velocity pressure:

$$h_e = F_e p_v \qquad (2.42)$$

where F_e is the hood entry loss *factor*. Combining Eqs. 2.40 and 2.41 gives

$$p_{s,h} = p_v(1 + F_e) \qquad (2.43)$$

The hood entry loss factor depends on the geometry of the entry. The coefficient of entry C_e is a dimensionless measure of this geometry and is calculated as the ratio of the actual flow observed in the duct at some hood static pressure to the maximum hypothetical flow possible at the same static pressure. This maximum flow would occur under the ideal conditions of no entry loss ($F_e = 0$, $h_e = 0$). In this case, all of the static pressure loss is converted into velocity pressure, so that $p_{s,h} = p_v$. In other words,

$$C_e = \frac{Q_{actual}}{Q_{ideal}} \qquad (2.44)$$

Recalling Eqs. 2.6, 2.17, and 2.44, and setting $F_e = 0$ for Q_{ideal}, we obtain

$$C_e = \frac{4000A\sqrt{p_v}}{4000A\sqrt{p_{s,h}}} \qquad (2.45)$$

$$= \sqrt{\frac{p_v}{p_{s,h}}} \qquad (2.46)$$

For $Q_{actual}, p_v = p_{s,h}/(1 + F_e)$,

$$C_e = \sqrt{\frac{p_{s,h}}{p_{s,h}(1 + F_e)}}$$

$$= \sqrt{\frac{1}{1 + F_e}} \qquad (2.47)$$

Solving for F_e yields

$$F_e = \frac{1 - C_e^2}{C_e^2} \qquad (2.48)$$

Typical values for C_e are provided in Fig. 2.12, ranging from 0.98 for a bell-mouthed inlet to 0.72 for a plain, unflanged duct opening.

In-Line Contractions. Contractions occurring within the duct can be abrupt (Fig. 2.11b) or gradual (Fig. 2.11c). The general form of Eq. 2.30 is used because the velocity pressure in the larger, upstream section is not zero and the static pressure in that section is no longer equivalent to ambient.

$$\Delta p_s = -(\Delta p_{loss} + \Delta p_v) \qquad (2.30)$$

The loss term for the tapered contraction is given as

$$\Delta p_{loss} = L\Delta p_v \qquad (2.49)$$

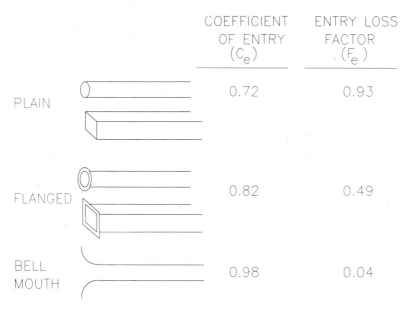

Figure 2.12 Some typical hood entries and the associated coefficients of entry and entry loss factors. More specific data can be found in the *Ventilation Manual* .

while for the abrupt contraction it is

$$\Delta p_{\text{loss}} = K p_{v,2} \qquad (2.50)$$

Values of L and K are given on Fig. 5–19 of the *Ventilation Manual*. The tapered loss factor L is a function of the taper angle, while the abrupt loss factor K is dependent on the ratio of the two areas. Any angle greater than 60° is considered an abrupt contraction. In designing a system, abrupt contractions should be avoided because of their inefficiency in converting static pressure into velocity pressure, primarily as a result of the flow separation and resultant eddy formation.

2.9.3 Elbows

The pressure loss encountered as the airstream passes through an elbow (Fig. 2.13) is approximated in ventilation system design as

$$\Delta p_{\text{loss}} = K_{90} \frac{\theta}{90} p_v \qquad (2.51)$$

where K_{90} is the loss factor for a 90° elbow and θ is the angle of the bend, in degrees. Because the loss is treated as being directly proportional to the angle (a 30° elbow exhibits one-third the loss of a 90° elbow), tables of K values are

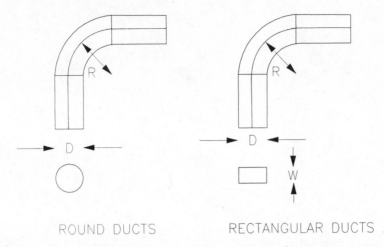

ROUND DUCTS   RECTANGULAR DUCTS

Figure 2.13   The frictional loss factor associated with elbows is dependent on the elbow aspect ratio for round ducts and both the elbow aspect ratio and the duct aspect ratio for rectangular ducts. The elbow aspect ratio is simply *R/D*, where *R* is the radius of curvature and *D* is the duct diameter. The duct aspect ratio is given by *W/D*, where *W* is the width of the duct. A table of elbow loss factors is provided as Fig. 5–16 of the *Ventilation Manual*.

usually provided only for right-angle elbows ($\theta = 90°$). For round ducts, K_{90} is a function solely of the elbow aspect ratio (R/D), as shown in Fig. 2.13. For rectangular ducts, both the elbow aspect ratio, R/D, and the duct aspect ratio, W/D, are important. Values of K_{90} for both round and retcangular ducts are provided on Fig. 5–16 of the *Ventilation Manual*.

Although the standard practice is to assume that the loss is proportional to θ, this is not completely accurate. However, because the actual loss depends on a number of other factors, including the length of straight run immediately prior to the elbow, this approximation is usually adequate.

2.9.4   Branch Entries (Junctions)

As the air in a branch enters a main duct (Fig. 2.14), there are turbulence losses which, in the design process, are assumed to occur in the branch and are therefore proportional to the velocity pressure there, not the velocity pressure in the main. The loss is calculated as

$$\Delta p_{\text{loss}} = K p_v \qquad (2.52)$$

with values of K provided in the *Ventilation Manual*, Fig. 5–17, for angles between 10 and 90°. As with elbows, higher values of K are associated with the wider angles. The loss is associated with the entering airstream and therefore the velocity pressure in the branch is used in the calculation. If there is a significant difference in velocities between the resulting combined airstream and either of the merging streams, the change in velocity pressure would be accompanied by a

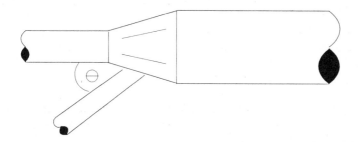

Figure 2.14 In a branch entry, the static pressure loss attributable to the fitting is attributed to the entering branch and is proportional to the velocity pressure there. The loss can be reduced by maintaining a small angle of entry.

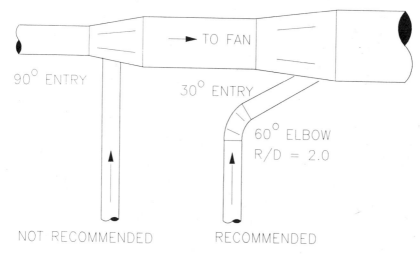

Figure 2.15 As a branch joins a perpendicular main, the use of a 90° branch entry is discouraged in favor of a 60° elbow combined with a 30° entry. The static pressure loss associated with the elbow–entry combination is substantially less than that of a 90° entry.

change in static pressure. In the design process, this difference is treated as an "other loss".

A frequent situation presented to a ventilation designer is the connection of a branch running perpendicular to the main duct (Fig. 2.15). A simple tee fitting, producing a right-angle entry, would induce a static pressure loss equal to $1.0p_v$ (*Ventilation Manual*, Fig. 6–12; $\theta = 90°$). If, however, the designer uses the combination of a 60° elbow and a 30° entry, the pressure loss is reduced to

$$\left(\frac{60°}{90°}\right)(0.27)p_v + 0.18p_v = 0.36p_v \qquad (2.53)$$

The designer should avoid entry angles of greater than 45°, with 30° being a better choice.

2.10   SUMMARY

This chapter should have provided the reader with an appreciation for the theory behind the behavior of moving air. With this background, some of the mystery surrounding the origins of the tables, nomograms, and procedures used to design these systems should be dispelled. An understanding of many of the relationships presented, especially those between velocity, static, and total pressure and between velocity pressure and velocity, is critical to the design and testing of effective ventilation systems.

LIST OF SYMBOLS

A	area
C_e	coefficient of entry
D	diameter
E_k	kinetic energy
E_p	potential energy
f	friction factor
F_e	hood entry loss factor
g	gravitational constant
h_e	hood entry loss
h_t	total head
K	loss coefficient
L	length, distance
m	mass
M	mass flow
p	pressure
p_s	static pressure
$p_{s,h}$	hood static pressure
p_t	total pressure
p_v	velocity pressure
Q	volume flow
r	ratio of cross-sectional areas
R	radius of curvature
R_a	gas constant for air
Re	Reynolds number
T	temperature
v	volume
V	average velocity
W	width (of rectangular duct)
z	elevation above reference datum
ρ	density (mass/volume)
θ	angle (of elbow, branch, entry)

3

Airflow Measurement Techniques

In this chapter we cover the methods available for the measurement of velocity, airflow rate, and certain system pressures. After a ventilation system is designed and installed, airflow measurements are required to determine if the system is performing according to the design; that is, whether the minimum critical velocities and airflows proposed in the design have been achieved in the installation (Fig. 3.1). There are other reasons why airflow measurements are made. In many cases, ventilation systems for the control of airborne contaminants must meet the minimum standards established by various regulatory agencies. Periodic airflow measurements also provide a maintenance history of the ventilation system and highlight conditions that require correction. Finally, the measurement and inventory of total exhaust rate are necessary to evaluate the adequacy of replacement air.

In most exhaust systems airflow is most conveniently measured by determining the average air velocity at a point in the system where the cross-sectional area is known. The airflow is then calculated as the product of the area and average velocity. As shown in Fig. 3.2, the common measurement locations in ventilation systems are the face of the hood or a suitable duct location. Methods for measuring velocity at these locations are given principal attention in this chapter.

A second method of airflow measurement utilizes total flow devices such as orifice and venturi meters. Since these devices have little application in measuring airflow in industrial ventilation systems and are covered adequately in other publications, they are given limited attention in this chapter.

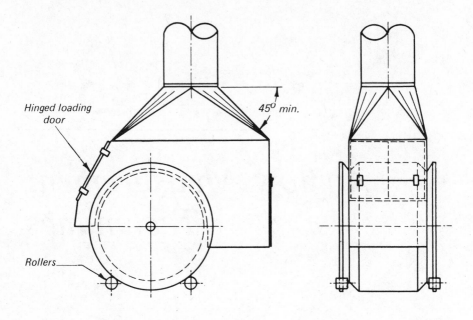

Hinged loading door

45° min.

Rollers

Q = 400 cfm/sq ft of opening

Duct velocity = 1000 fpm minimum or transport velocity for horizontal runs

Entry loss = 1.78 slot VP plus entry loss factor X duct VP

ELECTRIC ROCKING FURNACE

Figure 3.1 The design plates in the *Ventilation Manual* provide the basic design elements of the hood, including (1) geometry, (2) entry loss, (3) capture velocity, (4) exhaust rate, and (5) minimum duct velocity.

3.1 MEASUREMENT OF VELOCITY BY PITOT-STATIC TUBE

The relationship between velocity pressure and the velocity of air is

$$V = 4000 \sqrt{p_v} \qquad (2.17)$$

where velocity is expressed in feet per minute (fpm) and the velocity pressure is in in. H$_2$O. The velocity pressure can be measured as the difference between total pressure and static pressure at a fixed measurement location. This technique is used occasionally in industry for critical monitoring of airflow in a branch or main duct (Fig. 3.3.). However, for general measurement purposes, the Pitot-static tube is the most widely used device. An integral combination of Pitot tube for measurement of total pressure and a static pressure measurement element,

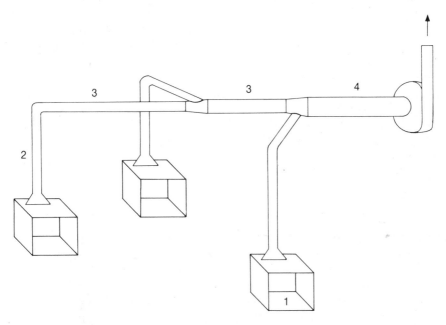

Figure 3.2 Ventilation measurements are commonly made (1) at the face of booth-type hoods, (2) just downstream of a hood at a branch location, (3) in the main to define hood exhaust rate by difference, and (4) in the main to define total system airflow.

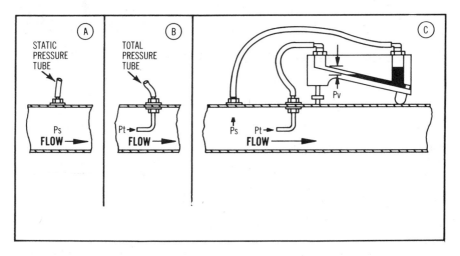

Figure 3.3 Fixed location total and static pressures are measured individually and velocity pressure is obtained by difference. (Courtesy of Dwyer Instruments, Inc.)

this device can be inserted into the duct at a convenient location for direct measurement of velocity pressure (Fig. 3.4).

A number of standard Pitot-static tube designs have evolved worldwide. Three widely accepted designs based on tapered, hemispherical, and ellipsoidal nose designs all have calibration factors of essentially 1.0 for the velocity range of interest in industrial ventilation work. For the wide range of conditions encountered in industrial ventilation practice, the error attributable to a well-designed Pitot-static tube used in the prescribed manner is less than 1%. Since the Pitot-static tube is based on fundamental principles and has a calibration of 1.0, it is considered a primary standard and is used to calibrate other flow-measuring devices.

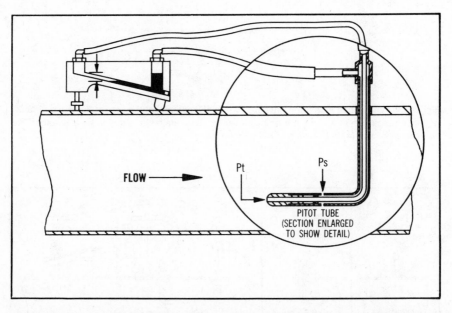

Figure 3.4   The total pressure probe of the Pitot-static tube consists of an inner tube that is positioned to point upstream and parallel to airflow. A second tube is sleeved over this tube to form an inner volume sealed on both ends and provided with a second manometer tap. This tube has a series of eight small holes drilled around its perimeter at a specified distance from the nose of the tube. The velocity pressure, or the difference between total and static pressures, can be measured with the Pitot-static tube by connecting the two Pitot-static taps to a manometer. A common problem with the Pitot-static tube in dusty industries is the plugging of the static and total pressure holes. This frequently occurs when the hole for the Pitot-static tube is drilled in the bottom of a duct conveying particles with inadequate transport velocity. The Pitot-static tube must be pushed through the settled dust and it frequently plugs. The holes in the tube are usually easily cleared by blowing through the tubing taps. If high dust loadings occur in the duct and frequent plugging occurs, the problem can be solved by using a special Pitot-static tube with larger holes. (Courtesy of Dwyer Instruments, Inc.)

3.1.1 Pressure Measurements

A critical element in the effective use of the Pitot-static tube in industrial ventilation work is the pressure-measuring device. There are many instruments available for measuring the low pressures encountered in industrial ventilation systems, each of which has its advantages and limitations. In most situations, the ventilation engineer or industrial hygienist will be measuring differential or relative pressure. Therefore, barometric instruments designed to determine absolute pressure will not be covered in this chapter; for information on these devices, the reader is referred to the ASHRAE Fundamentals (1985).

The simplest and most reliable pressure indicator is the vertical manometer, which consists of two connected columns of liquid, usually water, oil, or mercury (Fig. 3.5a). Since each column is connected to distinct atmospheres, the difference in heights of the two columns is a direct indicator of differential pressure between those atmospheres. The vertical manometer used with water can be read as low

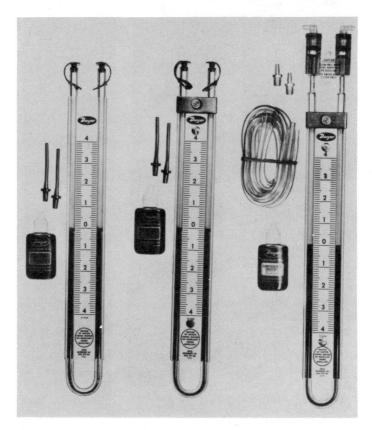

Figure 3.5 (*a*) Vertical manometers. If velocities are greater than 1270 fpm, the measured velocity pressure will be greater than 0.10 in. H_2O, a value that can be read on a vertical manometer using water as a fluid.

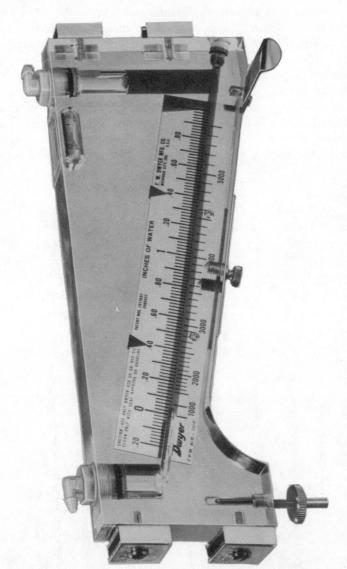

Figure 3.5 (*b*) Inclined manometer. At a velocity of 1000 fpm, the observed velocity pressure will be 0.07 in. H_2O, a pressure that cannot accurately be measured with a vertical manometer. In this case an inclined manometer, usually with a rise of 1 in. per 10 in. of run, will provide the necessary accuracy.

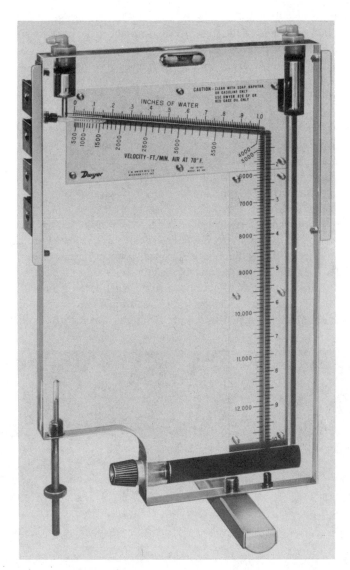

Figure 3.5 (c) Inclined/vertical manometer. Since static and velocity pressures will frequently exceed the range of the inclined manometer, a combination inclined/vertical manometer is most convenient for field work. Velocity pressures as low as 0.02 in. H_2O (566 fpm) may be read on the inclined scale; however, in this range accuracy is limited. Other total and static system pressures may be read up to 10 in. H_2O on the vertical leg. (Courtesy of Dwyer Instruments, Inc.)

as 0.10 in. H_2O, the velocity pressure equivalent of 1270 fpm. Frequently, duct velocities are below this velocity, so a more sensitive manometer must be used. The vertical manometer can be modified to improve readability at low pressures by tilting it at an angle so that a small pressure change will result in a longer deflection of the fluid meniscus. The inclined twin-leg manometer with a slope of

10:1 can be accurately read as low as 0.02 in. H_2O, corresponding to a velocity of 600 fpm. The advantage of the inclined manometer is its simplicity and the fact that it can be constructed in the laboratory from standard glass tubing and a wooden wedge. The difficulty encountered in its use is the problem in leveling the two legs and ensuring a 10:1 inclination plane.

The single-leg inclined manometer is a solution to the difficulties encountered with the double-leg manometer. In this unit, one leg is replaced by a fluid reservoir, while the reading leg is a takeoff from the reservoir (Fig. 3.5b). The reading leg may be positioned at a fixed inclination (i.e., 10:1) or it may be adjustable, affording a wide measurement range. In the United States, the most common manometer in industrial ventilation field work has a vertical and an inclined leg which can be used to measure both velocity pressures and system total and static pressures (Fig. 3.5c).

The discussion has referenced a water-filled manometer; in fact, the commercial manometers available in the United States use a special red gage oil with a specific gravity of 0.826. This oil will deflect approximately 20% farther than water under the same pressure, thus increasing the readability of the instrument. This fluid has an additional advantage in that a well-defined meniscus is formed and the fluid "clears" or seeks a stable reading quickly. For high-pressure measurements, for example, those made on high-velocity/low-volume systems

Figure 3.6 Hook gage micromanometer with electronic sensing that can be read to ± 0.00025 in. H_2O. (Courtesy of Dwyer Instruments, Inc.)

where total and static pressures are encountered in the range of 1 to 10 in. Hg, a vertical mercury manometer is indicated.

Any change in manometer fluid must be noted so that the subsequent pressure readings can be adjusted for density. In emergencies, water can be substituted in a manometer designed for another fluid. Unless circumstances dictate otherwise (i.e., excessive loss of fluid), the use of other fluids in an oil manometer is strongly discouraged. Under no circumstances should more than one liquid be used in a manometer.

For measurements as low as 0.001 in. H_2O (corresponding to a velocity of 130 fpm), a number of fluid-based micromanometers are available (Fig. 3.6). Although widely utilized in the laboratory, these devices are not suitable for field investigations. A description of a number of such devices is presented by Ower and Pankhurst (1977).

In addition to the liquid-filled manometers, both mechanical and electronic-based pressure sensors are now widely used in the United States. The most popular mechanical device is based on a diaphragm sensor which is displaced under differential pressure driving a linkage connected to a meter display (Fig. 3.7). Electronic micromanometers are now available with microprocessor-

Figure 3.7 Magnehelic devices are available for a range of pressures and are useful for measuring velocity pressures as low as 0.05 in. H_2O. These sensors are convenient for field use in that they do not require leveling, are easily zeroed, and do not use fluids. Since they are not primary standards they do require periodic calibration, but this is easily done with a water manometer. These devices are commonly used to monitor airflow continuously at individual hoods using the hood static suction measurement technique described in Section 3.5. (Courtesy of Dwyer Instruments, Inc.)

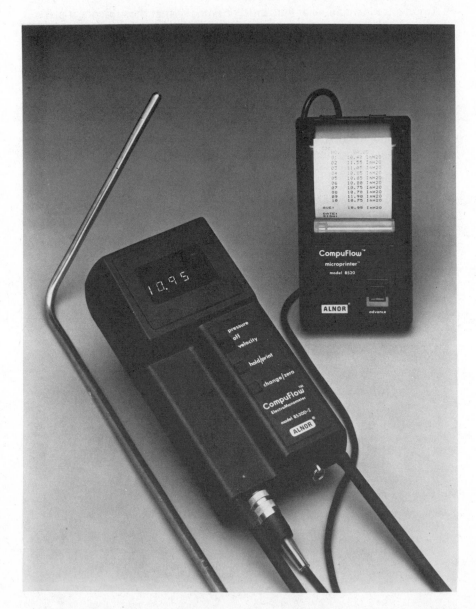

Figure 3.8 Microprocessor-based system for direct Pitot-static pressure and velocity measurements and averaging capability at the test location. A belt-mounted printer permits one person to conduct and record all measurements. (Courtesy of Alnor Instrument Company.)

based instruments for direct reading and averaging of pressure and velocity measurements in the field (Fig. 3.8). Devices of this type have definite advantages in the field, but frequent calibrations against a liquid manometer are important.

The most effective solution to the measurement of air velocities below 500 fpm may be the use of one of the heated devices to be described later in this chapter. A second solution to the measurement of low air velocities are special Pitot-static tubes which provide a reading that is some multiple of the velocity pressure. Since these devices are not primary standards, they must be calibrated under the conditions of use. As a result, these special designs are used in the laboratory but see little application in field measurements.

3.1.2 Velocity Profile in a Duct

The measurement of airflow in a branch, submain, or main duct is based on a careful measurement of both the cross-sectional area of the duct at the measurement location and the mean velocity of air moving through the duct at that point.

The velocity profile in a duct has been described in Chapter 2. As air moves through the duct, the velocity at the surface of the duct is zero. As one moves away from the duct wall, the velocity increases, first as a very thin laminar flow layer of air and then to a turbulent zone. The velocity gradient is at first very rapid and then it becomes asymptotic. The characteristic profile in a relatively smooth duct at constant temperature and pressure depends solely on Reynolds number. The relationship between the velocity at a given point and the centerline velocity (Ower and Pankhurst, 1977) is presented by the *power law*,

$$\frac{v_r}{v_c} = \left(1 - \frac{r}{a}\right)^m \tag{3.1}$$

where r = radius of point
a = internal radius of duct
v_r = velocity at radius, r
v_c = velocity at duct centerline
m = exponent that varies with the Reynolds number corresponding to the centerline velocity as shown in Table 3.1

The characteristic velocity profile represented by Eq. 3.1 is noted only in fully developed flow (Fig. 3.9).

It would be useful to determine the mean velocity from a single centerline velocity measurement. The relationship between mean and centerline velocity is useful in field investigations if an accuracy of 95% is sufficient (Fig. 3.10).

As noted in Fig. 3.9, the ideal fully developed velocity profile is observed only in a long duct run of undisturbed flow. This is an unusual situation in industry,

TABLE 3.1 Values of the Exponent *m* in Eq. 3.1

Reynolds Number at Duct Centerline	Exponent, *m*
1.27×10^4	0.168
2.49×10^4	0.156
6.10×10^4	0.141
1.20×10^5	0.130
2.37×10^5	0.120
5.84×10^5	0.109
1.16×10^6	0.101

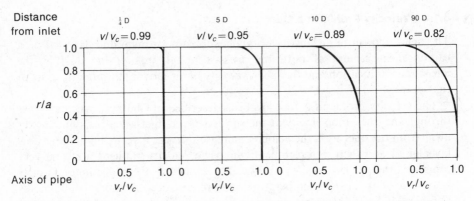

Figure 3.9 Velocity profile in a duct with progression to fully developed turbulent flow. V_r velocity at radius r, V_c velocity at axis; V, mean velocity; a, full internal radius. [From E. Ower and R. C. Pankhurst (1977), *The Measurement of Air Flow*. Used with permission of Pergamon Press.]

where long, straight runs are rarely encountered and system fittings such as elbows, entries, expansions, and contractions are numerous. To assist in choosing a proper location for the Pitot-static measurement the characteristic velocity profiles caused by common duct fittings are discussed.

The velocity profile downstream of an elbow is distorted due to the inertial "throw" of the air to the far wall as the air changes direction. The aspect ratio of the elbow and the duct velocity determine the observed profile. The characteristics of the velocity profile just downstream of the elbow are difficult to predict, as is the distance of straight pipe necessary to reestablish a reasonable profile. The progressive development of the velocity profile downstream of a representative elbow is shown in Fig. 3.11.

Branch entries also cause major disturbances of the velocity profile in the main. The effect of this disturbance for a representative branch with a 30° angle of entry and a branch air flow approximately 20% of the main flow is shown in Fig. 3.12. The distance downstream from the entry location to a suitable Pitot-static traverse location varies with the angle of entry, the ratio of the velocity in

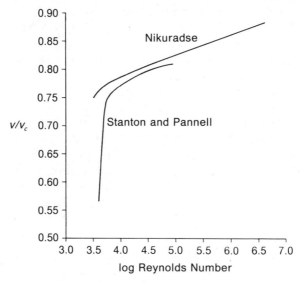

Figure 3.10   Relationship of average duct velocity to centerline velocity in smooth circular duct for various Reynolds numbers. *V*, mean velocity; V_c, velocity at axis. [From E. Ower and R. C. Pankhurst (1977), *The Measurement of Air Flow*. Used with permission of Pergamon Press].

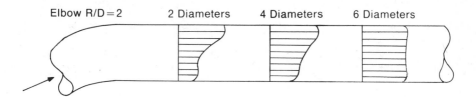

Figure 3.11   Velocity profile in a duct after an elbow. (Elbow $R/D=2$, $Q=1000$ cfm, 9-in. - diameter duct).

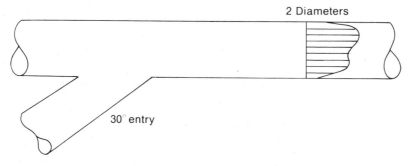

Figure 3.12   Effect of branch entry on the velocity profile in main duct. Main duct diameter, 9 in.; branch duct diameter, 4 in.; Q main, 1000 cfm; Q branch, 200 cfm.

the branch to the main duct, and the volumes flowing in the two ducts. In many field situations it takes a distance of 8 to 10 duct diameters before the profile is acceptable for a Pitot-static measurement. Frequently, the branch-to-branch inlet distance is much less than this. In such cases the flow through the individual branches are measured and added to the flow in the main duct to obtain the total system flow downstream of the branch entry.

Expansion and contraction fittings are less likely than other common duct fittings to complicate the choice of correct Pitot-static tube location. A low angle of expansion or contraction will present little velocity profile modification, and in many cases, a sharp contraction will improve the velocity profile.

No generalizations can be made about the velocity profiles at Pitot-static tube locations downstream of exhaust hoods since the velocity profile is defined by the geometry of the hood. The profile may be reasonably good a few diameters downstream of the hood if the hood has a symmetrical tapering design; however,

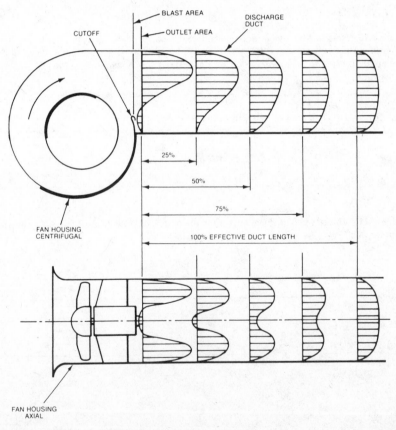

Figure 3.13 Velocity profiles at discharge from centrifugal blower (top) and from axial flow blower (bottom). (Reproduced through the courtesy of Air Movement and Control Association, Inc.)

the profile may be difficult to assess for a hood with a tangential delivery to the branch duct.

It is extremely difficult to obtain a reasonable assessment of the velocity profile immediately downstream of a fan. In systems served by a centrifugal blower the velocity gradient across the duct increases dramatically, due to the throw from the impeller tip (Fig. 3.13). If the fan discharges to a duct run and there is access at some distance downstream, a reasonable traverse is possible. If the fan discharge is vertical, access may be difficult. In such cases Pitot-static measurements are best made on the suction or negative-pressure side of the fan. The discharge from axial flow fans (especially roof exhausters) may be accessible for measurement, but the velocity profile across the discharge face is quite uneven (Fig. 3.13). With roof exhausters, the flow actually reverses near the centerline of the air mover. It is common practice to estimate the delivery of these low-static roof exhausters by noting the pressure at the inlet to the fan and then estimating airflow from the manufacturer's catalog as described in Chapter 10.

3.1.3 Pitot-Static Traverse

The discussion above should make it obvious that the mean duct velocity cannot be defined by taking one Pitot-static tube measurement; rather, multiple measurements must be made across the area of the duct. This so-called traversing technique is done in the United States by dividing round duct into annular rings of equal area (Fig. 3.14a) and determining the depths to which the Pitot-static tube must be immersed in this duct to position the nose of the tube at the center of each annular ring. Two traverses are made at right angles to define adequately the average velocity. A six-point traverse is used for ducts less than 6 in. in diameter; ducts between 6 and 40 in. in diameter are measured with a 10-point traverse and those greater than 40 in. with a 20-point traverse. The Pitot-static tube locations may be calculated from the relationship shown in Fig. 3.14a or directly from Tables 9.2 through 9.4 in the *Ventilation Manual*. The Pitot-static tube may be equipped with movable position clips; if not, the measurement points can be marked directly on the tube. The velocity pressure is noted from the manometer for each point of the traverse, each velocity is calculated from the velocity pressure, and the velocities are averaged to obtain the average duct velocity. A common error is the averaging of the velocity pressures.

The proper alignment of the Pitot-static tube with the duct axis can be checked by rotating the tube until the minimum reading is obtained. This is contrary to the widely accepted practice of turning the probe until the maximum velocity pressure is noted on the manometer (Fig. 3.15). Pitot-static tubes are available in 12- to 60-in. lengths, so a large range of ducts can be evaluated. If the duct diameter exceeds the length of the available Pitot-static tube, the entry can be made from both sides of the duct.

The British Standards Institute (BSI) traverse method is also based on a series of measurements across the duct diameter; however, the measurement locations differ from those used in the United States. The Pitot is placed not at the geometric center of the annular ring but at the location at which the predicted mean velocity occurs for that annular ring (Fig. 3.16a). This method is stated to ensure measurement errors of less than 1%. Other measurement methods have been proposed as a way of reducing the number of measurements; however, these methods have been validated only for fully developed flow with minimum velocity variation, and as a result they are not useful for general industrial ventilation practices.

Rectangular ducts present a different problem. One observes similar variations in velocity across the rectangular duct as noted with round duct, and

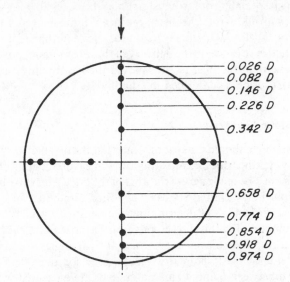

Pitot Traverse Calculation: Traverse on 10-in. Duct from Charge Port Hood on Reactor

Pitot Location	p_v (in. H₂O)	V (fpm)
1	0.92	3841
2	1.20	4387
3	1.50	4905
4	1.55	4986
5	1.45	4823
6	1.30	4566
7	1.30	4566
8	1.20	4387
9	1.15	4295
10	0.90	3799

Figure 3.14 (a) Technique for establishing measurement points for a Pitot-static traverse of a round duct commonly used in the United States and an example of calculations from such a traverse.

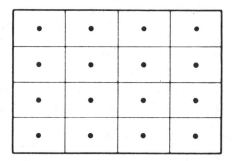

Figure 3.14 (**b**) Conventional procedure for establishing Pitot-static traverse points for rectangular duct in the United States.

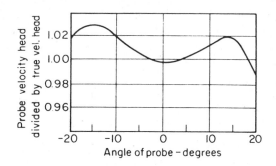

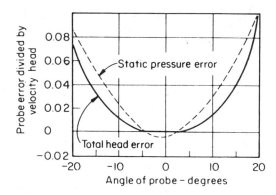

Figure 3.15 Error introduced in static pressure and velocity pressure readings due to misalignment of the Pitot probe in the airstream. Zero angle represents correct alignment of probe parallel to airstream.

multiple readings must be taken to get a reasonable measurement of mean velocity. The measurement technique used in the United States (Fig. 3.14*b*) requires that the duct be divided into equal areas and that measurements be taken at the center of these equal areas. The number of areas necessary to obtain a valid mean velocity depends on the type of upstream duct fittings, the distance between the last fitting and the measuring location, and the duct velocity. The distance between measuring points should not exceed 6 in. A different measurement grid for rectangular ducts is used in Great Britain to provide improved accuracy over the center-of-equal-area method used in the United States (Fig. 3.16*b*).

The conventional Pitot-static tube has a diameter of $\frac{5}{16}$ in.; since its cross-sectional area is small, it can be used to traverse ducts as small as 6 in. in diameter. For duct diameters below 6 in., a Pitot-static tube with a $\frac{1}{8}$-in. diameter is used to minimize disturbances. Since this small Pitot-static tube has the same design as the $\frac{5}{16}$-in. tube, it too is a primary standard and does not require calibration. The large Pitot-static tube requires a $\frac{3}{8}$-in. hole for insertion in the duct wall; the small tube requires a $\frac{3}{16}$-in. hole. Insulated duct requires a larger access hole depending on the thickness of the insulation.

The fate of the small holes drilled in the duct for insertion of the Pitot-static tube requires some discussion. If the measurements are made inside a building

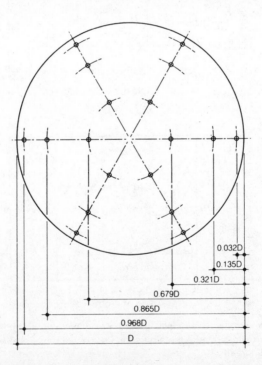

Figure 3.16 (*a*) British Standard Institute procedure (B.S. 1042: Section 2.1: 1983) for establishing Pitot traverse points for round duct.

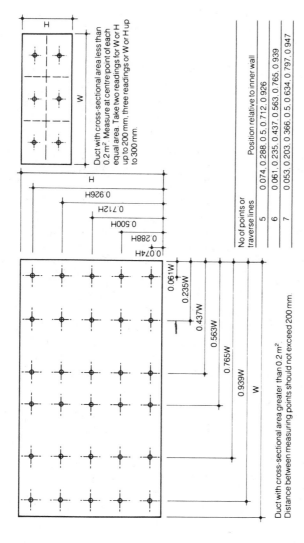

Duct with cross-sectional area less than 0.2 m². Measure at centre point of each equal area. Take two readings for W or H up to 200 mm, three readings or W or H up to 300 mm.

Duct with cross-sectional area greater than 0.2 m².
Distance between measuring points should not exceed 200 mm.

No of points or traverse lines	Position relative to inner wall
5	0.074, 0.288, 0.5, 0.712, 0.926
6	0.061, 0.235, 0.437, 0.563, 0.765, 0.939
7	0.053, 0.203, 0.366, 0.5, 0.634, 0.797, 0.947

Figure 3.16 (b) BSI technique for choosing traverse points for large and small rectangular ducts. (Courtesy of Airflow Developments Ltd.)

TABLE 3.2   Leak through 3/8 in. Pitot-Static Hole at Various Static Pressures

Static Pressure (in. H_2O)	Leak Rate (cfm)
1.0	0.39
2.0	0.57
4.0	0.84
6.0	1.01
7.8	1.23-

downstream of the fan, that is, on the positive-pressure side of the fan, it is obvious that the holes should be sealed to prevent air contaminants from escaping to the workplace. On the suction side of the fan, this is not necessary since only a small leak inward will occur (Table 3.2). However, plant personnel frequently insist that such holes be taped over or plugged.

3.1.4   Application

The major advantage of the Pitot-static tube is that it is a primary standard and does not require calibration; indeed, it is used to calibrate orifice and venturi meters and direct-reading instruments. The major disadvantage of the Pitot-static tube is that it cannot be used at velocities below 800 fpm unless special manometers are used.

3.2   MECHANICAL DEVICES

3.2.1   Rotating Vane Anemometers

The rotating vane anemometer (RVA) used widely in mining and to a lesser extent in general industry has proved useful in airflow measurements over the last five decades. Introduced in the 1920s, this device, shown in its simplest form in Fig. 3.17a, uses a lightweight rotating vane as the sensor. The windmill is supported by low-friction bearings which permit it to rotate at a speed proportional to air velocity. In the conventional RVA design shown in Fig. 3.17a, the impeller drives a mechanical transmission to provide a reading of the linear feet of air which have passed through the instrument in a given time. If the elapsed time of the observation is noted, one can easily calculate the velocity of the airstream by dividing the linear feet of air displayed on the meter face by the elapsed time. The vane design, type of bearings, and readout system define the operating characteristics of the device.

   In the last decade, a number of new RVA designs have been introduced which make this device more attractive to the practitioner. A digital readout and integral timer improve the ease of use in field measurement. Low-friction

Figure 3.17 (*a*) The conventional rotating vane instrument (Birams type) consists of an aluminum vane impeller with a mechanical gear train driving a series of three dials which read 100, 1000, and 10,000 ft. The instrument is equipped with a zero reset lever and a lever to activate the instrument. This unit has a 4-in.-diameter frame and has a useful range of 100 to 5000 fpm. It is available from several manufacturers in bronze and ball bearings (courtesy of Davis Instrument Manufacturing Company).

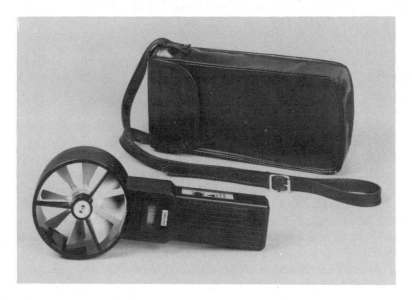

Figure 3.17 (*b*) Direct-reading rotating vane instrument with capacitance speed sensing and liquid crystal digital display, button reset, and a stop-and-start switch. Measuring range is 50 to 6000 fpm (courtesy of Airflow Developments Ltd.).

(c)

Figure 3.17 (c) Greater sensitivity is obtained with this rotating vane anemometer, which has a magnetic pickup and provides starting speeds as low as 20 to 30 fpm (courtesy of Airflow Developments Ltd.).

bearings and counters based on photoelectric or capacitance sensors have improved the low-velocity performance of the device (Fig. 3.17b and c).

These instruments are available in a range of sizes based on 1-, 3-, 4-, and 6-in.-diameter impellers. The size defines the operating range. The larger the diameter, the greater the available starting torque for a given velocity. Larger instruments are thus more sensitive and able to measure lower velocities, although they have a limited upper range. The small-diameter instruments are useful at high velocities. The 3- and 4-in. devices are by far the most common with a useful velocity range of approximately 100 to 5000 fpm.

The performance characteristics of several 4-in.-diameter rotating vane anemometers are shown in Fig. 3.18. Minimum starting speed is in the range of 50 fpm for a conventional 4-in.-diameter device. The response is not linear until approximately 100 to 125 fpm, when the calibration does becomes linear for the balance of the useful range of the instrument. The manufacturer normally provides a calibration for each instrument. The response of a newer device with ball bearings and nonmechanical counters exhibits improved instrument sensitivity at low velocities.

The instrument is commonly used to measure velocity at mine roadways, hood faces, exhaust grilles, and supply registers. To minimize errors, the cross-sectional area of the anemometer must be small compared to the area of the opening. In general, it is not useful for duct measurements. If used to measure the velocity at a duct opening, the diameter of the duct should be at least six times the diameter of the instrument to maintain errors less than 1%.

The RVA measures the velocity at the axis of the instrument. Therefore, if a 4-in. RVA is to be used to measure mean velocity by a traverse across the duct face, the minimum ratio of duct diameter to RVA diameter is 15 for a six-point traverse and 26 for a 10-point traverse (Ower and Pankhurst, 1977).

The instrument can be used in either a fixed-point survey or a hand traverse mode (Fig. 3.19). The timing of the RVA reading is conducted in two ways. The most accurate method is to position the instrument at the measurement point and allow it to come up to speed. A stopwatch is used to note the time for a given number of revolutions on the dial or for a counter interval on the newer digital-based instruments. The linear speed can then be easily calculated. A second technique, which although not as accurate, is probably suitable for most field measurements, involves positioning the instrument, permitting it to come to speed and then actuating the counter for a given period, usually 2 to 4 minutes. At the end of this interval, the counter is turned off, the linear feet passed through the RVA is noted, and the velocity is calculated. In any case, no reading should be taken for less than 100 seconds or an instrument reading of less than 160 ft, and two readings should be taken at each location (Ower and Pankhurst, 1977).

Teale (1958) has identified the following sources of error in using the rotating vane anemometer.

- The conditions under which the field tests were conducted do not represent those under which the device was calibrated.
- Errors in averaging in the fixed-point survey may be due to errors in the measurement of the small areas. In the traverse method, errors may result from failure to cover equal distances in the same time interval.
- Measuring the cross-sectional area being evaluated is subject to error.
- The presence of the observer in the test field may influence the readings.
- Variations in velocity across the measurement plane during the period of measurement can affect the results.

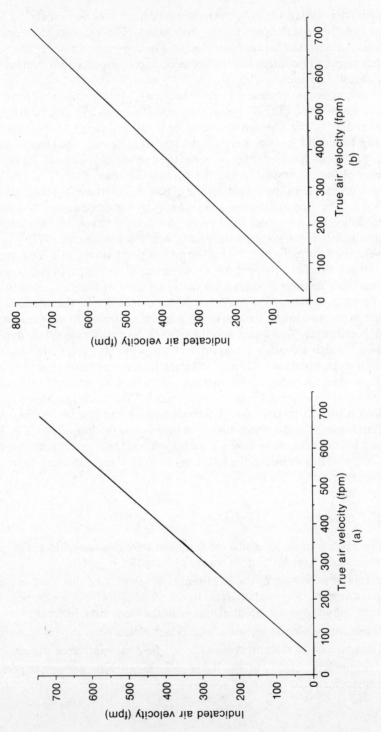

Figure 3.18 (*a*) and (*b*)

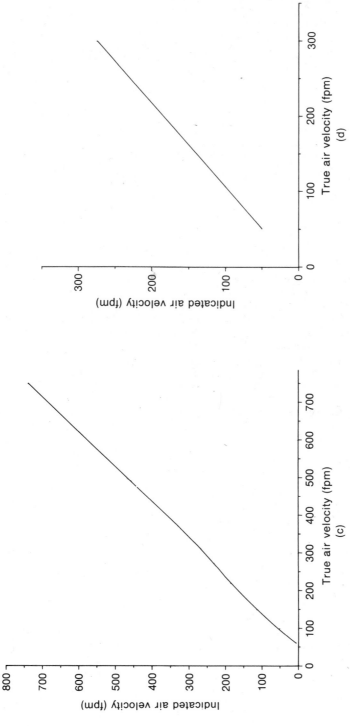

Figure 3.18 Indicated versus true velocity for four conventional rotating vane anemometers with a variety of bearings and readout systems. To identify the useful lower range of the instrument, starting and stopping speeds can be defined. Starting speed is defined by placing the instrument in a calibration tunnel and slowly increasing the air-velocity until the vane starts to rotate. To define stopping speed, the tunnel speed is slowly reduced until the vane ceases to turn. The starting and stopping speeds are indicated for all models. (*a*) RVA 1: Gear train readout; bronze bearing; starting speed 63 fpm, stopping speed 54 fpm. (*b*) RVA 2: Gear train readout; ball bearings; starting speed 23 fpm, stopping speed 18 fpm. (*c*) RVA 3: Gear train readout; jewel bearings; starting speed 61 fpm, stopping speed 67 fpm. (*d*) RVA 4: Magnetic pickup; ball bearings; starting speed 60 fpm, stopping speed 40 fpm. [From Partell (1979).]

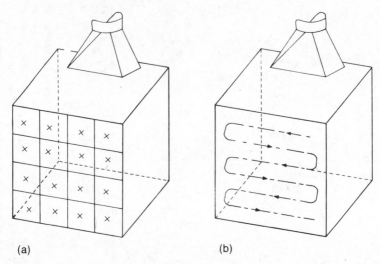

(a)   (b)

Figure 3.19   Two techniques are used to conduct velocity measurements with the rotating vane anemometer. (a) The cross-sectional area to be measured can be divided into equal areas and the measurement taken at the center of each area. The velocities are then averaged to obtain the average velocity at the plane of the opening. This fixed-point technique requires an assessment of the area geometry and assignment of measurement locations. It also requires a series of individual timed measurements; for a large opening this may be rather time consuming. (b) The second technique is to conduct a running traverse at the opening for a given period of time. This procedure requires some judgment in pacing the traverse, and it may be necessary to make more than one traverse to achieve a traverse speed that will permit equal velocity weighing over the entire area and completion at an assigned time. The traverse procedure is, in general, more convenient and faster than the fixed-point technique. Comparison of the two techniques have shown the traverse technique to be the more accurate while the fixed-point method is more precise.

Detailed investigations of the systematic errors in using conventional rotating vane anemometers with mechanical readout have been performed by Ower and Pankhurst (1977), Swirles and Hinsley (1954), and Teale (1958). Since the vane anemometer is directional, it must be aligned with the plane of the opening being measured or an error will occur (Fig. 3.20). The magnitude of this error depends on the angle of incidence or *yaw*. It is evident that one must exceed an angle of 10° before a 1% error in indicated velocity is noted; a 30° yaw will result in a 10% error.

A second systemic error may occur if the traversing velocity is large compared to the true stream velocity being measured. Again, although the Teale study is restricted to early designs of rotating vane anemometers, it is obvious from Fig. 3.21 that if one maintains a traversing velocity less than 20% of the true stream velocity, the measurement error will be less than 5%. The 4-in. RVA has limited application below 150 fpm, so that a traverse velocity of 30 fpm or less ensures minimal error due to traverse speed.

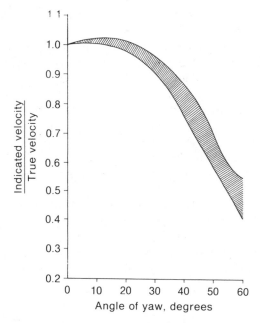

Figure 3.20 Effect of misalignment of rotating vane anemometer on accuracy. If the rotating vane is not aligned normal to the direction of airflow, an error may be noted. Ower has shown this error to be insignificant up to an angle of 20°. The effect of four early RVA designs is depicted here. Data are not available on the new instruments shown in Fig. 3.17; however, it is likely that similar results will be observed. [Adapted from Teale (1958).]

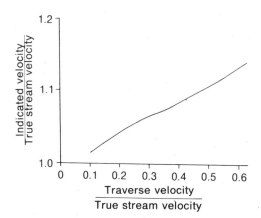

Figure 3.21 Movement of the rotating vane device while traversing a hood opening to determine the average face velocity results in an error. For the conventional Biram-type instrument the error is approximately 10% if the traversing speed is one-half the true air velocity. [Adapted from Teale (1958).]

Occasionally, an airflow stream is encountered that pulsates; that is, the air velocity varies with time. Since the RVA has a poor time response due to the inertia of the vane sensor, the correct mean velocity under pulsating flow may not be obtained. Under conditions of sinusoidal flow the instrument over-estimates the true velocity. The error is shown in Fig. 3.22 for two conventional instruments; if the amplitude of pulsation is $\pm 20\%$ of the true mean velocity, the instrument will overestimate the true velocity by less than 1%.

The modified airflow pattern created by the presence of the observer may be significant with the RVA. This effect can be minimized by mounting the instrument on a rod using the threaded mounting surface on the instrument. The required separation distance has not been thoroughly investigated; however, a minimum distance of at least 2 ft will usually minimize this interference.

If used properly, the rotating vane anemometer will permit direct measure-ment of air velocity in field investigations with an accuracy of $\pm 5\%$ of the true velocity. The RVA does have a bearing suspension that is subject to damage; the aluminum vanes may also be corroded by air contaminants encountered in the field. For these reasons the device should be calibrated periodically. A major

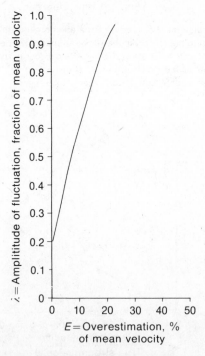

Figure 3.22 Teale has shown that a vane anemometer positioned in a fluctuating stream will give readings higher than the mean speed. The performance of the Biram-type instrument shown in this figure indicates that the amplitude of velocity fluctuations must exceed 50% before errors of 8 to 10% are encountered. This range of fluctuation is rarely encountered in industrial ventilation systems. [Adapted from Teale (1958).]

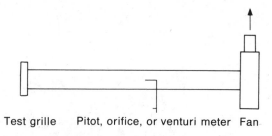

Test grille   Pitot, orifice, or venturi meter   Fan

Figure 3.23 If a rotating vane anemometer is used to measure velocity at an HVAC supply or exhaust grille, a correction factor must be applied when calculating airflow. The correction factors available from the manufacturer apply to pressure and suction openings with specific geometry and net area. Frequently, grilles or registers are encountered which do not meet these specifications. If a number of locations must be measured, the correction factor should be evaluated in the laboratory. A Pitot tube traverse can be used to measure a set airflow or an orifice or venturi meter can be utilized. The product of the velocity at the face of the grille or register measured with the RVA and the gross area is the indicated airflow. The ratio of the actual airflow measured by Pitot-static tube, or flowmeter, and the indicated airflow from the RVA is equal to the correction factor applicable to this specific grille.

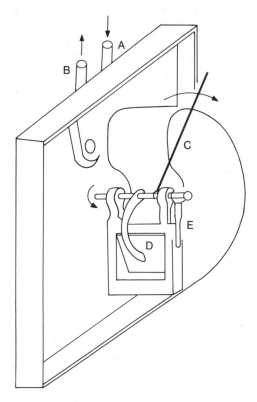

Figure 3.24 Construction of the swinging vane anemometer showing the sensing element. The two probe connections *A* and *B* penetrate the instrument case. Probe *A* is connected to the sensor tunnel casing *E* and probe *B* communicates to the interior of the case. Air moving through probe A strikes vane *D*, causing a rotation of the counterweighted needle *C* which displays the air velocity on the meter face. The air passing through the sensor tunnel exits through probe *B*.

67

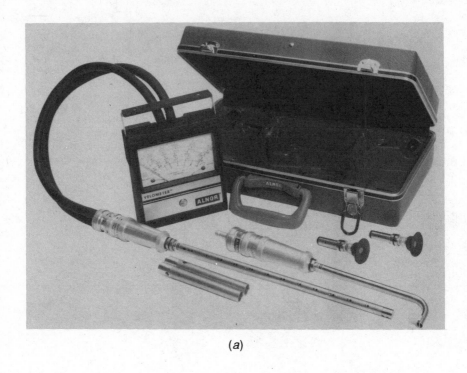

(a)

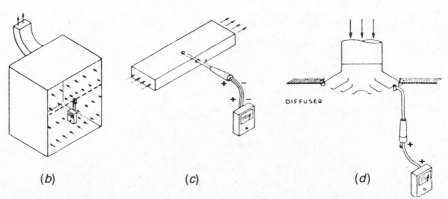

(b)   (c)   (d)

Figure 3.25   (a) The basic swinging vane instrument (Alnor Model 6000). Applications include (b) a low-velocity probe for air-velocity measurements in plane of hood opening; or (c) a Pitot probe with a diameter of 1/2 in. Since it does not have the 90° bend of the pitot, it requires only a slightly oversized hole. The advantage of this probe is that the velocity may be read directly from the meter. The device is not a primary standard and requires calibration by a standard Pitot-static tube. Also shown is (d) a probe for measuring delivery from diffusers. By measuring the diffusers throat velocity the delivery can be calculated using the manufacturer's k factor. (Courtesy of Alnor Instrument Company.)

fault of the RVA is its poor response at low velocities. If the device is used for measurement of supply and exhaust openings, corrections must be applied for the effect of the grille. These corrections should be validated for the particular grille or register in a laboratory test as shown in Fig. 3.23.

3.2.2 Deflecting Vane Anemometers (Velometer®)

The sensor for the deflecting or swinging vane anemometer (Velometer) consists of a lightweight vane positioned in a rectangular duct and mounted on a balanced taut-band suspension (Fig. 3.24). Air from the probe passes through the instrument, striking the vane and causing an angular deflection that is proportional to air velocity. The cross-sectional area of the tunnel in which the vane moves expands, providing increased clearance at high velocities.

The original instrument design had a series of probes or fittings which were attached to the inlet side of the instrument and restricted the airflow by varying degrees. Each fitting had a separate calibration and the velocity scale for each probe was identified on the dial face. This rather confusing arrangement was changed in the 1970s to a new design with a basic series of three probes, all of which are read on the same scale (Fig. 3.25a). With the new design it is not necessary to calibrate each probe for a given instrument.

The principal application of this device is to measure air velocity in the plane of large hood openings using the low-velocity probe (Fig. 3.25b). In this configuration the meter has a range of 30 to 300 fpm. The manufacturer claims an accuracy of $\pm 5\%$ of full scale. With the low-velocity probe, the cross-sectional area of this instrument is 0.12 ft^2. It cannot be used for small openings due to the change in effective hood area which occurs and the possible change in calibration due to the airflow geometry.

Other special probes are available for this instrument. A double-opening Pitot-static tube provides the same pressure output for a given velocity as does the standard pitot tube (Fig. 3.25c). In effect, the manufacturer has forced the instrument to respond to the basic flow equation of the Pitot-static tube. A probe is also available for evaluating flow from HVAC air supply diffusers as shown in Fig. 3.25d. The diffuser throat velocity is measured with this probe, and the noted velocity is multiplied by the diffuser manufacturer's calibration factor to obtain air flow. The proper placement of the probe in the throat is critical for accurate measurements. For extensive balancing of airflow through supply diffusers, the use of the swinging vane anemometer with the canopy device shown in Fig. 3.26 appears to be a more accurate technique. Two simple adaptations of the vertical deflecting vane anemometer concept are shown in Fig. 3.27 and a simple horizontal vane deflecting device is shown in Fig. 3.28.

Few data are available on the accuracy and precision of the swinging vane anemometer. The response of the standard instrument has been investigated by Purtell (1979), and these studies indicate that the accuracy below 100 fpm is tenuous at best (Fig. 3.29). When used with the low-velocity probe, precise alignment with direction of airflow is not necessary; the deviation from the true

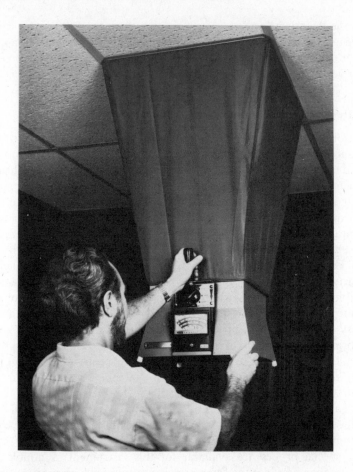

Figure 3.26 Canopy device which can be positioned over the HVAC diffuser and directs the supply air past an Alnor instrument. The device is calibrated directly in cfm. This procedure is more accurate than the probe technique, which measures diffuser throat velocity. (Courtesy of Shortridge Instruments, Inc.)

velocity is minimal up to 20°. The instrument has a visually fast response time for field application; indeed, one must average readings in pulsating flow fields. The effect of flow pulsation on accuracy of measurement has not been explored. Special correction factors must be utilized for calculation of flow through grilles and registers.

3.2.3 Bridled Vane Anemometer

The sensor of this air-velocity device (Fig. 3.30) is a constrained windmill-type impeller whose angle of rotation is proportional to air velocity. The range of the

(a)

(b)

Figure 3.27 Two simple and inexpensive adaptations of the swinging vane anemometer. Both devices are designed to monitor face velocity at simple hood openings. (a) A scaled-down version of the standard Alnor unit. The instrument in (b) utilizes a simple, replaceable vane sensor equipped with a pin that fits in a pivot. Its modest cost and ease of use make it useful for laboratories and production personnel or for continuous monitoring of critical hood applications. (Courtesy of Alnor Instrument Company and Dwyer Instruments, Inc.)

instrument (50 to 1000 fpm) is determined by the spring used to establish the counter-rotational torque on the impeller. The device is equipped with a brake mechanism that fixes the impeller at its maximum angular position to provide an instantaneous reading. The indicated velocity is read directly as the position of a pointer on a scale on the perimeter of the instrument housing. The performance

Figure 3.28 Horizontal deflecting vane anemometer. (Courtesy of Bacharach, Inc.)

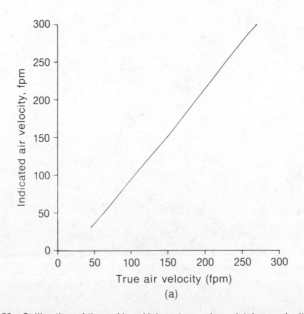

(a)

Figure 3.29 Calibration of three Alnor Velometer probes: (*a*) Low-velocity probe.

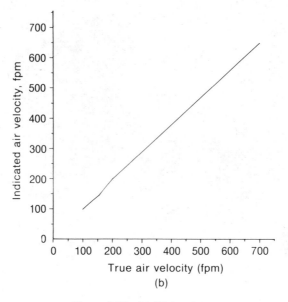

Figure 3.29 (*b*) Pitot probe.

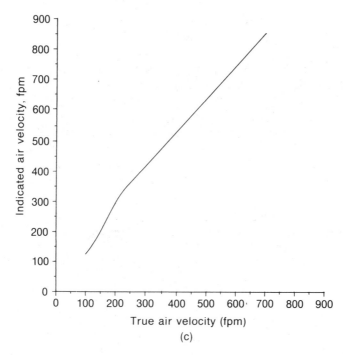

Figure 3.29 (*c*) Diffuser probe. As defined by NBS, the minimum operating speed is the air velocity at which the device under test indicates a speed of one-half the lowest marked division on the meter face above zero. [From Purtell (1976).]

Figure 3.30 The bridled vane anemometer is a convenient device for remote measurement of velocity at grille and register openings using a reach rod. (Courtesy of Bacharach, Inc.)

of this instrument has not been studied; however, it is a convenient device for approximate measurements of HVAC grilles and registers.

3.3 HEATED-ELEMENT ANEMOMETERS

The large number of heated-element air-velocity sensors available commercially in the 1980s is based on the original hot-wire anemometry work by King (1914). These devices operate by virtue of the fact that a heated element placed in a flowing airstream will be cooled and the rate of cooling is proportional to the velocity of air movement. The introduction of a variety of heated sensors, including wires, thermistors, thermocouples, and films, coupled with advances in methods of evaluating the cooling rate, have added an important dimension to airflow measurement. These heated-element anemometers have attractive performance at low velocities, a range that is difficult to measure with the mechanical devices discussed in Section 3.2.

The basic design and operating features of the family of heated devices can be described by reviewing the hot-wire device. King showed that the heat loss from electrically heated wires in a moving fluid stream can be described by the following equation:

$$H = kT + (2\pi k S \rho dv)^{0.5} T \tag{3.2}$$

where H = rate of heat loss per unit length of wire
d = diameter of wire
T = temperature of wire above ambient air temperature
k = thermal conductivity of fluid

S = specific heat at constant volume of fluid
ρ = density of fluid
v = velocity of fluid

For a given wire operating at the same elevated temperature in the same fluid, this equation becomes

$$H = A + Bv^{0.45} \qquad (3.3)$$

where A and B are constants for a given instrument. For a more detailed discussion of this phenomenon, the reader is referred to Ower and Pankhurst (1977).

The heated-wire sensor is normally a fine platinum or nichrome wire less than 0.01 in. in diameter with an operating temperature of 250 to 700°F, depending on the velocity to be measured. In the original instrument the cooling of the element in the velocity field was measured as a change in resistance of the heated element. In later designs the heated element is maintained at a constant temperature. As heat is lost from the element, additional current is required to maintain it at a constant temperature. The current is monitored as a measure of air velocity, with the normal response shown in Fig. 3.31. To permit the application of these devices over a wide temperature range, a reference sensor shielded from air movement is mounted in the probe.

The simple heated-wire sensor described above provides a powerful tool for velocity studies and is still in wide use today. However, the heated-wire element is fragile and the calibration of the instrument changes due to the aging of the wire. A more suitable instrument for industrial field studies is the shielded-element device. In this device the wire or other heated element is placed inside a small, thin-walled tubing, or a fine wire coil is coated with a rigid plastic coating. In both cases the sensor is rugged and quite acceptable for field applications. Due to

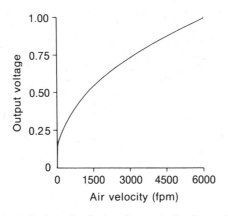

Figure 3.31 Output of a hot-wire device, demonstrating its nonlinear response.

its mass and shielding, these devices have poor response time compared to the simple heated-wire device. Although they are not suitable for resolving rapid changes in velocity required in studies of turbulence, they are quite suitable for industrial ventilation work.

In addition to heated wires, other sensors have been applied to industrial anemometry (Fig. 3.32). One popular approach is a sensor based on the deposition of a metallic coating, often platinum, on an insulating form. The performance of this sensor follows the same basic law described by King and it has a response time similar to the shielded heated wire. Other instruments utilize thermistors or thermocouples as the sensing agent in a constant-current operating mode. The thermistor sensor heated to 200 to 400°F is especially attractive since it has a much greater change in resistance per unit change in temperature than other sensors. These instruments also utilize a reference junction so that reasonable excursions in air temperature will not introduce an error (+20 to 150°F).

The heated airflow sensors have excellent response at low velocities. With the exception of the heated wire the probes are fairly rugged, although all may be damaged by exposure to corrosive chemicals. The probe output is not linear and field instruments thus have good sensitivity at low velocities and rather poor sensitivity at high velocities (Fig. 3.33). Linearizing electronics may be obtained, but this is not usually necessary in industrial ventilation field work. Short battery life is the chronic difficulty with all battery-powered systems, including heated anemometers. The probes may or may not be directional, depending on their

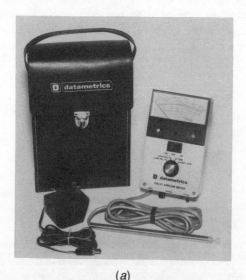

(*a*)

Figure 3.32 Various heated-element anemometers: (*a*) hot wire (courtesy of Datametrics).

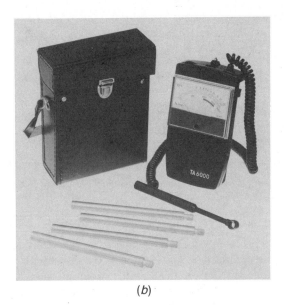

(b)

Figure 3.32 (b) Heated thermistor (courtesy of Airflow Developments Ltd.).

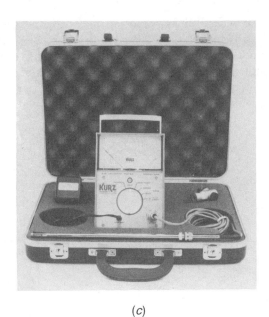

(c)

Figure 3.32 (c) Heated film (courtesy of Kurz Instruments, Inc.).

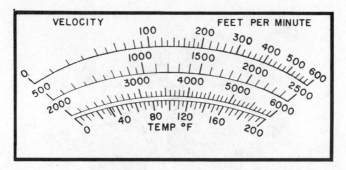

Figure 3.33 Nonlinearity of a heated element anemometer as shown by the meter display of a standard instrument.

geometry. As with all airflow-measuring devices other than the Pitot-static tube, these devices also require periodic maintenance and calibration.

Given the limitations above, the heated sensor instruments are extremely valuable instruments and are useful in studies of local exhaust ventilation, comfort, heat stress, and HVAC systems.

3.4 OTHER DEVICES

3.4.1 Vortex Shedding Anemometers

If an obstruction is placed in an airflow path, a discrete vortex forms just downstream of the obstruction (Fig. 3.34). The pressure pulse that is formed by the new vortex at one side of the obstruction prevents formation of a vortex on the other side until the initial vortex is downstream. A series of vortices is generated and moves down the so-called "vortex street." The frequency of vortex generation is a function of air velocity and can be used as the basis of an air-velocity meter.

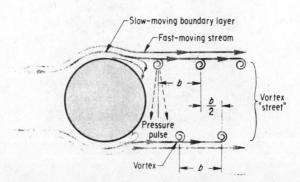

Figure 3.34 Basis of vortex-shedding air-velocity meter.

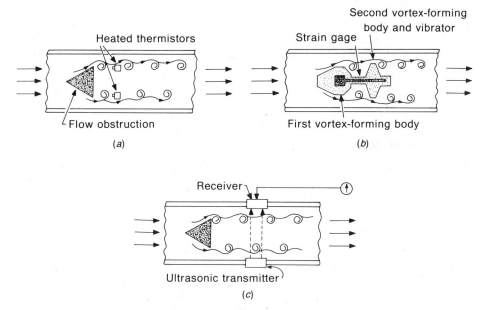

Figure 3.35 Various vortex sensing systems used in anemometers: (*a*) thermistor array senses individual vortex; (*b*) a strain gage on second obstruction notes vortex-induced displacement; (*c*) ultrasonic sensing at the vortex.

In one such device (Fig. 3.35*a*) the vortex shedding rate is sensed by a thermistor array downstream of the obstruction. Each vortex causes rapid cooling of the heated thermistor, which results in an instantaneous change in resistance with a pulsed signal whose frequency is directly proportional to air velocity. In another design (Fig. 3.35*b*) the vortex shedding phenomena is stabilized by placing another obstruction directly after the first. The vibrations of the second obstruction caused by each individual vortex are sensed by a strain gage in the link between the two obstructions. Again, the vortex frequency is directly proportional to air velocity.

A third technique uses an ultrasonic sensing system as shown in Fig. 3.35*c*. This instrument couples excellent sensitivity with a linear response. Two models are available for velocity ranges of 70 to 3600 fpm and 150 to 7000 fpm; the instrument provides both an instantaneous digital readout and a time-weighted average. This instrument and its calibration are shown in Fig. 3.36.

3.4.2 Orifice Meters

The orifice meter (Fig. 3.37) is designed as a sharp-plate restriction in the duct and provides a means of accelerating flow and creating a static pressure drop between the upstream static tap and the downstream tap at the vena contracta. The flow through the orifice is proportional to the square root of the orifice pressure differential. The orifice plate can be sized to provide a given pressure drop at the anticipated flow. Unfortunately, the pressure loss across an orifice meter may be 40 to 90% of the orifice pressure differential. Particles and corrosive chemicals transported in many industrial exhaust systems may erode the plate necessitating frequent orifice replacement and calibration. For these reasons the orifice meter is rarely used for flow measurements in general industry exhaust systems.

3.4.3 Venturi Meters

In a venturi meter, the entrance contraction and the exit expansion are defined by the geometry shown in Fig. 3.38. Due to the gradual contraction and

(a)

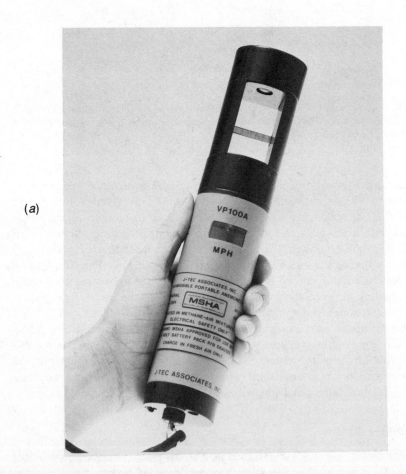

Figure 3.36 Vortex shedding instrument (*a*) based on an ultrasonic sensing system and its calibration (*b*). (Courtesy of Davis Instrument Manufacturing Company.)

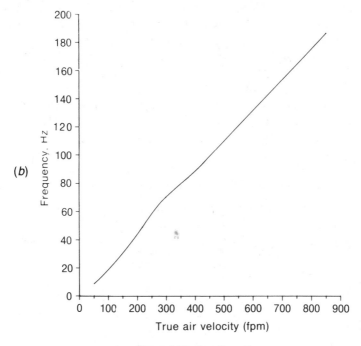

(b)

Figure 3.36 (*continued.*)

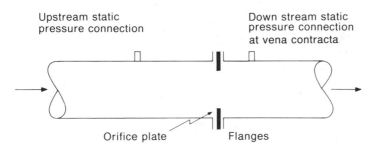

Figure 3.37 Orifice meter for measurement of airflow quantity.

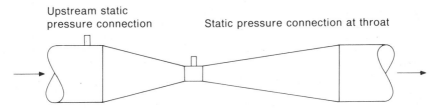

Figure 3.38 Venturi meter for measurement of airflow quantity.

subsequent reenlargement, the conversion from static pressure to velocity pressure and then back to static pressure is accomplished more efficiently than in the orifice meter and the pressure loss is reduced significantly. The venturi meter, although more efficient than the orifice meter, still has a loss of 10 to 20% of the venturi differential pressure and is rarely used in local exhaust ventilation systems, due to its initial cost, maintenance, and operating cost. If either the orifice or venturi meters are considered for field application, they should be constructed following the design of the American Society of Mechanical Engineers (1959).

3.5 HOOD STATIC PRESSURE METHOD

As described in Chapter 2, one of the major energy losses in a local exhaust ventilation system is the entry loss as the air moves into the hood and duct. The hood static pressure $p_{s,h}$ measured just downstream of the hood (Fig. 3.39) can be expressed as

$$p_{s,h} = h_e + p_v \tag{2.41}$$

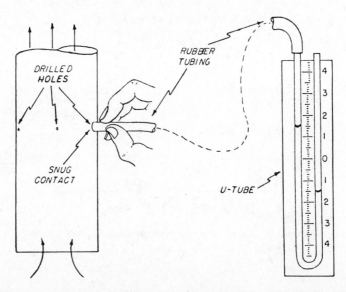

Figure 3.39 Hood static pressure method. In using this technique in the field, care must be given to the correct placement of the static tap. A deburring tool should be used to prepare the tap for measurement. The use of screens at the hood face to prevent debris from entering the hood confuse the measurement. The screen presents an element whose resistance varies as the screen plugs. Since this resistance is not a part of the entry loss factor, it is not included in the coefficient of entry. If the screen cannot be cleaned, it should be removed. Frequently, there is an elbow or other fitting directly after the hood. This fitting may be a part of the normal hood geometry, and if so, it is included in the published coefficient of entry. If not, a correction must be made in the flow rate calculations to account for this added resistance to flow.

where p_v is the velocity pressure in the duct and h_e represents the losses attributable to turbulence in the hood and transition (hood entry loss).

If the hood geometry were ideal, turbulent losses would not occur and the hood static pressure would be totally converted to velocity pressure. However, for even the best aerodynamic form possible, there are still turbulent losses. The aerodynamic quality of a hood is identified by the coefficient of entry C_e defined as the ratio of the actual airflow at a given static pressure at the hood to the theoretical airflow that would occur if the total static pressure were available as velocity pressure,

$$C_e = \frac{Q_{\text{actual}}}{Q_{\text{ideal}}} \tag{2.44}$$

$$= \frac{4000\,A\,\sqrt{p_v}}{4000\,A\,\sqrt{p_{s,\,h}}} \tag{2.45}$$

$$= \sqrt{\frac{p_v}{p_{s,\,h}}} \tag{2.46}$$

and rearranging Eq. 2.44 and 2.45 yields

$$Q_{\text{actual}} = 4000\,A_h\sqrt{p_{s,\,h}}/C_l \tag{3.4}$$

Values of C_e have been developed for many hoods, as shown in the *Ventilation Manual*. It follows that the airflow through a standard hood with an assigned coefficient of entry can be calculated by the simple measurement of the duct area and the static pressure in the duct downstream of the hood. This technique is reasonably accurate ($\pm 10\%$) for standard hoods and can be accomplished in a fraction of the time of other techniques. It is especially useful for measuring airflow through a group of identical hoods, such as a line of grinding or buffing wheels.

The principal difficulty with this system is the accuracy of the C_e values available to the field engineer. The published values for simple hoods such as a sharp entry or a flanged opening are quite accurate and can be used without reservation. For more complex hoods, such as grinding wheel hoods, small differences in construction may modify published entry loss values. In such cases, if the application warrants it, the hood may be tested in the laboratory or the field to validate the C_e value.

3.6 CALIBRATION OF INSTRUMENTS

With the exception of the Pitot-static tube, all air-velocity measuring devices discussed in this chapter require periodic calibration. Instruments should be checked when received from the manufacturer, calibrated at regular intervals, and recalibrated when abused in the field.

The required frequency of calibration depends on both the history of instrument use and the criticality of the application. Shipment of an instrument from a central laboratory to various plants may subject the instrument to serious shock in handling. Instruments shipped as luggage are also subjected to vibration and shock. Industrial applications in hostile chemical environments may damage the basic suspension of a mechanical device or the probe and associated electronics of heated devices. Dusty environments such as foundries or mineral crushing plants are especially harmful to mechanical devices. If the instruments are used by several persons, the variation in the knowledge and experience of the practitioner must be considered in establishing a time schedule for calibration.

Given this range of applications, how often should such devices be calibrated? If the device is used infrequently (e.g., once a month) and by a single trained person, annual calibration is probably adequate. If the device is used frequently (e.g., once a week), if the environment is harsh, if the instrument is used by several persons, or if the device is used for critical applications, monthly calibration may be required.

It is difficult for the occasional user of an instrument working in an isolated industrial setting to calibrate anemometers. The devices may be returned to the manufacturer or a consulting laboratory, but this takes time and the instrument may be needed at the plant site. A backup instrument may be the answer to this dilemma.

If the instrument is routinely used for critical applications, the plant may wish to set up its own calibration facility. A number of systems have been developed for evaluating all types of instruments over a range of air velocities. In general, these systems use either a variable-speed or fixed-speed fan with a bypass to move air through a large-area test tunnel and then a small-area duct to an orifice or a venturi calibrated with a Pitot tube. The tunnel and duct sections are chosen to provide a range of velocities and will handle both a large instrument that must be placed directly in the stream and a small sensor such as a heated element that is used as a duct probe. Such a calibration system can also be used for other purposes, such as determining the calibration factor for diffusers and determining the coefficient of entry for hoods. Design information for such systems is included in the *Ventilation Manual.* Manufacturers of heated devices also provide bench top flow calibrators which are convenient for field use.

Every effort should be made to measure air velocity accurately. Manufacturer claims of accuracy of $\pm 2\%$ of full scale are probably unrealistic for field instruments. Since design calculations are usually not better than $\pm 5\%$, an accuracy of $\pm 5\%$ is sufficient for a field instrument.

LIST OF SYMBOLS

a	duct radius
C_e	coefficient of entry

D	duct diameter
d	wire diameter
H	heat loss
h_e	hood entry loss
$p_{s,h}$	hood static pressure
p_v	velocity pressure
Q	airflow
r	radius at measurement point
S	specific heat of fluid
T	temperature above ambient
V	average velocity
v_r	velocity at radius, r
v_c	velocity at duct centerline
ρ	density (mass/volume)

REFERENCES

Airflow Developments Ltd. (1983), "Airflow Measurements in Buildings," Airflow Developments Ltd., High Wycombe, Buckinghire, England.

American Society of Mechanical Engineers (1959), *Fluid Meters, Their Theory and Application*, ASME, New York.

British Standards Institute (1983), "Methods for the Measurement of Fluid Flow in Closed Conduits—Methods Using Pitot-Static Tubes," B. S. 1042, Section 2.1, BSI, London. (Identical to ISO 3966–1977.)

Hitchin, E. R., and C. B. Wilson (1967), "A Review of Experimental Techniques for the Investigation of Natural Ventilation in Buildings," *Build. Sci.*, **2**: 59–82.

Jorgensen, R., ed. (1983), *Fan Engineering*, Buffalo Forge Company, Buffalo, NY.

King, L. V. (1914), "On The Convection of Heat From Small Cylinders in a Stream of Fluid." *Philos. Trans. R. Soc.* London, **A214**:373.

Purtell, L. P. (1979), "Low Velocity Performance of Anemometers," *National Bureau of Standards*, Report No. NBSIR 79–1759.

Stanton, T. E., and J. R. Pannell (1914), "Similarity of Motion in Relation to the Surface Friction of Fluids," *Philos. Trans. R. Soc.* London, **A214**:199.

Swirles, J., and F. B. Hinsley, (1954), "The Use of Vane Anemometers in the Measurement of Air Flow," *Trans. Inst. Min. Eng. London*, **113**: 895.

Teale, R. (1958), "The Accuracy of Vane Anemometers," *Colliery Eng.*, 239–246, June.

ADDITIONAL READINGS

American Society of Heating, Refrigerating and Air Conditioning Engineers (1985), *ASHRAE Handbook—1985 Fundamentals*, ASHRAE, Atlanta, GA.

Ower, E., and R. C. Pankhurst (1977), *The Measurement of Air Flow*, Pergamon Press, Oxford.

MANUFACTURERS OF AIRFLOW MEASURING INSTRUMENTS

Airflow Developments (Canada) Ltd., 1281 Matheson Boulevard, Mississauga, Ontario, Canada L4W1R1 and Airflow Developments Ltd., Lancaster Road, High Wycombe, Buckingshire HP 12 3QP, England.

Alnor Instrument Company, 7555 North Linder Avenue, Skokie, IL 60077.

Bacharach, Inc., 625 Alpha Drive, Pittsburgh, PA 15238.

Datametrics, 340 Fordham Road, Wilmington, MA 01887.

Davis Instrument Manufacturing Company, 519 East 36th Street, Baltimore, MD 21218.

Dwyer Instruments, Inc., P. O. Box 373, Michigan City, IN 46360.

Kurz Instruments, Inc., 2411 Garden Road, Monterey, CA 93940.

Neotronics N. A., Inc., P. O. Box 370, 411 Bradford Street, NW, Gainesville, GA 30503.

Shortridge Instruments, Inc., 14609 North Scottsdale Road, Scottsdale, AZ 85260.

Sierra Instruments, Inc., P. O. Box 909, Carmel Valley, CA 93924.

TSI, Inc., 500 Cardigan Road, P. O. Box 64394, St. Paul, MN 55164.

4

General Exhaust Ventilation

General exhaust ventilation (GEV), also known as *dilution ventilation*, differs from *local exhaust ventilation* (LEV) in the manner in which the contaminants are removed from the workplace as well as the equipment employed. As discussed in Chapter 1, LEV is based on the principle of contaminant collection and removal at the emission source, thereby preventing it from entering the internal building environment. In practice, complete capture and removal is rarely achieved with LEV, resulting in partial release into the workplace air. General ventilation relies on *dilution* to minimize contaminant concentrations. As workplace air is removed by mechanical systems, it is replaced by outside air entering the building, either through replacement-air systems or simply through openings in the building. As this clean air enters the workplace, it mixes with and dilutes the contaminated air, resulting in lowered concentrations. In designing or evaluating a GEV system, it is assumed that all the material released from the process will enter the workplace air. Therefore, a sufficient quantity of clean air must be introduced to dilute the concentration to an acceptable level.

The primary element in a general ventilation system is an exhaust fan or blower, usually a propeller-type unit mounted in the ceiling or wall. A GEV system does not include hoods and extensive *exhaust* ducting, although the system for supplying replacement air may be quite sophisticated. Replacement air is often provided by *natural ventilation,* which is simply air movement through openings in the building shell (windows, cracks, or roof ventilators) driven by pressure and thermal differences as well as by wind. Rather than relying on the uncontrolled infiltration of replacement air, equipment capable of providing clean, conditioned air through properly designed and positioned ducts and diffusers should be included. The design of appropriate replacement air systems is covered in Chapter 12.

The basic concept governing general exhaust ventilation design is that sufficient air must be provided to prevent the contaminant concentration from exceeding a specified "unsafe" level at any location at which a person could be exposed. This design must incorporate knowledge of

- Physical properties of the contaminant
- Contaminant generation rate
- Health hazard guidelines (e.g., threshold limit value)
- Relative positions of contaminant generation points, work areas, and air supply and exhaust points
- Existing ventilation (natural and mechanical)

One common use of general exhaust ventilation is to prevent the accumulation of flammable gases or vapors at concentrations that would represent a hazardous condition. For these situations, the "safe" concentration is usually specified as 10% of the *lower flammable limit* (LFL) of the vapor in question. Because this concentration (10% of LFL) is nearly always well above any health hazard guidelines (e.g. threshold limit values, permissible exposure limits, maximum allowable concentrations), this approach is restricted to areas where worker exposure would be brief and intermittent, such as drying ovens and storage tanks. The method can also be employed in storage rooms as well as production areas to prevent vapor buildup in the event of a leak or spill.

4.1 LIMITATIONS OF APPLICATION

General exhaust ventilation should be considered as an acceptable method of environmental workplace control only after gaining a thorough understanding of both the situation to be controlled and the limitations of GEV. Often, one or more factors will preclude the use of general ventilation. An initial consideration is the *physical nature* of the material. Is it present as a gas, vapor, or particulate (dust, mist, or fume)? General ventilation is usually not considered appropriate for the control of particulate contaminants, for a number of reasons. Because air velocities associated with general exhaust are low, large particles will not be transported efficiently from their point of generation to an exhaust point; they are likely to settle out and present a housekeeping problem. A second limitation is that the proper design of a GEV system requires reasonable knowledge of the contaminant generation rate, a parameter that is often difficult to determine for processes producing particulate emissions. Third, many processes that generate particulate emissions do so on a sporadic basis, so that the generation rate, even if known, is not constant. Unless the GEV system is designed to handle the peak emissions, occasional excursions above the target concentration will occur. The final and most important reason is that particulate contaminants are often produced and released in a small area, making it much more desirable, for both health and economic reasons, to use local exhaust ventilation.

These arguments do not necessarily apply to gaseous contaminants. Nearly all gases and vapors disperse homogeneously in air and so move with the air through the workplace; they do not "settle out" at low air velocities. Second, the contaminant generation rate can often be determined using simple mass-balance calculations. In a number of processes, particularly continuous ones, vapors are generated at known and fairly constant rates. Such processes might also involve contaminant release over a large area. For example, a material applied to a product at one point in the manufacturing process may continue to be a source of vapors after removal from the point of application. A solvent-based paint sprayed on a workpiece will continue to dry after it is removed from the locally exhausted spray booth. The evaporating solvents from such processes may require additional control by general exhaust ventilation if they are not controlled adequately by an LEV system in the drying area.

One method for estimating contaminant release rates is to determine the "loss" of material during a time period and assume that all of this loss reflects evaporation into the workplace. Consider, for example, an electroplating shop in which $\frac{1}{2}$ gallon of trichloroethylene (TCE; density $= 1.5$ g/ml) must be added to a continuously operating wet degreaser every 8-hour shift to account for solvent loss. In this shop, local exhaust ventilation is not used to control the degreaser emissions. Because concentrations are most often expressed in metric units, the first task is to convert gallons of TCE into milligrams:

$$m = \left(\frac{1}{2} \text{ gal}\right)\left(3785 \frac{\text{ml}}{\text{gal}}\right)\left(1.5 \frac{\text{g}}{\text{ml}}\right)\left(1000 \frac{\text{mg}}{\text{g}}\right)$$
$$= 2.8 \times 10^6 \text{ mg}$$

Next, determine the loss rate by averaging the loss over the 8-hour work shift:

$$\frac{m}{t} = \frac{2.8 \times 10^6 \text{ mg}}{480 \text{ min}} = 5800 \frac{\text{mg}}{\text{min}}$$

For this degreaser, operating continuously, the average generation rate of TCE would be 5800 mg/min.

Given that the contaminant can be efficiently conveyed by slow-moving air, that the generation rate is known and stable, and that the area over which the material is potentially released is quite large and thus ill-suited to LEV, the toxicity of the material in relation to its generation rate remains to be considered. How much air will be necessary to reduce the concentration to a safe level? Is it feasible to remove, and supply, this quantity of air? These questions can be answered with the following quantitative approach to predicting airborne concentrations.

4.2 EQUATIONS FOR GENERAL EXHAUST VENTILATION

To facilitate calculations using equations for predicting GEV performance, several simplifying assumptions must be made.

- There is perfect mixing in the room.
- The generation rate is constant.
- The dilution air contains negligible amounts of the contaminant.
- The contaminant is introduced into the workplace solely through process generation.
- The contaminant is removed from the workplace solely through general exhaust ventilation.

The first assumption, that of perfect mixing, applies both spatially and temporally. Perfect mixing means that when the contaminant is released, it completely and immediately mixes throughout the room. It means that the replacement air completely and immediately mixes, as well. Although this, as well as the other assumptions, can never be satisfied in real-life conditions, appropriate adjustments can be made to the equations describing the ideal case to accommodate most deviations. In practice, while the problem of imperfect mixing creates dilemmas for the designer in calculating the system capacity, the concentration gradients can actually be used to optimize the effectiveness of the GEV system.

To calculate the airborne concentration C of a substance as a function of time t, the generation rate G, the ventilation rate Q, and the room volume V (Fig. 4.1) must be known or estimated. Any change in concentration ΔC during a time period Δt is due to addition of material to the room air through process

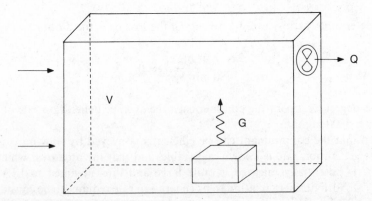

Figure 4.1 The workplace is represented as a simple box, with a single process generating contaminants at a mass rate G. The volume of the room is V and air is being removed with the general exhaust system at a volumetric rate Q. No replacement-air system is shown, so it is assumed that natural infiltration is providing the replacement air.

generation and/or removal of material by exhaust ventilation. The amount of material introduced during the time period Δt is equal to $G \Delta t$, while the amount removed is $QC \Delta t$. Thus the net change in the total amount of contaminant present in the space ΔM is

$$\Delta M = G \Delta t - QC \Delta t \tag{4.1}$$

This change can also be expressed in terms of a change in the air concentration ΔC by dividing both sides of the equation by the volume V:

$$\Delta C = \frac{\Delta M}{V}$$

$$= \frac{G \Delta t}{V} - \frac{QC \Delta t}{V} \tag{4.2}$$

Expressed in the form of a differential equation, we have

$$dC = \frac{G}{V} dt - \frac{QC}{V} dt$$

$$= \frac{G - QC}{V} dt \tag{4.3}$$

or

$$\frac{V}{G - QC} dC = dt \tag{4.4}$$

The change in concentration from any time t_1 to a later time t_2 can be obtained by integration:

$$V \int_{C_1}^{C_2} \frac{1}{G - QC} dC = \int_{t_1}^{t_2} dt \tag{4.5}$$

Integrating yields

$$\frac{-V}{Q} \ln\left(\frac{G}{V} - \frac{QC_2}{V}\right) - \frac{-V}{Q} \ln\left(\frac{G}{V} - \frac{QC_1}{V}\right) = t_2 - t_1$$

$$\ln\frac{G - QC_2}{G - QC_1} = -\frac{Q}{V}(t_2 - t_1)$$

$$\frac{G - QC_2}{G - QC_1} = \exp\left[-\frac{Q}{V}(t_2 - t_1)\right]$$

$$C_2 = \frac{1}{Q}\left\{G - (G - QC_1)\right.$$

$$\left.\exp\left[-\frac{Q}{V}(t_2 - t_1)\right]\right\} \tag{4.6}$$

Equation 4.6 can be used to calculate the concentration C_2 at any time t_2 if the concentration C_1 at an earlier time t_1 is known. This is the general equation describing concentration change (either buildup or decay) as a function of time, generation rate, room volume, and ventilation rate. Several specific cases produce simplified forms of the general equation.

Case I ($C_1=0$ at $t_1=0$, $G>0$). If the initial concentration is assumed to be zero, the resulting concentration at any time t is given by

$$C = \frac{G}{Q}\left[1 - \exp\left(\frac{-Qt}{V}\right) \right] \tag{4.7}$$

where the contaminant is being generated at a rate G for the time t and the air is being exhausted at a rate Q.

Case II ($t \gg t_1$, $G>0$). As seen in Fig. 4.2, when contaminant generation first starts, the concentration rises rapidly and then levels off. After sufficient time, the exponential term of Eq. 4.7, $\exp(-Qt/V)$, approaches zero and the concentration asymptotically approaches a maximum steady-state concentration $C_{\max}$ given by

$$C_{\max} = \frac{G}{Q} \tag{4.8}$$

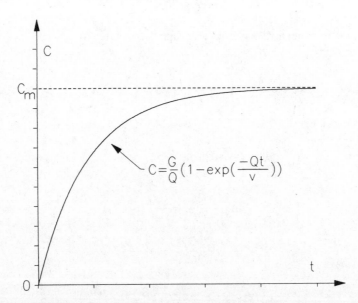

Figure 4.2 The general shape of the curve corresponding to Eq. 4.7 is shown. The concentration C asymptotically reaches a steady-state concentration C_m. This graph assumes that the concentration was zero at $t=0$.

At this point, the material is being removed from the room at the same rate as it is being introduced. This can also be seen by setting the net mass change ΔM equal to zero in Eq. 4.1, since steady state means no change in concentration. Rearranging Eq. 4.1 for this special case will also produce Eq. 4.8. The steady-state equation does not depend on the room volume. Room size will affect how quickly the steady-state concentration is achieved, but not the magnitude of the final concentration.

Case III (G=0).   For intermittent sources, the decrease in concentration when generation stops can be calculated by setting $G=0$ in Eq. 4.6,

$$C_2 = C_1 \exp\left[\frac{-Q}{V}(t_2 - t_1)\right] \tag{4.9}$$

As shown in Fig. 4.3, the concentration decreases exponentially after generation stops, approaching zero asymptotically. If generation resumes again at time t_2, Eq. 4.6 can be used to calculate subsequent concentrations.

  Purge time, defined as the time necessary to reduce the concentration to a specified level upon cessation of the generation source, can also be calculated. The purge time is often measured in *half-lives* $(t_{1/2})$ referring to the amount of time required to reduce the concentration by 50% $(C_2 = 0.5C_1)$ once the

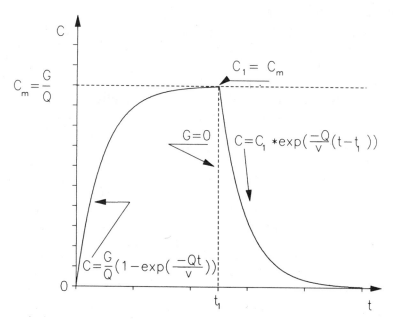

Figure 4.3 If the generation rate ceases $(G=0)$ at time t_1, the concentration declines according to Eq. 4.9.

generation of contaminant has stopped ($G=0$). After 2 half-lives, the concentration has been reduced to 25% of the original level, after 3 half-lives it has been reduced to 12.5%, and so on. The half-life is a function of the room volume and the ventilation rate. Let $t_2 - t_1 = t_{1/2}$ in Eq. 4.9:

$$\frac{C_2}{C_1} = \exp\left(\frac{-Qt_{1/2}}{V}\right) \tag{4.10}$$

After time $t_{1/2}$, $C_2 = 0.5\,C_1$:

$$\frac{0.5\,C_1}{C_1} = \exp\left(-\frac{Qt_{1/2}}{V}\right)$$

$$t_{1/2} = -\frac{V}{Q}\log_e 0.5$$

$$= 0.693\,\frac{V}{Q} \tag{4.11}$$

There are five useful equations for describing the general case and the special situations. The general equation describes the change in concentration with time (Eq. 4.6). The special cases are:

(a) Concentration buildup from an initial value of zero (Eq. 4.7)
(b) Steady-state concentration (Eq. 4.8)
(c) Rate of concentration decrease after generation cessation (Eq. 4.9)
(d) Rate of purging, or half-life (Eq. 4.11)

If the degreaser described previously is used intermittently during the work shift, these equations can be used to construct a temporal concentration profile. For example, consider a process which operates for the first 45 minutes of every hour during the first 4 hours of the 8-hour shift, for the next 2 hours continuously, and then not at all for the last 2 hours. When the degreaser is not in use, the employees have been instructed to cover it. It is located in a 30,000-ft^3 room that has a wall-mounted fan exhausting 3000 cfm. The concentration is negligible ($C=0$) at the start of the shift.

To determine the concentration at any time during the shift, Eqs. 4.6, 4.7, and 4.9 are used. Equation 4.7 is used to predict concentrations during the first 45 minutes; Eq. 4.9 is used for the next 15 minutes, when the generation stops. Equation 4.6, the general equation, must be used to calculate the next period, because the initial concentration is not zero as it was at the start of the shift. The concentrations for the remainder of the shift can be calculated using Eq. 4.6 while TCE is being released and Eq. 4.9 when it is not. The resulting profile is shown in Fig. 4.4. The steady-state concentration can be determined graphically from Fig.

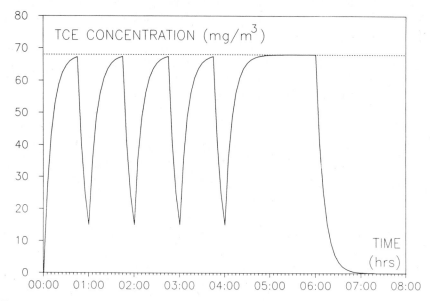

Figure 4.4 For the degreaser operating on an intermittent basis, the concentration profile will change over time. To determine the increase in concentration when the emission begins, Eq. 4.7 is used when the initial concentration is zero and Eq. 4.6 is used when it is not. To determine the decay curve, Eq. 4.9 is used. The concentration is seen to approach the steady-state concentration, 68 mg/m³, after approximately 45 min.

4.4 or by using Eq. 4.8. First, airflow is changed to metric units:

$$Q = \left(3000 \, \frac{\text{ft}^3}{\text{min}} \right) \left(0.0283 \, \frac{\text{m}^3}{\text{ft}^3} \right)$$

$$= 85 \, \text{m}^3/\text{min}$$

Then, solving for the concentration C_{max} given

$$C_{\text{max}} = \frac{G}{Q}$$

$$= \frac{5800 \, \text{mg/min}}{85 \, \text{m}^3/\text{min}}$$

$$= 68 \, \text{mg/m}^3 \qquad (4.8)$$

The half-life $t_{1/2}$ is useful for predicting the decay rate upon cessation of generation. From Eq. 4.11, each half-life is

$$t_{1/2} = 0.693 \, \frac{V}{Q} = 0.693 \, \frac{30,000 \, \text{ft}^3}{3000 \, \text{ft}^3/\text{min}} \approx 7 \, \text{min}$$

For example, after 14 minutes (2 half-lives), the concentration would be

$$68 \text{ mg/m}^3 \times \tfrac{1}{2} \times \tfrac{1}{2} = 17 \text{ mg/m}^3$$

From the steady-state concentration equation, the minimum ventilation rate Q_{min} necessary to maintain the concentration at a target level C_{max} can be calculated as

$$Q_{min} = \frac{G}{C_{max}} \tag{4.12}$$

If the target concentration C_{max} were 80 mg/m^3, then

$$Q_{min} = \frac{5800 \text{ mg/min}}{80 \text{ mg/m}^3}$$

$$= 72.5 \frac{\text{m}^3}{\text{min}} \times \frac{35.3 \text{ ft}^3}{\text{m}^3}$$

$$= 2560 \text{ ft}^3/\text{min}$$

As stated at the beginning of this section, the derivation of these formulas relied on several assumptions. Because these assumptions are rarely met, this design flow Q_{min} must be adjusted to accommodate variations in generation rate, inadequate mixing, worker position, and other appropriate safety factors, as discussed below.

Often, the rate of air exchange E in general ventilation systems is measured in terms of room air changes per hour (ACH) with a unit of inverse hours (h^{-1}). The number of ACH is obtained by dividing the volumetric flow Q by the volume of the ventilated work space V:

$$E = \frac{Q}{V} \tag{4.13}$$

The inverse hour unit for E arises when the flow unit is m^3/h (or ft^3/h) and the volume unit is m^3 (or ft^3).

The air exchange rate is a ventilation rate that has been normalized by the room volume. Air exchange rates when expressed in this manner are often used (and misused) to compare ventilation rates of workplaces with different volumes. A comparison is appropriate when the equation describing the concentration can be simplified by eliminating the individual Q and V terms, and replacing them with E. This is only feasible for the purge rate (Eqs. 4.9 and 4.11). In these cases, the (Q/V) term can be replaced with E, leaving no dependence on Q or V alone. However, the steady-state concentration (Eq. 4.8) is a function of the generation and ventilation rates and is independent of the room volume. Therefore, the volume-normalized unit would be inappropriate for purposes of comparison. To compare anticipated steady-state concentrations, only the generation and volumetric rates for the two situations need be known. The room volumes are unimportant. Similarly, when considering the concentration build-up $(G > 0)$, the equations describing the change in concentration contain not only a (Q/V) term, but also contain the flow term Q, alone. Attempting to simplify Eq. 4.6 or 4.7 by replacing the (Q/V) term with E would not help, since the flow must still be known.

4.3 VARIATIONS IN GENERATION RATE

If the variations in the contaminant generation rate are less than 20%, they are usually ignored. Similarly, if the periodicity of the variation is small, those variations can be safely overlooked. However, when lengthy, pronounced duty cycles are present, the possibility of transient excess exposures must be consider-ed. For example, if a process emits a contaminant at a rate G for one hour out of every two, with no emissions during the second hour, it would usually be inappropriate to use an average emission rate of $0.5G$ because this would provide inadequate control during the periods of peak emission.

If the system cannot be designed to provide higher exhaust rates during periods of emission, the most prudent approach to this problem is to assume, for design purposes, that the generation rate is constant at a level equal to the maximum rate observed during the cycle. Although this will result in more than adequate ventilation during most of the work shift, it will assure that the intermittent emissions will not produce concentrations above the target level. If the period of contaminant release is brief compared to the period of nonrelease, as might be the case with some batch processes, excursions above the target level might be allowable. This would depend on the toxicity of the material, the frequency of release, the duration of the release, and the projected exposure concentrations, both during the peak emission episodes as well as the intervening periods.

In some situations, the distance between the point of release and the employees' work stations can be used effectively to "average out" the peaks. Unfortunately, in many operations the release of emissions is associated directly with the presence of the workers. For example, emission of solvent vapors from a degreasing tank is often caused by the operator-attended removal of wet parts. Similarly, the opening of a dryer-oven door may produce a brief, yet intense material release. During this phase of the process, an operator standing nearby often receives a considerable exposure. However, if a process can be modified to avoid human involvement during peak emissions, the amount of air being handled by the general exhaust system might safely be reduced.

4.4 MIXING

The mechanisms governing dispersal of airborne material in buildings are not well understood. The equations developed in Section 4.2 assume that the mixing of both the contaminant and the clean replacement air into the workplace environment was complete and instantaneous, although this is never attained. Two approaches have been taken in attempts to accommodate, in the theoretical treatments, the inherent lack of ideal mixing encountered in the real world.

The industrial ventilation engineers traditionally incorporate into the equations a multipurpose safety factor K which addresses the imperfect mixing as well as other variables such as material toxicity, seasonal changes in natural ventilation, reduced operating efficiencies of ventilation systems, process cycle and duration, worker location, number of workers exposed, location and number of contaminant release points, and other circumstances affecting worker exposure (Soule, 1978). This K factor is introduced at the last stage of the steady-state calculation (Eq. 4.8), so that the desired amount of dilution air is now provided as

$$Q_a = KQ_i = K\frac{G}{C} \qquad (4.14)$$

where Q_i = air exhaust rate under ideal conditions
$\quad Q_a$ = air exhaust rate under actual conditions
$\quad K$ = multipurpose safety factor (dimensionless)

The range of values traditionally cited for K is 3 to 10, although experimental validation is lacking. Moreover, a rigorous method to choose the proper value has yet to be developed.

As noted above, it is conventional to aggregate all of the various factors into a single K factor. However, it is more appropriate to examine individually each of the reasons for adjusting the design ventilation rate. At the least, the physical mixing should be considered apart from the safety and health issues mentioned above. The adequacy of the mixing would be included in a mixing factor K_m while a safety factor K_s would incorporate the toxicity of the material, the number of workers, and so on. The multipurpose factor K would be the product of K_m and K_s. The mixing factor K_m could be used to calculate expected concentrations, given knowledge of generation rates, recirculation rates, air cleaner removal efficiencies, and outdoor concentrations. However, virtually no work has been conducted to characterize the parameters affecting and describing mixing rates in large, well-ventilated spaces. Given this lack of knowledge, the ability to choose and assign an appropriate mixing and safety factor remains a very uncertain art, not a science.

4.5 INLET/OUTLET LOCATIONS

Because the physical mixing factor K_m is dependent on the locations of the exhaust outlets and supply inlets, these must be specified in the design process, bearing in mind that judicious selection will result in both a cleaner workplace and cost savings. In the derivation of the equations describing steady-state conditions, it was assumed that the concentration was homogeneous throughout the workplace; that is, the contaminant was evenly distributed. Of course, there will always be a concentration gradient. The concentration will be higher near the emission sources and lower near the replacement air inlets, a fact of some importance in the design phase for two reasons. First, the location of workers relative to the source must be considered. The closer the workers are to the point of generation, the greater their exposure is likely to be. This should be addressed in selection of the safety factor K_s. Second, the differences in concentration can be used to provide more effective removal of the contaminant by adapting certain concepts from local exhaust ventilation. If the point of exhaust is located as close to the point of release as possible, the rate at which material is removed from the workplace QC will actually increase, thus providing lower overall concentrations throughout the workplace. This premise rests, of course, on the reasonable assumption that the highest concentrations are encountered near the source.

Consider the simplified case illustrated in Fig. 4.5, where one source emits material at a generation rate G. The concentration near the source C_s will be higher than the background room concentration C_r. There is an intermediate zone, with concentration C_i. Although this approach will assume only these three, somewhat distinct, zones, in real life there are no clear delineations between them. Rather, there would be continuous concentration gradient as one moved farther from the source. For illustrative purposes, however, one can consider these three regions in which $C_s > C_i > C_r$.

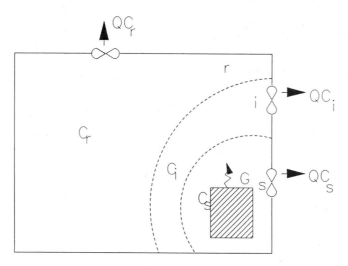

Figure 4.5 The three regions (source, s; intermediate, i; and room, r) are shown. If the exhaust fan pulls air from the general room region, the mass rate of contaminant removal is given as QC_r. If the fan is located in the intermediate zone, the pollutant is removed at a faster rate QC_i, since $C_i > C_r$. Even more effective removal can be achieved by locating the exhaust fan near the source, so that the removal rate is QC_s, which is greater than QC_r.

One common approach to locating the exhaust outlet is to place the fan in the ceiling, with little or no regard for its position relative to the three regions (source, intermediate, or room). Therefore, it is usually exhausting air from the room region, removing the contaminant at a rate QC_r. Note, however, that by locating the fan in the intermediate region where the concentration is higher than the background concentration, more effective contaminant removal is achieved, since the material is now being removed at a rate QC_i, which is greater than QC_r. It follows that moving the exhaust point into the source region will provide yet more effective removal. "Spot" ventilation, as this is frequently termed, is a hybrid, with attributes of both general and local ventilation systems. Although the equations describing general exhaust no longer apply, the desirability of LEV-like systems is clearly apparent. It is seen, therefore, that the effectiveness of a GEV system can be increased by locating the exhaust point as close as possible to the point of contaminant release.

4.6 OTHER FACTORS

A consideration mentioned earlier is the location of the workers. Obviously, personal exposures should be minimized. By locating the exhaust and replacement air points strategically, it is possible to establish an airflow pattern that will draw contaminated air away from the employees. The practice of moving contaminated air through populated regions of the plant should be avoided if at all possible. Rather, the system should be designed to move air from clean areas into dirty areas. If this is not possible, the suitability of GEV should be reconsidered.

The locations of the supply air inlets are dictated by logic similar to that used to locate the exhaust outlets. Given that the tempered replacement air is cleaner than the general room air, this air should be delivered directly into the area populated by employees before passing through the contaminated zone. This has a twofold effect. Not only does it provide clean, fresh air for the workers but also discourages intrusion of dirty air into this zone. This helps to establish and maintain a pattern of flow within the plant, as shown in Fig. 4.6. This supply air requires conditioning and a system for delivery (i.e., fans, ducts, diffusers). Guidelines for replacement air systems are provided in Chapter 12.

The provision of adequate replacement air is a critical aspect of GEV system design. Propeller-type fans, which are often used as wall or ceiling exhaust fans, suffer substantial degradation in performance at even small negative static pressures (see Chapter 10). Therefore, if a sufficient quantity of replacement air is not provided, the building will come under negative pressure (with respect to the outdoors) and the exhaust fans will not perform as expected. Natural ventilation patterns in the building should also be considered. Open doors, windows, roof ventilators, and so on, will affect both the quantity and patterns of airflow. Temperature differences in the plant may also induce air movement. Natural dilution ventilation is highly dependent on seasonal characteristics as well as

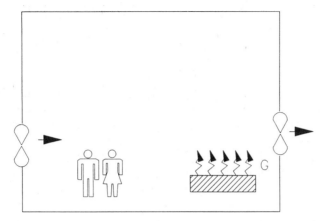

Figure 4.6 Providing clean, tempered air to the populated areas of the workplace and exhausting dirty air directly from the process area is good practice. This discourages intrusion of the contaminated air into the populated zone. The supply air should be properly conditioned and a system for delivery (i.e., fans, ducts, diffusers) should be designed.

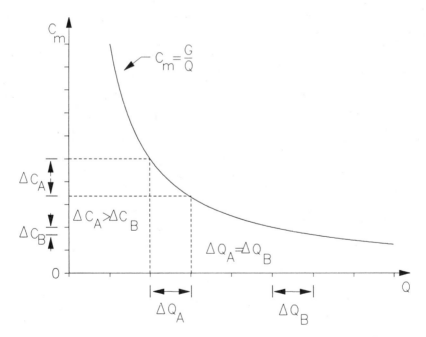

Figure 4.7 The steady-state concentration C_m is shown as a function of the exhaust rate Q assuming a constant generation rate G. The extent of concentration reduction to a lower steady-state level is a function of the change in Q *and* the existing Q. At low exhaust rates, small increases can produce substantial decreases in concentration. For example, ΔQ_A produced a large concentration decrease ΔC_A. However, at a higher Q, the same increase in Q will produce only minor decreases in C. Even though ΔQ_B is the same as ΔQ_A, the change in concentration attributable to the former ΔC_B is much less than ΔC_A. The efficiency of additional general ventilation is highly dependent on existing ventilation.

other factors such as window opening and door propping. The potential for interaction between natural and mechanical ventilation must be considered in the design phase, as well as during testing after installation and routinely thereafter.

If possible, the amount of natural ventilation should be measured. This can be accomplished by using tracer-gas decay techniques. It is important to know the extent of natural ventilation because the reduction in concentration is not linearly proportional with increasing ventilation rate, as illustrated in Fig. 4.7, and by the steady-state equation (Eq. 4.8). At low ventilation rates, a small increase in Q will have a marked effect on the concentration. If a building has little existing ventilation (either naturally or from other ventilation systems), the introduction of a mechanical GEV system is likely to have real benefits in lowering in-plant concentrations. However, if the present ventilation rates are high, any further decrease in concentration will require substantial amounts of exhaust air.

Consider a warehouse with propane forklift trucks producing indoor workplace carbon monoxide concentrations of 100 ppm, in which a concentration reduction to 40 ppm is to be accomplished with the introduction of a new GEV system. The generation rate G and mixing factor K are assumed to be constant. Rearranging Eq. 4.14 yields

$$C_1 = K\frac{G}{Q_1}$$
$$C_2 = K\frac{G}{Q_1 + Q_2}$$

where $C_1 = 100$ ppm CO
$\quad C_2 = 40$ ppm CO
$\quad Q_1 =$ actual natural (existing) ventilation rate
$\quad Q_2 =$ actual mechanical (additional) ventilation rate

Rearranging and combining gives us

$$C_1 Q_1 = C_2(Q_1 + Q_2) \tag{4.15}$$
$$= C_2 Q_2 + C_2 Q_1$$

$$Q_2 = \frac{C_1 - C_2}{C_2} Q_1$$

$$= \frac{60}{40} Q_1$$
$$= 1.5 Q_1 \tag{4.16}$$

If there was an initial airflow of 10,000 cfm through the warehouse, the GEV system should be designed to move an additional 15,000 cfm. However, if Q_1 was already 50,000 cfm, the system would be required to handle 75,000 cfm, a fivefold increase over the 15,000-cfm figure. Thus, without knowing the contaminant generation rate G for a process, the amount of air required to reduce the concentration to a target level from an existing concentration can be calculated using Eq. 4.16 if the existing ventilation rate is known.

4.7 COMPARISON OF GENERAL AND LOCAL EXHAUST

Although this chapter has emphasized the theory and design of GEV systems, the problems associated with this form of environmental control need discussion. Although it does have valid applications, GEV is too often used as a quick and dirty alternative to LEV, based on its low capital cost, simple design, and easy installation. There are several serious deficiencies associated with GEV, however.

There may be adverse health consequences of GEV compared with alternative control methods, primarily LEV. By allowing the contaminants to be released into the workplace air, personal exposure to the contaminant will always occur, albeit at a nominally "safe" level. Local exhaust ventilation, on the other hand, has the *potential* of capturing the contaminants at the point of release, thereby eliminating worker exposure. Second, the amount of air handled by a GEV system will always be greater than that of a LEV system designed to control the same process (see Chapter 6). Since all of the exhausted air should be replaced with clean, tempered air, the costs of operating a replacement air system for the GEV system will be substantially more than for a LEV system. Third, the problem of cleaning the exhaust air must be addressed. It is easier and less expensive to clean the small volume of highly contaminated air produced by a LEV system than a large volume of less-contaminated air resulting from a GEV system. In general, capital and operating costs for air-cleaning devices are proportional to the quantity of air being handled, rather than the total amount of contaminant being removed.

In summary, a general exhaust ventilation system should be considered as a primary means of environmental control only if:

- The workers will not be exposed to excessive concentrations.
- A LEV system is not feasible or is inappropriate.
- The contaminant is not highly toxic.
- The contaminant will be transported effectively to the exhaust point.
- The costs of supplying replacement air will not be excessive.
- Existing ventilation rates are relatively low.

This generally limits the use of GEV to widely distributed sources of gaseous or vapor-phase contaminants such as solvent vapors.

LIST OF SYMBOLS

C	concentration
E	air exchange rate
G	generation rate
K	multipurpose safety factor
K_m	mixing factor
K_s	safety factor
m, M	mass
Q	volume flow
t	time
$t_{1/2}$	half-life
V	volume

REFERENCES

American Conference of Governmental Industrial Hygienists, Committee on Industrial Ventilation (1988), *Industrial Ventilation, A Manual of Recommended Practice* 20th ed., ACGIH, Lansing, MI, p. 2–1.

Soule, R. D. (1978), "Industrial Hygiene Engineering Controls," in *Patty's Industrial Hygiene and Toxicology*, 3rd ed., Vol. 1, G. D. Clayton and F. E. Clayton, eds., Wiley, New York, 786–787.

5

Hood Design

The key concept underlying local exhaust ventilation is the capture of airborne contaminants at the source. The system should be designed to capture air containing the contaminants given off by the process being ventilated. In many instances this is an extremely difficult task, and the success or failure depends on the exhaust hood design.

An exhaust hood is defined to be a suction source in an exhaust system through which air moves from the ambient environment, carrying with it any contaminant generated or released from the process or piece of equipment served by that suction source. Hoods can range in design from a plain opening at the end of a duct to a complex enclosure surrounding an entire process. Past studies and common sense tell us that the specific design of an exhaust hood is the most important element in the success or failure of any exhaust system.

The critical elements of any exhaust hood are its geometry (which determines the airflow patterns in the vicinity of the hood), its location relative to the process, and the amount of air being moved through it. These factors determine *how much* air is being removed from the vicinity of the process and *from where* the air is drawn. If a sufficient quantity of air is being removed from regions where the contaminant is released, those contaminants will be captured by the exhaust system with high efficiency. If any of the three elements are deficient, however, exhaust flow will not be provided at the point of generation in sufficient quantity to ensure contaminant capture.

A considerable portion of the *Ventilation Manual* (1988) is devoted to the design of exhaust hoods. Detailed guidelines concerning hood shape, airflow, and hood energy loss are presented for a wide range of specific operations; these are useful, but specific designs should always be used with care. Their routine application may lead to installed systems which have poor contaminant capture efficiencies. These potential deficiencies can be traced to many factors, such as shortcomings in the design plates, excessive cross-draft conditions, and processes that differ from the design assumed in the plate. Circumstances under which the guidelines must be tempered with sound engineering judgement are discussed in this chapter. Hood designs for several specific processes are discussed in Chapter 6.

5.1 CLASSIFICATION OF HOOD TYPES

Hemeon (1963) was perhaps the first to describe a simple scheme to categorize hood designs. He concluded that all local exhaust hoods fall into three categories: enclosures, exterior hoods, and receiving hoods. The classification of a hood into one of these categories is useful, since the design procedure to be followed is different for each. The key element in classifying hoods is the *location* of the hood relative to the point of contaminant generation or escape. The relative location of hood with respect to source helps to determine the amount of airflow needed for effective capture. The design process for each hood type is discussed in the following sections.

5.1.1 Enclosures

If a hood is designed so that the contaminants are released from the process *inside* the hood, it is classified as an *enclosure* (Fig. 5.1). This type of hood generally is the most desirable from the ventilation engineer's point of view, since

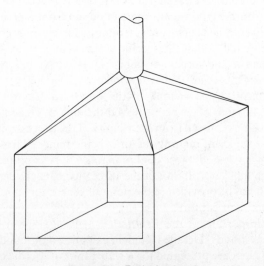

Figure 5.1 Enclosing hood.

the contaminants are always contained inside the boundaries of the hood. Effective control of emissions is much easier in this case than for exterior-type hoods (to be described in Section 5.1.2), where the exhaust airflow must "reach out" and capture the contaminated air as the contaminant is generated.

Enclosing hoods can be further categorized, depending on the completeness of the enclosure provided. Complete enclosures are typified by glove boxes (Fig. 5.2a); the process/work is conducted entirely inside the exhaust hood. This type of hood offers the greatest protection to the user, since the possibility of worker exposure to air contaminants in a well-designed hood during normal operation is negligible. It also conserves energy, since of the three basic hood types this design requires the least amount of airflow to ensure adequate control. An exhaust airflow is chosen sufficient to create a slight negative pressure inside the enclosure; this ensures that contaminated air will not escape from the enclosure.

Generally, glove boxes and other similar enclosures are provided with a designed air inlet so that a small continuous airflow is maintained in the absence of leaks. This airflow can be filtered at the inlet if the interior of the box must be maintained in a clean condition. Glove boxes frequently are specified when handling highly toxic chemicals. In such cases it is usually necessary to provide a high-efficiency air-cleaning system for the exhaust air (see Chapter 11).

The principal problem with using complete enclosures is the lack of access to the process. For this reason, complete enclosures are used only for the most hazardous exposures, where product cleanliness must be assured, or where the process is completely automated and no worker access is required. An example of the latter is the flat deck screen hood pictured in Fig. 5.2b; obviously, instances where complete enclosures are feasible will be limited in most industries.

Booths, the second subcategory of enclosures, are simply enclosures with one side partially or completely open to provide access. Common examples are hoods used for abrasive cutoff sawing (Fig. 5.3a) and spray painting (Fig. 5.3b). Booths are classified as enclosures because the contaminant release takes place inside the hood itself, as it does with glove boxes and other complete enclosures. Although the contaminants are generated inside the hood, the presence of one open side gives rise to the possibility of contaminant escape through this opening. To prevent this, a sufficiently high air velocity must be maintained through the open area (called the hood face) to preclude the loss of any of the contaminants from the hood due to internally generated air currents or external eddies. The specification of this *minimum face velocity* is a major part of the design process for booths.

Tunnels (Fig. 5.4), the third type of enclosing hood, are similar to booths, except that they have two open faces for process flow or access. The contaminants are still given off inside the hood, but now two routes of escape exist. As with booths, it is the face velocity of the air drawn in through the openings that determines the hood capture efficiency.

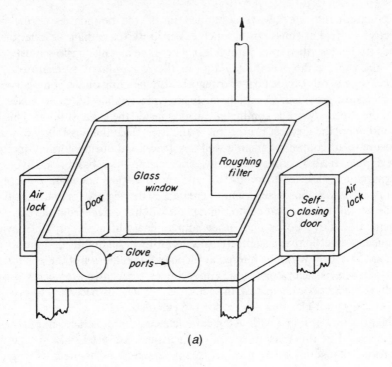

(a)

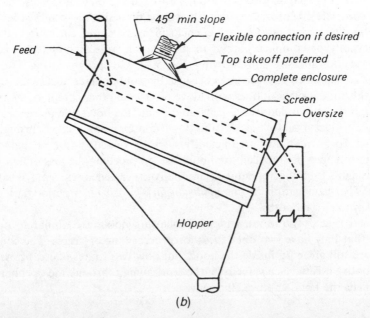

(b)

Figure 5.2 Examples of enclosing hoods: (a) glove box; (b) flat deck screen hood. [After Hagopian and Bastress (1976).]

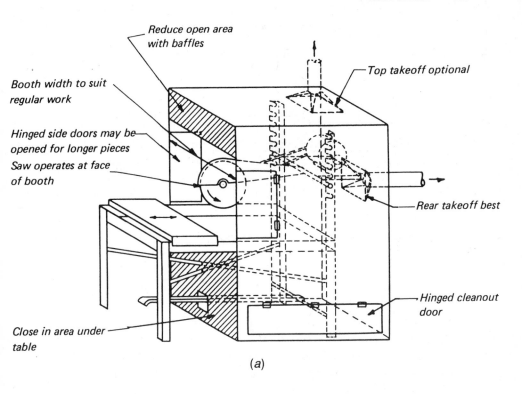

Reduce open area with baffles

Top takeoff optional

Booth width to suit regular work

Hinged side doors may be opened for longer pieces

Saw operates at face of booth

Rear takeoff best

Hinged cleanout door

Close in area under table

(a)

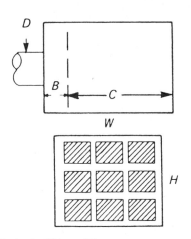

D

B

C

W

H

1. *Split Baffle or Filters*

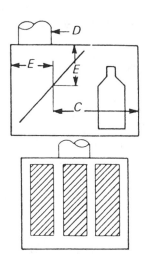

D

E

E

C

2. *Angular Baffle*

(b)

Figure 5.3 Examples of booth-type enclosures: (a) abrasive cutoff saw hood; (b) spray painting hood. [After Hagopian and Bastress (1976).]

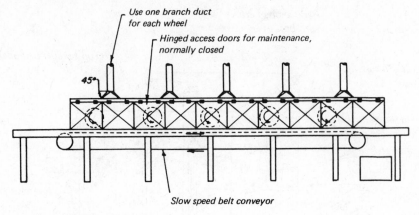

Figure 5.4 Straight-line multiple buffing, an example of a tunnel-type enclosure. [After Hagopian and Bastress (1976).]

5.1.2 Exterior Hoods

The basic difference between an enclosing hood and an exterior hood is that the latter does not surround the contaminant at the point of release. The contaminant is given off *outside* the hood and must be *captured* by the hood rather than allowed to escape to the general plant atmosphere. A typical exterior hood is the portable welding hood (Fig. 5.5).

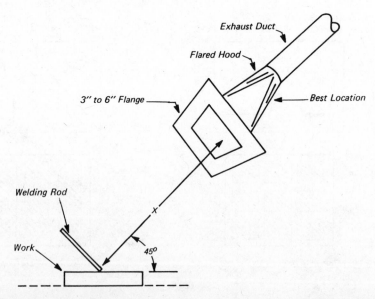

Figure 5.5 Movable welding hood, an example of an exterior hood. [After Hagopian and Bastress (1976).]

With an exterior hood the contaminant is captured by the movement of air past the point of generation into the hood. This air movement is caused by the exhaust system fan, which draws air from the duct connecting the hood to the fan. This creates a partial vacuum at the hood, causing air to flow from outside the hood to replace the air that has been drawn into the duct system. If the contaminants are given off into the air that enters the hood, they will be captured by the exhaust system.

In many cases process access requirements necessitate the use of exterior hoods, which inherently are less desirable than enclosing hoods. With enclosing hoods, contaminants are given off directly into the hood and not into the general workplace atmosphere. The process creating the contaminants and the exhaust system can be thought of as a closed system, and if this system is working properly the worker is never exposed. The exterior hood suffers greatly in comparison to such a closed system, since the contaminant is first *released* into the air surrounding the process and then the air containing the contaminant is *captured* by the exhaust system. Unfortunately, many adverse effects can occur between generation and capture.

External hoods are susceptible to extraneous air movements which can disrupt the airflow patterns between the point of contaminant release and the exhaust hood. Such air currents are ubiquitous; almost any room experiences random air movements with velocities of at least 30 to 50 fpm when no overt sources of air movement are present (as a point of reference, 100 fpm is approximately equal to 1 mile per hour). In industrial facilities, where numerous sources of air movement exist, severe cross-draft conditions may completely disrupt the performance of exterior hoods. Cross-drafts are discussed extensively in the following section covering exterior hood design.

Exterior hoods have other problems in addition to cross-drafts. Since the contaminants are released into the air surrounding the process and then

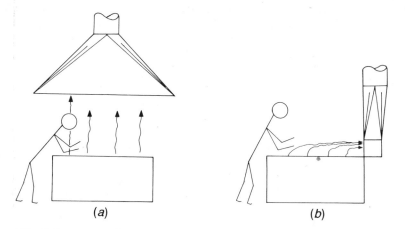

Figure 5.6 Airflow patterns for hoods used on an open-surface tank: (*a*) canopy hood; (*b*) slot hood.

captured, it is possible for a worker to be in the path between the contaminant source and the hood, and be exposed to the contaminants before they are captured. Consider a worker at an open-surface tank, which is exhausted by an overhead, or canopy, hood, as shown in Fig. 5.6a. As part of the job the worker must lean over the tank to insert and remove objects, and in so doing is exposed to the contaminants being generated. In this case, even if the hood is capturing 100% of the contaminants being released, the worker would still be exposed. A better design for the exhaust hood on this process might be the slot hood shown in Fig. 5.6b, which creates an airflow pattern that bypasses the worker's breathing zone. This concept is discussed in detail later in this chapter.

5.1.3 Receiving Hoods

Receiving hoods are exterior hoods designed to take advantage of some aspect of the contaminant generation process to assist in moving the contaminant from the source to the hood. If the contaminant is released in a preferential direction, the exhaust hood is positioned to take advantage of this preferential motion and "receive" the contaminant. Most exterior hoods have to "reach out" and direct the contaminant into the hood with exhaust air; in receiving hoods the contaminants "ride with" the exhaust air into the hood.

The two properties of contaminant release that are commonly utilized in receiving hoods are particle momentum and thermal updrafts. Certain processes, such as the grinding shown in Fig. 5.7, propel particles into the air with a considerable amount of momentum. Grinding wheel exhaust hoods (Fig. 5.7a) are designed to intercept the particles as they are thrown off; if the hood cannot be placed in the path of particle travel (Fig. 5.7b), it is much more difficult to "turn" the particle by air movement and direct it into the hood. A detailed discussion of the design of grinding wheel hoods is included in Section 5.4.2.

Thermal drafts can be powerful forces directing contaminants upward. This can often be used to advantage when designing exhaust hoods for hot processes, such as canopy-type hoods (Fig. 5.8). The hood in Fig. 5.8a is exhausting a room-temperature process, while the hood in Fig. 5.8b is exhausting a hot process. Since there are no external forces directing contaminants into the exhaust hood for the room-temperature process, this hood is an *exterior* hood and as such

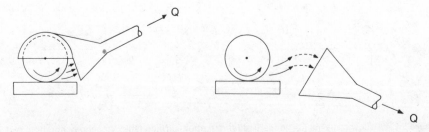

Figure 5.7 Grinding wheel hood types: (a) receiving hood; (b) exterior hood.

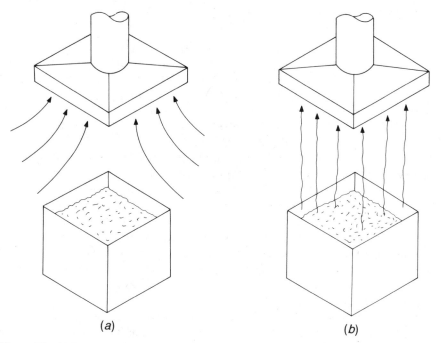

Figure 5.8 Airflow into a canopy hood: (*a*) hood over a room-temperature process; (*b*) hood over a heated process.

must create sufficient air velocity by *exhaust air alone* to capture the contaminants evolved at the tank surface. The heated process, on the other hand, creates a thermal updraft that *carries* the contaminants toward the hood. In this case the canopy hood acts as a receiving hood. The discussion in Section 5.4.1 will demonstrate that canopy-type receiving hoods may be effective, but canopy hoods are poor exterior hoods.

5.1.4 Summary

The discussion above should convince the reader of the desirability of using enclosing-type hoods whenever possible. The basic idea when designing an exhaust system is to use a hood that encloses each process as completely as possible, consistent with necessary access by personnel and materials. In many cases this hood selection and design process relies on the engineering judgement and common sense of the ventilation designer. The process to be ventilated should be observed closely whenever possible to determine what access is required by the worker, material transport, and the process equipment. The worker should be consulted and the possible interactions between the proposed exhaust system and the process should be discussed. An exhaust hood can then be designed which encloses the process as completely as possible but which is not looked upon by the worker as an interference with effective task performance.

5.2 DESIGN OF ENCLOSING HOODS

With an enclosing hood the contaminant, by definition, is given off *inside* the hood. The important considerations in designing such a hood are to ensure that the contaminant is actually given off inside the hood and to *keep* it inside the hood at all times. The process creating the contaminant and the exhaust system should be thought of as a closed system that contains the contaminant at all times; the worker is positioned outside this closed system and thus has no possible exposure to the contaminant.

In most practical applications a complete enclosure is not possible. The contaminant may be given off inside the physical dimensions of the hood, but openings exist in the hood to allow necessary access to the process. In these cases sufficient exhaust must be provided so that adequate inward velocity is created through all such hood openings to prevent the escape of any contaminant.

The abrasive blasting cabinet (VS-101.1), one example of a true enclosure, illustrates the limitations entailed by requiring an exhaust system to enclose a process completely. The cabinet is extremely effective in controlling worker exposure to airborne contaminants, but imposes severe limitations on the worker's interaction with that process.

Most enclosing-type hoods fall into one of two subcategories, booths and tunnels. A booth (Fig. 5.3) is an enclosure with one open side, while a tunnel (Fig. 5.4) has two open sides. Each meet the definition of an enclosure, since the contaminants are still released within the confines of the hood itself. Booths are probably the most common type of enclosure found in industry; examples include the paint spray booth (VS-601, 603, 604), the swing grinder booth (VS-414), and the laboratory hood (VS-203). Tunnels are used typically to enclose material transfer or conveying operations such as conveyor belts (VS-306), screens (VS-307), and straight-line automatic buffing (VS-405).

In designing enclosures, the important consideration is *face velocity*, in this case defined as the average velocity through all hood openings that penetrate the enclosure. Thus, in looking at the simple booth of Fig. 5.3, the face velocity V_f (in fpm) is given as

$$V_f = \frac{Q}{A_f} \tag{5.1}$$

where Q is the total airflow through the hood (in cfm) and A_f is the total cross-sectional area of the hood openings (in ft^2). The design of enclosing-type hoods must concentrate on (1) fitting the hood to the process so that the enclosure is as complete as possible, and (2) selecting an airflow through the hood which will result in a face velocity of sufficient magnitude to ensure that contaminants cannot escape the hood.

The first of these steps, the fitting of the hood itself, is probably the most important consideration for ensuring good control. The design process to be followed cannot, unfortunately, be described in detail, since it depends largely on the specific process being controlled. General guidelines for designing enclosures are available, but the specific details are best left to the ventilation designer. The following basic rules apply:

1. Make the enclosure as complete as possible. This reduces the total area of the openings through which exhaust air must be drawn, thus minimizing the airflow required for contaminant control.
2. Include design features to ensure that airflow is evenly distributed across the openings. This is especially important with large open areas, such as those associated with paint spray booths. If such booths are designed to be very shallow, as shown in Fig. 5.9a, the average face velocity (Eq. 5.1) may be adequate but the airflow will be poorly distributed. Since much of the air will flow in near the center of the booth face, which is the path of least resistance, the velocity near the edges of the booth face will be inadequate.

To ensure an even distribution of exhaust air, steps must be taken to equalize the resistance to airflow across all areas of the hood face. To accomplish this, the following design features, singly or in combination, can be applied:

1. Make the booth deeper (Fig. 5.9b). The *Ventilation Manual* (VS-603, 604) recommends that the depth of the booth should be at least 75% of the booth face height or width, whichever is larger.
2. Use a baffle. Examples of a solid baffle, an angular baffle, and a split baffle are shown in Fig. 5.9c. The dimensions for such baffles are given in VS-603 for large hoods and VS-604 for small hoods. Baffle design is also pictured in Fig. 4–16 of the *Ventilation Manual.*
3. Use filters. Filters or other air-cleaning devices are commonly used in paint spray and other operations for air pollution control purposes. When placed at the rear of the booth (Fig. 5.9d), they provide relatively high resistance to airflow and thus serve to even out the air distribution. In fact, such devices, unless properly designed and maintained, are more effective as air distributors than as air cleaners (see Chapter 11).

Once a basic enclosing hood configuration is selected, all that remains is to calculate the airflow Q required to ensure contaminant control. The important quantity for system design purposes is Q, but the important quantity for control is the average face velocity V_f. In most applications, face velocities in the range 100 to 200 fpm are sufficient. It is important to realize that velocities which are too high can be just as detrimental to good hood performance as velocities which are too low. For example, velocities greater than 150 fpm at the face of laboratory hoods have been known to create serious problems (Chapter 7). High air velocities can disrupt experiments being carried out in the hood (as, for example, when a Bunsen burner or other combustion source is being used).

The presence of the operator introduces additional difficulties. Standing in front of the hood, the operator creates an obstruction to airflow; if the air flows

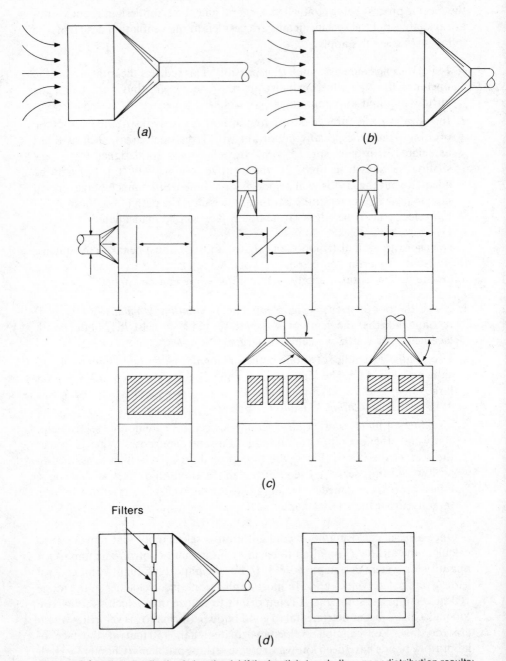

Figure 5.9 ´ Airflow distribution in booths: (a) if the booth is too shallow, poor distribution results; (b) distribution can be improved by making the booth deeper; (c) baffles can be used to help distribute airflow [after Hagopian and Bastress (1976)]; (d) filters are also effective.

past the operator at too great a velocity, eddies can be created between the operator and the hood opening. Such turbulent eddies can entrap contaminants given off inside the hood and carry them out of the hood. These types of problems can be minimized by the selection of the proper fume hood face velocity, as discussed in Chapter 7.

Recommended face velocities for enclosures used on many processes are given in the *Ventilation Manual*. In many cases, a range of velocities is specified. For these cases, and where a standard design is not available, the designer must pick a specific velocity within the recommended range. The following factors should be considered in making such a selection:

1. The presence of disruptive cross-drafts requires the use of higher face velocities. Cross-drafts are usually thought of as being disruptive to the performance of exterior hoods, but they can also affect the performance of booths and tunnels.
2. The collection of highly toxic materials requires a more conservative design to ensure positive control. A lower permissible exposure limit for the material being collected would indicate a higher face velocity (within the recommended range).
3. Material-handling systems can cause induced airflows which must be overcome by increased hood face velocities. This is discussed in Chapter 6.

5.3 DESIGN OF EXTERIOR HOODS

As stated earlier, exterior hoods are inherently less satisfactory than enclosing hoods as collection devices. Since such hoods must "reach out" and capture contaminated air beyond the boundaries of the hood, it is extremely important that they be designed properly. Poor design invariably ensures that contaminants will escape in significant quantities into the workplace, causing worker exposure.

The three elements of critical importance in designing exterior hoods are the capture velocity to be created at the point of contaminant release, the airflow through the hood required to create the capture velocity outside the hood, and the physical shape and location of the hood itself. Each is discussed separately.

5.3.1 Determination of Capture Velocity

Capture velocity is defined to be the air velocity at the point of contaminant release sufficient to "capture" the contaminants and carry them into the exhaust hood. The proper selection of the capture velocity is of critical importance in the design of exterior hoods. The air velocities created by an exterior hood are imposed on a complex airflow pattern which is always present due to other sources of air movement, such as replacement air, perimeter infiltration, and process-induced airflow. The capture velocity must be large enough to overcome

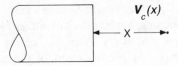

Figure 5.10   Unflanged round exterior hood used to capture contaminants at distance x.

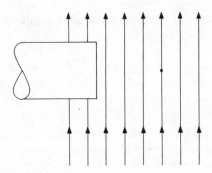

Figure 5.11   Unflanged round exterior hood in a uniform cross-draft parallel to the hood face.

any other air-velocity patterns at the point of contaminant release so that the resultant airflow is from the release point to the exhaust hood.

These points can be illustrated by the simple example shown in Fig. 5.10. A plain round duct exhausts a process at a distance x directly in front of the opening. For simplicity the contaminant is assumed to be released as a point source. The exhaust induces an airflow at point x with a velocity $V_c(x)$ (the relationship between the amount of air flowing Q and the velocity induced at x is developed in the next section).

Under the assumption that no other air movement exists in the vicinity of x, the air is completely still when the local exhaust hood is not operating. Under these circumstances the operation of the exhaust hood will induce an airflow at point x which will be directed into the exhaust hood. The capture velocity $V_c(x)$ can take any value at all; in the absence of other air currents, the actual velocity is irrelevant since all air passing point x will eventually be drawn into the exhaust hood.

Such a condition is not realistic, of course. In any practical situation extraneous airflows will exist at point x. The simplest such situation is a uniform cross-draft parallel to the hood face, as shown in Fig. 5.11. If the cross-draft has a velocity $V_d(x)$ at point x and the exhaust system is operating, the total velocity of the air at point x is now the vector sum of the capture velocity and the cross-draft velocity:

$$V(x) = V_c(x) + V_d(x) \tag{5.2}$$

In this case, whether or not the contaminant is drawn into the exhaust hood depends on the relative magnitudes of V_c and V_d and the distance x from the release point to the hood opening.

This example illustrates the critical importance of cross-drafts to the performance of exterior hoods. In the complete absence of cross-drafts, any exterior hood will capture all contaminants given off in its vicinity. It is the action of cross-drafts and other air motions in the workplace that degrade the performance of exterior hoods and make their design particularly difficult. For each operation being exhausted, a capture velocity must be selected to overcome the action of expected extraneous airflow patterns and direct the resultant velocity vector into the exhaust hood.

The performance of exterior hoods in the presence of crossdrafts is the subject of considerable recent research (Conroy et al., 1988; Conroy and Ellenbecker, 1988; Dalrymple, 1986; Ellenbecker et al., 1983; Fletcher and Johnson, 1986; Flynn and Ellenbecker, 1985, 1986, 1987; Jansson, 1980, 1985; Regnier et al., 1986). For example, Flynn and Ellenbecker have recently developed a method to predict the airflow patterns into flanged circular hoods in the presence of uniform cross-drafts. As an example, the airflow pattern into a 6-in. flanged circular hood exhausting 100 cfm in the presence of a 100-fpm cross-draft is shown in Fig. 5.12. The zone of influence of the hood is defined by the pattern of streamlines present. Any contaminants given off in an area where streamlines converge on the exhaust hood will be carried by the exhaust air into the hood

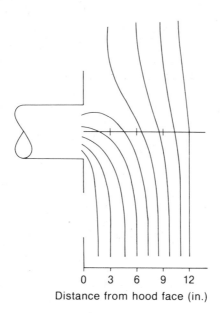

0   3   6   9   12
Distance from hood face (in.)

Figure 5.12 Airflow patterns in the vicinity of a 6-in. flanged circular hood exhausting 100 cfm in the presence of a 100-fpm cross-draft.

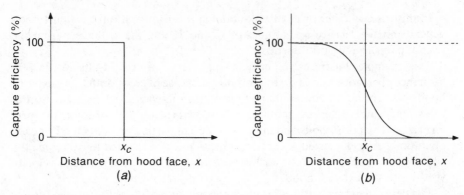

Figure 5.13 Hood capture efficiency versus distance from hood face: (*a*) theoretical, no turbulence; (*b*) actual, with turbulence.

and captured; any contaminants given off outside this area will likewise escape the hood. For the case shown in Fig. 5.12, the capture efficiency of the hood along the x axis might be expected to follow a step function (Fig. 5.13*a*). In actual practice, turbulent diffusion will act on the contaminant as it travels toward the hood, and the capture efficiency would be expected to follow a sigmoidal curve (Fig. 5.13*b*). Capture efficiency is discussed in more detail in Chapter 13.

Given the overriding importance of cross-drafts, the novice ventilation designer might think that the selection of a capture velocity for a specific application is relatively straightforward; one simply measures or estimates the cross-draft and selects a capture velocity sufficient to overcome it. It is unfortunate that the "real world" does not operate in such a straightforward manner. Experimental and theoretical work to quantify the effect of cross-drafts on hood capture efficiency, of the type described above and in Chapter 13, is in its infancy. Until considerably more work is done, the selection of capture velocities sufficient to overcome extraneous airflows will remain an art rather than a science. Our inability to quantify existing airflow patterns and the effect of a new exhaust system on those patterns limits the estimation of hood performance to an approximation. Given this, current design procedures result in capture velocities specified in a manner that acknowledges, either implicitly or explicitly, the uncertainties in the design process.

The selection of a capture velocity is covered in less than one page in the *Ventilation Manual*. The essential information is given in Table 3-1 of that volume, reproduced here in its entirety as Table 5.1. This table was adapted directly from Brandt (1947), and has seen little change in the last 40 years. Four categories, or ranges, of capture velocity are given. A particular category is selected based on a qualitative estimate of both cross-draft magnitude and velocity of contaminant release; no quantitative guidelines are given to help the designer select a category.

Within each category, a range of capture velocities is specified. Rules of thumb to assist in selecting a value from that range describe how much *room for error*

TABLE 5.1 Range of Capture Velocities

Condition of Dispersion of Contaminant	Examples	Capture Velocity (fpm)
Released with practically no velocity into quiet air	Evaporation from tanks; degreasing, etc.	50–100
Released at low velocity into moderately still air	Spray booths; intermittent container filling; low-speed conveyor transfers; welding; plating; pickling	100–200
Active generation into zone of rapid air motion	Spray painting in shallow booths; barrel filling; conveyor loading; crushers	200–500
Released at high initial velocity into zone of very rapid air motion	Grinding; abrasive blasting, tumbling	500–2000

In each category above, a range of capture velocity is shown. The proper choice of values depends on several factors:

Lower End of Range
1. Room air currents minimal or favorable to capture
2. Contaminants of low toxicity or of nuisance value only
3. Intermittent, low production
4. Large hood—large air mass in motion

Upper End of Range
1. Disturbing room air currents
2. Contaminants of high toxicity
3. High production, heavy use
4. Small hood—local control only

Source: ACGIH *Industrial Ventilation*, 20th Edition.

should be allowed when the exhaust hood is operated; in other words, they are an attempt to estimate the importance of ensuring complete contaminant capture for the operation in question. Selection of capture velocities at or near the lower end of each range in essence means that some escape of contaminants will be allowed, while selection from the upper end is an attempt to ensure complete capture.

Some difficulties in using this table should be obvious. The range of capture velocities presented is enormous, with the highest velocity being a factor of 40 greater than the lowest. Several critical terms meant to help in the selection of a value within this wide range are qualitative and almost cryptic. For example, the system designer must decide whether the contaminants are released into "quiet" air, "moderately still" air, air with "rapid air motion," or air with "very rapid air motion"; none of these terms are defined or quantified. Many other critical terms (e.g., size of hood, toxicity, active generation, room air currents favorable to capture) are similarly undefined. What if the process releases contaminants at high velocity into still air, or at low velocity into rapidly moving air? No hint is given as to the value to be selected under these conditions.

Inexperienced ventilation designers may thus find it difficult to use this table properly in the selection of capture velocity for exterior hoods. The designer has two approaches to making this critical design decision. Design plates for selected operations are specified in the *Ventilation Manual*. This approach has the effect of allowing the author of the design plate to select the capture velocity. The second approach is to choose what is thought to be a conservatively high value of capture velocity.

Both approaches have disadvantages. The indiscriminate use of design plates may not properly take into account specific characteristics of the application, such as cross-drafts, which are greater than the design plate was meant to overcome. An attempt to use an overly conservative capture velocity introduces two distinct possibilities. The value may not be conservative enough, in which case the hood capture efficiency may be low, or the capture velocity may be much higher than actually required. In the latter case the hood will have high capture efficiency, but the amount of air exhausted will be excessive and energy will be wasted. These possibilities are discussed in more detail in Chapter 13. In any case, the selection of a capture velocity remains the weakest part of the design procedure for an exterior hood.

5.3.2 Determination of Hood Airflow

The next step in designing an exterior hood is to select an airflow through the hood which will give the desired capture velocity at the furthest point of contaminant generation. To do this it is necessary to understand the airflow patterns that exist around an exhaust hood of the type being used.

Comparison of Blowing and Exhausting Air. A look at the different velocity patterns of blowing and exhausting air will illustrate the difficulties encountered in trying to use an exhaust hood to "reach out" and capture contaminants. The simple exhaust system shown in Fig. 5.14 will be used for illustration; it consists of a fan and plain rectangular outlets and inlets. The air being blown from a plain duct opening forms a jet that persists for a long distance from the opening.

The *ASHRAE Fundamentals Handbook* (1985) describes four distinct zones in a jet. The first two, an initial zone and a transition zone, are relatively short, extending about 12 duct diameters from the outlet. The third zone, called the zone of fully established turbulent flow, extends quite far, from 25 to 100 duct diameters. In this zone the velocity of a jet of air at a distance x from a plain round opening is

$$V(x) = \frac{KQ}{x\sqrt{A_f}} \tag{5.3}$$

where $K =$ a dimensionless constant
$Q =$ airflow at the opening, cfm
$A_f =$ cross-sectional area of the opening, ft^2

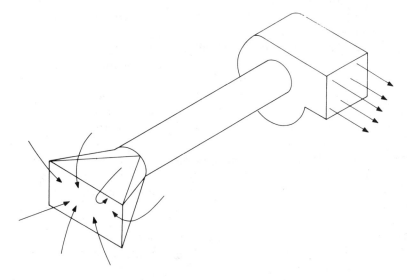

Figure 5.14   Simple exhaust system, showing both blowing and exhausting air.

For a round opening, Hemeon states that K is equal to 2. Solving Eq. 5.1 for Q and substituting for Q and K in Eq. 5.3 gives an expression for the velocity at a distance x as a fraction of the face velocity V_f:

$$\frac{V(x)}{V_f} = \frac{2\sqrt{A_f}}{x} \qquad (5.4)$$

Substituting the diameter of the opening D for the area gives the following equation:

$$\frac{V(x)}{V_f} = \frac{1.8D}{x} \qquad (5.5)$$

In the fully established turbulent zone, the velocity falls off linearly in front of a jet. The influence of the jet can be felt at some distance from the opening. From Eq. 5.5, the jet will still have 10% of its face velocity $[V(x)=0.1\ V_f]$ at a point 18 duct diameters downstream $(x=18D)$.

The characteristics of a jet can be contrasted with those of a source of suction, such as an exhaust hood. In the simple exhaust system shown in Fig. 5.14, the blower is the energy source that causes air to move; it does this by transferring air molecules from the downstream duct to the upstream duct. The act of physically removing air molecules from the downstream duct causes a partial vacuum to be created inside the duct. Air molecules move from outside the duct to fill this partial vacuum, and it is this motion that constitutes the airflow into the hood.

The ventilation system designer intends to create a velocity at some point *in front of* the hood opening in order to capture contaminants. Unfortunately, the air that moves into the hood opening to fill the partial vacuum does so from *all* directions, not only from directly in front. Some of the air actually enters the hood from *behind* the opening (Fig. 5.14). This means that the velocity of the entrained air falls off very rapidly with distance from the hood face. It would be desirable for the airflow into the hood to be highly directional, as in a jet, but the aerodynamic behavior of a suction source does not conform to this wish.

Theoretical Considerations for Exhausting Air. The relationship between velocity, airflow, and distance from the source of suction must be known if a hood is to be designed to attain a desired capture velocity at a particular location in front of the hood. This relationship can best be understood by first considering a hood with vanishingly small dimensions, so that it becomes a point source of suction (Fig. 5.15). This point source will draw air equally from all directions; if the amount of air being exhausted is Q, the velocity of the air at any distance x will be given by the simple formula

$$V(x) = \frac{Q}{A} \tag{2.6}$$

where A is the area through which the air is flowing. In this case the area is simply the surface of a sphere with radius x, so that

$$V(x) = \frac{Q}{4\pi x^2} \tag{5.6}$$

In theory, the velocity is inversely proportional to the square of the distance from the hood.

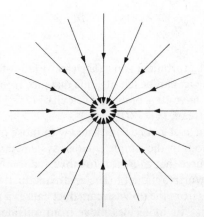

Figure 5.15   Streamlines into a point source of suction.

Experimental Determination of Capture Velocities. The simple inverse-square relationship demonstrated in Eq. 5.6 is valid only for (hypothetical) point sources of suction. Actual sources of suction vary from this model in two ways. The suction is applied over finite area (the hood face area A_f), and the hood and duct take up part of the physical space surrounding the point of suction, so the suction cannot operate over the surface of an entire sphere. Real-world considerations such as these have hindered the development of theoretical predictions of capture velocities in front of exhaust hoods.

This question was first investigated empirically in a series of experiments performed at the Harvard School of Public Health by Joseph DallaValle (1930), who measured the velocity contours and streamlines created in front of several basic hood shapes operated under suction. He studied flanged and unflanged square hoods, round hoods, and rectangular hoods with aspect ratios (i.e., hood width divided by hood length) ranging from 0.75 to 0.33. This work generated a large quantity of data concerning the nature of airflow patterns in front of these simple hood shapes; typical examples of his results are shown in Fig. 5.16, taken from his thesis. Shown are the measured velocity contours and streamlines in front of a plain 8-in. square opening (Fig. 5.16a) and a similar plot for the same hood with a 5-in. flange installed (Fig. 5.16b).

The centerline velocity for the unflanged hood decreases rapidly with distance from the hood face. The velocity is reduced to 50% of the face velocity at $x = 2.25$ in., and falls to 10% of the face velocity at $x = 8$ in., or one duct diameter. The reduction in velocity with distance is much more severe at locations off the centerline; for example, along the edge of the duct the velocity falls to the 50% value at a distance of 1.25 in. in front of the hood and falls to 10% at 6.75 in. Note also the flow of air from *behind* the hood.

The improvement in performance gained by the addition of the flange is apparent in Fig. 5.16b. Air is now prevented from flowing from behind the hood; since the same total amount of air is flowing it must now all enter from in front of the hood. This improves the "reach" of the hood, so that the centerline velocity falls to 10% of the face velocity at a distance of 9 in. rather than 8 in. as occurred without the flange. In general, flanges should be used on all exterior-type exhaust hoods to gain maximum benefit from the air being exhausted.

The data described above, which were collected by DallaValle over 50 years ago, still form the basis for the design of exterior exhaust hoods. Empirical formulas were developed that fit the data collected along the centerline of each basic hood type. When all the data for unflanged hoods were examined, it was found (DallaValle, 1952) that the best fit was obtained by using the formula

$$V(x) = \frac{Q}{10x^2 + A_f} \qquad (5.7)$$

The velocity of air into an unflanged exhaust hood at any distance x along the centerline in front of the hood, for a given airflow Q and hood face area A_f can be

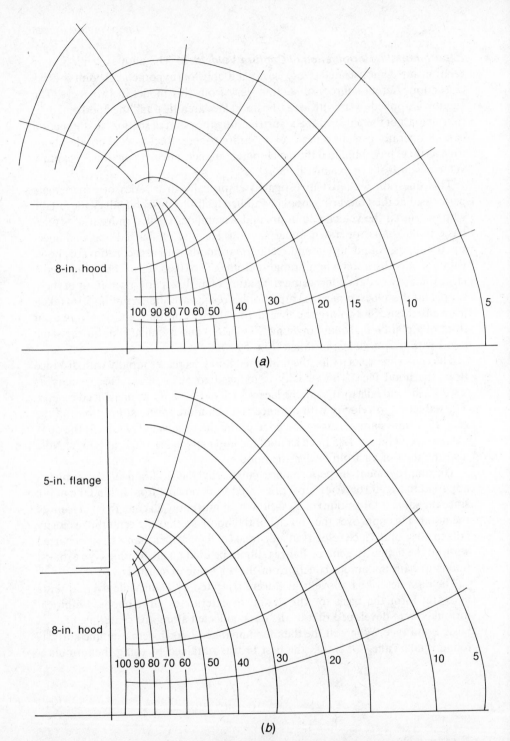

Figure 5.16 Typical equal velocity contours and streamlines measured by DallaValle (1930): (*a*) unflanged 8-in. square opening; (*b*) same hood with a 5-in. flange installed.

126

calculated with Eq. 5.7. This equation is valid for round hoods, square hoods, and rectangular hoods with aspect (i.e., width to length) ratios greater than 0.2.

This empirical formula is similar to the theoretical equation developed for a point source of suction (Eq. 5.6). Both include the inverse-square relationship between velocity and distance, with slightly differing constants (10 versus 4π); the empirical formula also includes the effect of the hood area, which, of course, is zero for the idealized point source.

To obtain the normalized velocity at any distance in front of the opening, Eqs. 5.1 and 5.7 can be combined:

$$V(x) = \frac{V_f A}{10x^2 + A_f}$$

Rearranging gives us

$$\frac{V(x)}{V_f} = \frac{A}{10x^2 + A_f} \tag{5.8}$$

The normalized velocity equations for blowing air (Eq. 5.4) and exhausting air (Eq. 5.6) are compared in Fig. 5.17 for a 1-ft-diameter round duct. The data have been plotted on log-log paper because of the drastic difference in performance at the two ends of the exhaust system. The blowing air forms a jet which still has 10% of its initial velocity at a distance of almost 20 duct diameters from the duct face. In contrast, the velocity of the air being drawn into the duct falls off very rapidly, reaching 10% of the face velocity in less than one duct diameter and 1% in about three duct diameters.

An equation similar to Eq. 5.8 has been developed to take into account the effect of the flange on the velocity characteristics in front of an exterior hood. DallaValle (1952) found that the following equation best fit his data for flanged hoods:

$$V(x) = \frac{Q}{0.75(10x^2 + A_f)} \tag{5.9}$$

In normalized form,

$$\frac{V(x)}{V_f} = \frac{A}{0.75(10x^2 + A_f)} \tag{5.10}$$

These equations illustrate the effect the flange has in increasing the "reach" of the hood; for a given value of airflow Q, the capture velocity at any distance x is increased by a factor of $1/0.75$, or 1.33. It should be emphasized that these are empirical formulas, not ones developed from theory. Since DallaValle, several investigators have conducted further studies into the performance of round, square, and rectangular hoods. Garrison (1981, 1983a, 1983b; Garrison and Byers, 1980a, 1980b) measured the centerline velocities in front of a large number of hoods operated at high face velocities (so-called low-volume/high-velocity

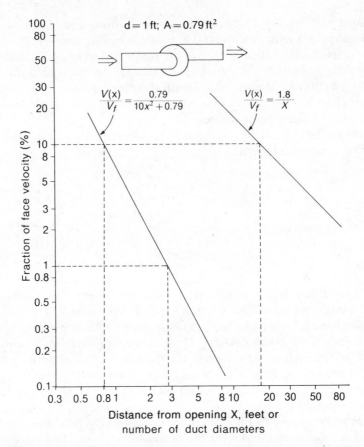

Figure 5.17 Comparison of air velocity patterns for blowing and exhausting air for a round duct 1 ft in diameter.

hoods) and developed empirical formulas relating V/V_f as a function of the relative distance x/D or x/W, where D is the hood diameter for round hoods and W is the width of square, rectangular, and slot hoods. He studied both flanged and unflanged hoods, and developed separate equations for each hood type.

Fletcher (1977, 1978, 1982; Fletcher and Johnson, 1982) performed similar studies but only on unflanged hoods. He developed a single empirical equation which accounts for hood shape by incorporating the hood width-to-length ratio.

Flynn and Ellenbecker (1985, 1986, 1987) developed equations for airflow into flanged circular hoods. Their approach differed from previous work in two respects: (1) the equations were derived theoretically from fluid mechanics rather than from empirical correlations; and (2) the equations describe the entire flow field, not just velocity along the hood centerline.

Two different models were developed. One, the so-called exact solution, rests on the assumption that potential flow theory describes the flow field in front of an exhaust hood. This assumption is valid at all points except near the hood face, where the model does not predict velocities well. The second model, the so-called approximate solution, modifies the first to improve prediction near the hood face. Figure 5.12 illustrates a typical flow field predicted by the approximate solution model.

The equations used in all of the above models to predict centerline velocity for round, square, and rectangular hoods, both flanged and unflanged, are presented in the Appendix to this chapter. The equations are given in dimensionless form for ready comparison with each other and with empirical data. One such comparison is shown in Fig. 5.18, which compares the centerline velocities predicted by the various models with DallaValle's data for a round 8-in. flanged slot hood exhausting 100 cfm. All of the models are seen to predict centerline velocities very well; Flynn and Ellenbecker's theoretical model predicts velocity slightly better than the empirical models at intermediate distances $(0.4 < V/V_f < 0.1)$ and has the additional advantage of predicting the entire flow field and not just centerline velocities.

Capture Velocities in Front of Slot Hoods. Slot hoods (Fig. 5.19*a*) are rectangular hoods with aspect ratios less than 0.2. Such hoods were studied by Leslie Silverman (1941, 1942*a*, 1942*b*, 1943) at the Harvard School of Public Health in the early 1940s as a follow-on to DallaValle's work. The behavior of slot hoods is fundamentally different from that of rectangular or round openings; this difference can be demonstrated both theoretically and experimentally.

A slot hood can be modeled in the ideal case by a line source of suction (Fig. 5.19*b*). Here the velocity falls off with distance as a function of the surface area of a cylinder. Ignoring the cylinder ends, the area of a cylinder is given by

$$A_c = 2\pi Lx \tag{5.11}$$

where L is the length of the cylinder and x is the radius. If the hypothetical line source of suction is inducing an airflow Q, the velocity at a distance x is then

$$V(x) = \frac{Q}{A_c}$$

$$= \frac{Q}{2\pi Lx} \tag{5.12}$$

Velocity is thus inversely proportional to distance for a line source of suction; this is an improvement over the point source (Eq. 5.6), where velocity falls off as the square of the distance. For a constant Q, however, the velocity at a point in front of a line source of suction is also a function of the length of the source L. Such sources of suction are useful only for contaminant sources which have an appreciable linear dimension, such as open-surface tanks.

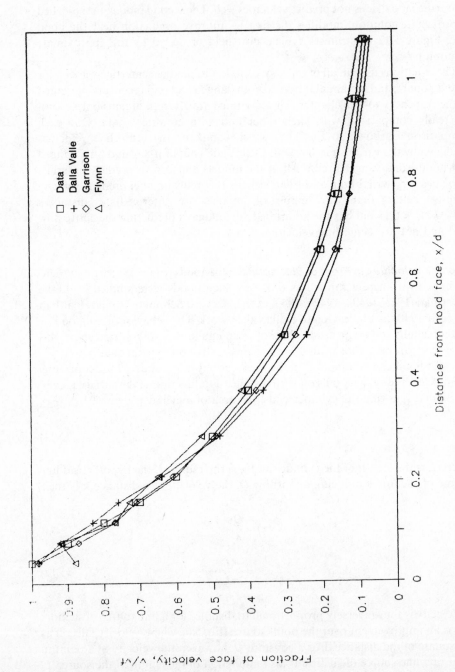

Figure 5.18 Comparison of models predicting centerline velocity as a function of distance for an 8-in. flanged circular hood. [Data from DallaValle (1930).]

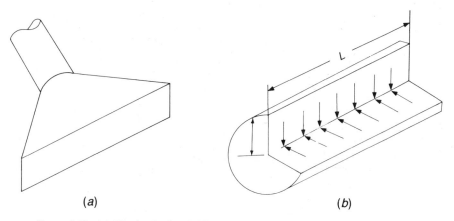

Figure 5.19   (a) Simple slot hood; (b) modeling a slot as a line source of suction.

Silverman (1943) measured velocity contours in front of several configurations of slot hoods, and derived empirical formulas relating capture velocity to airflow and distance. For unflanged slot hoods, the data were best fit by the following formula:

$$V(x) = \frac{Q}{3.7Lx} \qquad (5.13)$$

and for flanged hoods,

$$V(x) = \frac{Q}{0.75(3.7Lx)}$$

$$= \frac{Q}{2.8Lx} \qquad (5.14)$$

These equations are similar to those predicted by theory, differing only by a constant factor (i.e., 3.7 versus 2π).

In designing a slot hood, it is desirable to use a slot which is relatively narrow, so that the airflow entering the hood is distributed relatively evenly over its length. The principles involved are illustrated in Fig. 5.20. If the slot on this hood is too *wide*, the air will enter the hood primarily through the *center* of the slot, near the transition from the plenum to duct. Narrowing the slot will place a high-resistance element in series with the low-resistance plenum; since the pressure drop across the opening must be equal at all locations, the high resistance of a narrow slot will tend to even out the airflow along the length, ensuring uniform flow across the tank being ventilated.

It is possible, of course, to make the slot *too* narrow, causing an excessive pressure drop. Practical experience has shown that a slot sized to give a slot velocity of 1000 to 2000 fpm will provide good airflow distribution while

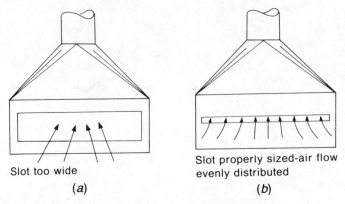

Slot too wide

(a)

Slot properly sized-air flow evenly distributed

(b)

Figure 5.20   Principles of slot design: (*a*) if the slot is too wide, most of the air will flow through the center of the slot; (*b*) narrowing the slot causes the airflow to distribute evenly.

avoiding excessive pressure drop. This is the velocity range prescribed in the design sections of the *Ventilation Manual.*

Recognizing that the average slot velocity V_s and width w are fixed by design considerations, Eqs. 5.13 and 5.14 can be rewritten to obtain normalized forms. In this case,

$$Q = V_s A_s$$
$$= V_s L w \tag{5.15}$$

where A_s is the slot area selected to give the desired slot velocity. Substituting and rearranging yields

$$V(x) = \frac{V_s L w}{3.7 L x} \tag{5.13}$$

$$\frac{V(x)}{V_s} = \frac{w}{3.7 x} \tag{5.16}$$

and for flanged hoods,

$$\frac{V(x)}{V_s} = \frac{w}{2.8 x} \tag{5.17}$$

Recent investigators have also studied slot hoods. The work of Fletcher and Garrison described previously also included slot hoods, and Conroy et al. (1987) have extended the work of Flynn and Ellenbecker to this subject. Centerline velocity equations for each slot model are also presented in the Appendix. Silverman's model predicts an infinite velocity at the hood face, and so is not accurate in this area. For x/w values greater than 1, all the models predict similar velocities. Fletcher's model can only be used for unflanged hoods, while that of

Conroy et al. can only be used for flanged hoods but has the advantage of predicting the entire flow field.

5.3.3 Exterior Hood Shape and Location

The basic principle behind shaping and locating an exterior hood is to make it as much like an enclosing hood as is physically possible. A complete enclosure is always the preferred design. Deviations from this ideal are caused by "real-world" considerations, such as worker or material access. Frequently, the designer must be satisfied with a partial enclosure, which is of course superior to a purely exterior hood. Examples of partial enclosures abound; they can be as simple as the inclusion of baffles at the ends of a welding bench (Fig. 5.21*a*, VS-416), or as complicated as a double side-draft hood around a foundry shakeout (Fig. 5.21*b*, VS-111).

Partially enclosing an exterior hood reduces the effects of cross-drafts and directs the exhaust airflow from the point of contaminant release toward the exhaust hood. To the extent that an exterior hood can be partially enclosed, the amount of airflow needed to capture all contaminants will be reduced.

5.4 DESIGN OF RECEIVING HOODS

Most receiving hoods fall into one of two categories, canopy hoods for heated processes and hoods for grinding and other material finishing operations. Each is a special category of exterior hood, where the contaminant capture is assisted by a property of the contaminated gas stream. The recommended design procedure is different for each of these hood types, so they will be described separately.

5.4.1 Canopy Hoods for Heated Processes

The design of ventilation systems for heated processes was studied and described extensively by Hemeon (1963). Goodfellow (1985) also discusses the subject in detail and presents design data for several heated processes. General principles for the design of ventilation systems for heated processes are *not* presented in the *Ventilation Manual*; the subject is covered only peripherally through several specific hood design plates. Basic principles for design are presented here; the reader who needs more information is referred to Hemeon (1963) and Goodfellow (1985).

Thermal Updrafts from Heated Sources. The principle underlying the operation of canopy receiving hoods is the capture of the thermal updraft created by a heated process; the contaminants are presumed to be contained in this heated airstream, so the capture of the air transported by the chimney effect or thermal head will ensure the efficient capture of the contaminants. A canopy hood used

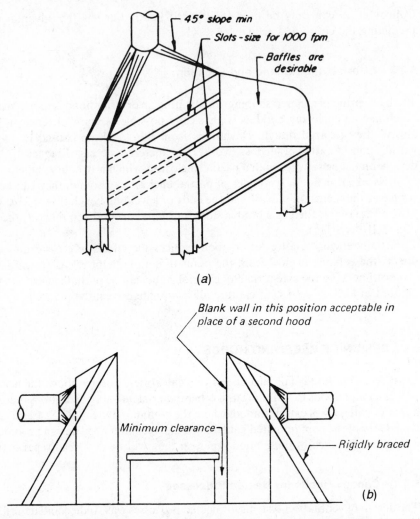

45° slope min

Slots - size for 1000 fpm

Baffles are desirable

(a)

Blank wall in this position acceptable in place of a second hood

Minimum clearance

Rigidly braced

(b)

Figure 5.21 Partially enclosing exterior hoods: (a) welding bench with end baffles; (b) double side-draft foundry shakeout hood. [After Hagopian and Bastress (1976).]

over a cold process, on the other hand, is an exterior hood; since the contaminants are not carried into the hood by the thermal head, the contaminants must be drawn into the hood by creating a sufficient capture velocity.

The difference between heated and unheated processes is illustrated in Fig. 5.22. For a room-temperature process (Fig. 5.22a), the canopy hood can be considered to be a booth with four open faces of dimension xy surrounding the process. Sufficient inward velocity must be developed across each face to ensure that the contaminants are drawn into the hood. The contaminants given off by a heated process are carried into the hood by the buoyancy of the exhaust gas (Fig. 5.22b). The required exhaust airflow in this case is governed by the rate at which

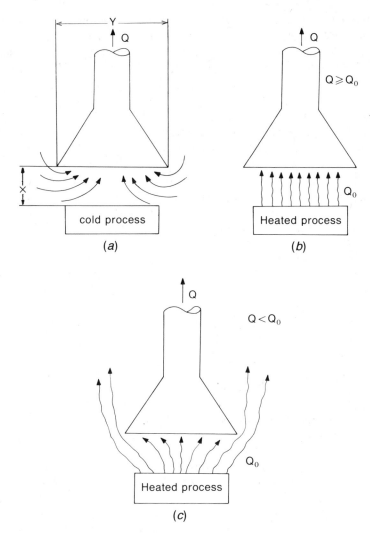

Figure 5.22 Comparison of canopy hoods used for heated and cold processes: (a) airflow patterns for a cold process, where the hood acts as an exterior hood; (b) airflow patterns for a heated process, where the induced airflow Q_h causes the hood to act as a receiving hood; (c) spillover for a heated process with $Q < Q_h$.

heated air is given off by the process; if the hood is properly sized and positioned, the exhaust airflow Q must be selected to be greater than the rate Q_h at which hot gas is generated. If this condition is not met, some of the heated air will "spill over" the edges of the canopy hood and escape (Fig. 5.22c).

Required Airflow for Heated Processes. The principal design consideration for canopy hoods, then, is the determination of Q_h. Hemeon (1963) and

Goodfellow (1985) have considered this problem in detail. The amount of airflow set in motion by a heated body Q_h is a function of the body's heat loss rate, which in turn is a function of the body's size, surface characteristics, and surface temperature relative to ambient temperature. The induced airflow rises and mixes turbulently with the surrounding air; it is also affected by any cross-drafts that might be present. The relative importance of these effects depends on the receiving hood location.

Hemeon distinguishes between low canopy hoods, which are located within about 3 ft of the heated surface, and high canopy hoods, which are located higher. The design of low hoods is much simpler since turbulent mixing and cross-draft interaction are less extensive than with high hoods. Low hoods are generally preferred because they can capture a higher percentage of the contaminants with a much lower required airflow than high hoods. If high hoods must be used, Goodfellow presents design equations and examples of proper design.

The design and use of slot hoods for heated sources is much more difficult than for room-temperature sources, due to the necessity of calculating the required capture velocity to overcome the buoyant plume. The design procedure for open-surface tanks discussed in Chapter 6 includes provisions for elevated liquid temperature; in general, the exhaust volumes specified by this procedure are much higher than those required for a canopy hood.

5.4.2 Hoods for Grinding Operations

Receiving hoods used in grinding and other finishing operations are quite different in concept from the canopy hoods described above. The property assisting in the transport of the contaminants to the hood is the momentum imparted to the particles by the grinding process. The hood is positioned and sized to "catch" the particles created by the grinding process, which are "thrown" with some momentum toward the hood (Fig. 5.23). Since the hood does not have to "reach" as far to capture the contaminant, the capture velocity can be supplied at some distance which is less than x, the separation between the wheel and the hood. The airflow required, using DallaValle's equations, would be less than if the particle throw were not taken into account. If these equations are used, the designer must estimate the new distance x' at which the capture velocity must be created (Fig. 5.23). This distance is given by

$$x' = x - S \tag{5.18}$$

where S is the particle stopping distance, defined as the distance a particle ejected into still air at an initial velocity will travel in decelerating to rest due to drag forces.

Hinds (1982) has calculated stopping distances for different-sized particles for the case of an initial velocity of 2000 fpm (10 m/s) (Table 5.2). These data illustrate the difficulty in "throwing" even fairly large particles an appreciable

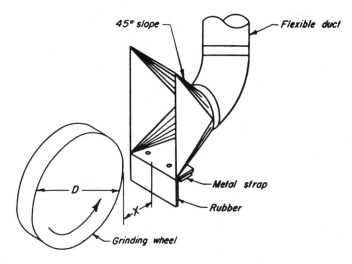

Figure 5.23 Receiving hood used for grinding. [After Hagopian and Bastress (1976).]

TABLE 5.2 Particle Stopping Distances for an Initial Velocity of 2000 fpm (10 m/s)

Particle Diameter (μm)	Stopping Distance (cm)
0.01	2.0×10^{-6}
0.1	6.8×10^{-5}
1.0	3.6×10^{-3}
10	0.23
100	12.7

distance in still air. Any particles that are in the inhalable range (i.e., $d_p < 10\ \mu$m) should be considered immovable in still air, so for industrial hygiene purposes it should always be assumed that $x = x'$. Larger particles, which may present a housekeeping problem if not captured, may travel appreciable distances and hood designs for such processes need to take this into account.

Although aerosol particles cannot travel a considerable distance in still air, the high-velocity projection of such particles into still air can *induce* an airflow similar to that caused by falling particles (Chapter 6). Hemeon (1963) investigated this phenomenon and concluded that significant amounts of air would be entrained only for large particles ($d_p > 1000\ \mu$m).

As part of a study of ventilation requirements for grinding operations (Bastress et al., 1974), detailed particle trajectories around these operations were modeled. A typical result of this effort is shown in Fig. 5.24. The behavior of particles can be divided into three categories, depending on particle diameter. Small particles

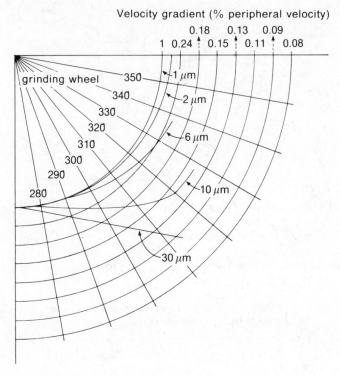

Figure 5.24 Particle trajectories around grinding operations.

$(d_p < 3\ \mu\text{m})$ travel around the periphery of the wheel and thus are not "thrown" toward the exhaust hood. Large particles $(d_p > 30\ \mu\text{m})$ are thrown in a nearly straight trajectory toward the exhaust hood, and thus are easily captured. Particles in the intermediate size range $(3\ \mu\text{m} < d_p < 30\ \mu\text{m})$ follow an intermediate trajectory which carries them some distance toward the hood and facilitates capture.

It is apparent from this analysis that the grinding wheel hood does act as a receiving hood for large particles but not for small, respirable ones. Such particles must be collected by creating a sufficiently high capture velocity near the surface of the grinding wheel. Bastress et al. (1974) found that respirable particles were not captured efficiently by typical grinding wheel exhaust systems and escaped to the workplace in the vicinity of the worker's breathing zone. For example, the study found that about 20% of the respirable particles generated when grinding mild steel escaped the exhaust system even when large airflows were employed (Fig. 5.25). Nonetheless, the standard hood designs recommended in the *Ventilation Manual*, while not 100% efficient at capturing respirable particles, were sufficient to provide worker protection at or below the threshold limit values for total and respirable inert dust.

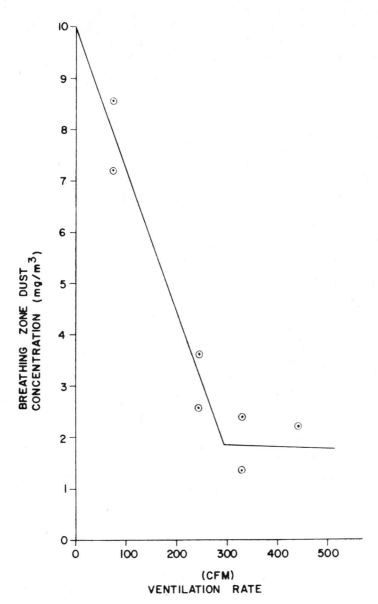

Figure 5.25 Typical performance characteristics of a ventilation system used in conjunction with a pedestal grinder.

5.5 EVALUATION OF HOOD PERFORMANCE

After installation the performance of exhaust hoods should be evaluated to confirm that (1) the airflow through the hood is equal to the design airflow, and

(2) the hood is capturing the contaminants given off by the process it is controlling. Because these issues more accurately assess the performance of the entire exhaust system rather than the exhaust hood alone, they are addressed in detail in Chapter 13.

LIST OF SYMBOLS

A	area
A_c	cylinder surface area
A_f	hood face cross-sectional area
A_s	slot area
D	duct diameter
d_p	particle diameter
K	constant in Eq. 5.3
L	cylinder length
Q	airflow
Q_h	airflow induced by a heated process
V	velocity
S	particle stopping distance
V_c	capture velocity
V_d	cross-draft velocity
V_f	hood face velocity
V_s	slot velocity
W	hood width
w	slot width
x	distance, length
y	distance, length

REFERENCES

American Conference of Governmental Industrial Hygienists, Committee on Industrial Ventilation (1988), *Industrial Ventilation*, 20th ed., ACGIH, Lansing, MI.

American Society of Heating, Refrigerating and Air Conditioning Engineers (1985), *ASHRAE Handbook—1985 Fundamentals*, ASHRAE, Atlanta, GA.

Bastress, E., J. Niedzwecki, and A. Nugent (1974), "Ventilation Requirements for Grinding, Buffing, and Polishing Operations," U.S. Department of Health, Education, and Welfare, National Institute for Occupational Safety and Health, Publication No. 75–107, Washington, DC.

Brandt, A. (1947), *Industrial Health Engineering*, Wiley, New York.

Conroy, L., M. Ellenbecker, and M. Flynn (1988), "Prediction and Measurement of Velocity into Flanged Slot Hoods," *Am. Ind. Hyg. Assoc. J.*, **49**(5): 226–234.

Conroy, L., and M. Ellenbecker (1988), "Capture Efficiency of Flanged Slot Hoods under the Influence of a Uniform Crossdraft," Doctoral thesis, Harvard School of Public Health, Boston.

DallaValle, J. (1930), "Studies in the Design of Local Exhaust Hoods," Doctoral thesis, Harvard School of Public Health, Boston.

DallaValle, J. (1952), *Exhaust Hoods*, Industrial Press, New York.

Dalrymple, H. L. (1986), "Development and Use of a Local Exhaust Ventilation System to Control Fume from Hand Held Soldering Irons," in *Proceedings of the 1st International Symposium on Ventilation for Contaminant Control*, October 1–3, 1985, Toronto, Canada, H. G. Goodfellow, ed., Elsevier, New York.

Ellenbecker, M., R. Gempel, and W. Burgess (1983), "Capture Efficiency of Local Exhaust Ventilation Systems," *Am. Ind. Hyg. Assoc. J.*, **44**(10): 752–755.

Esman, N., D. Weyel, and F. McGuigan (1986), "Aerodynamic Properties of Exhaust Hoods," *Am. Ind. Hyg. Assoc. J.*, **47**(8): 448–454.

Fletcher, B. (1977), "Centreline Velocity Characteristics of Rectangular Unflanged Hoods and Slots under Suction," *Ann. Occup. Hyg.*, **20**: 141–146.

Fletcher, B. (1978), "Effect of Flanges on the Velocity in Front of Exhaust Ventilation Hoods," *Ann. Occup. Hyg.*, **21**: 265–269.

Fletcher, B. (1982), "Centreline Velocity Characteristics of Local Exhaust Ventilation Hoods," *Am. Ind. Hyg. Assoc. J.* **43**(8): 626–627.

Fletcher, B., and A. Johnson (1982), "Velocity Profiles around Hoods and Slots and the Effects of an Adjacent Plane," *Ann. Occup. Hyg.*, **25**(4): 365–372.

Fletcher, B., and A. Johnson (1986), "The Capture Efficiency of Local Exhaust Ventilation Hoods and the Role of Capture Velocity," in *Proceedings of the 1st International Symposium on Ventilation for Contaminant Control*, October 1–3, 1985, Toronto, Canada, H. G. Goodfellow, ed., Elsevier, New York.

Flynn, M., and M. Ellenbecker (1985), "The Potential Flow Solution for Air Flow into a Flanged Circular Hood," *Am. Ind. Hyg. Assoc. J.*, **46**(6): 318–322.

Flynn, M., and M. Ellenbecker (1986), "Capture Efficiency of Flanged Circular Local Exhaust Hoods," *Ann. Occup. Hyg.*, **30**(4): 497–513.

Flynn, M., and M. Ellenbecker (1987), "Empirical Validation of Theoretical Velocity Fields into Flanged Circular Hoods," *Am. Ind. Hyg. Assoc. J.*, **48**(4): 380–389.

Garrison, R. (1981), "Centerline Velocity Gradients for Plain and Flanged Local Exhaust Inlets," *Am. Ind. Hyg. Assoc. J.*, **42**(10): 739–746.

Garrison, R. (1983a), "Velocity Calculation for Local Exhaust Inlets—Empirical Design Equations," *Am. Ind. Hyg. Assoc. J.*, **44**(12): 937–940.

Garrison, R. (1983b), "Velocity Calculation for Local Exhaust Inlets—Graphical Design Concepts," *Am. Ind. Hyg. Assoc. J.*, **44**(12): 941–947.

Garrison, R., and D. Byers (1980a), "Static Pressure and Velocity Characteristics of Circular Nozzles for High Velocity/Low Volume Exhaust Ventilation," *Am. Ind. Hyg. Assoc. J.*, **41**(11): 803–811.

Garrison, R., and D. Byers (1980b), "Static Pressure, Velocity, and Noise Characteristics of Rectangular Nozzles for High Velocity/Low Volume Exhaust Ventilation," *Am. Ind. Hyg. Assoc. J.*, **41**(12): 855–863.

Goodfellow, H. (1985), *Advanced Design of Ventilation Systems for Contaminant Control*, Elsevier, New York.

Hagopian, J., and E. Bastress (1976), "Recommended Industrial Ventilation Guidelines," Arthur D. Little, Inc., Contract No. CDC-99-74-33, U.S. Department of Health, Education, and Welfare, Public Health Service, Centers for Disease Control, National Institute for Occupational Safety and Health, Cincinnati, OH, January.

Hemeon, W. (1963), *Plant and Process Ventilation*, 2nd ed., Industrial Press, New York.

Hinds, W. (1982), *Aerosol Technology: Properties, Behavior, and Measurement of Airborne Particles*, Wiley, New York.

Jansson, A. (1980), "Capture Efficiencies of Local Exhausts for Hand Grinding, Drilling and Welding," *Staub-Reinhalt. Luft* (English ed.), **43**(3): 111–113.

Jansson, A. (1985), "Description of Air Movement Outside a Circular Exhaust Opening without Flange" (in Swedish), Undersökningsrapport No. 36.

Regnier, R., R. Braconnier, and G. Aubertin (1986), "Study of Capture Devices Integrated into Portable Machine-Tools," in *Proceedings of the 1st International Symposium on Ventilation for Contaminant Control*, October 1–3, 1985, Toronto, Canada, H. G. Goodfellow, ed., Elsevier, New York.

Silverman, L. (1941), "Fundamental Factors in the Design of Lateral Exhaust Hoods for Industrial Tanks," *J. Ind. Hyg. Toxicol.*, **23**(5): 187–266.

Silverman, L. (1942a), "Centerline Velocity Characteristics of Round Openings under Suction," *J. Ind. Hyg. Toxicol.*, **24**(9): 259–266.

Silverman, L. (1942b), "Velocity Characteristics of Narrow Exhaust Slots," *J. Ind. Hyg. Toxicol.*, **24**(9): 267–276.

Silverman, L. (1943), "Fundamental Factors in the Design of Exhaust Hoods," Doctoral thesis, Harvard School of Public Health, Boston.

APPENDIX: EXTERIOR HOOD CENTERLINE VELOCITY MODELS

Applicable Range	Equation	Equation Number
Round Hoods, Unflanged		
DALLAVALLE $$0 < \frac{x}{D} < \infty$$	$$\frac{V}{V_f} = \frac{1}{12.7\,(x/D)^2 + 1}$$	(5.19)
FLETCHER $$0 < \frac{x}{D} < \infty$$	$$\frac{V}{V_f} = \frac{1}{10.9(x/D)^2 + 1}$$	(5.20)
GARRISON $$0 < \frac{x}{D} < 0.5$$	$$\frac{V}{V_f} = 1.1\,(0.06)^{x/D}$$	(5.21)
$$0.5 \leqslant \frac{x}{D} \leqslant 1.5$$	$$\frac{V}{V_f} = 0.08\left(\frac{x}{D}\right)^{-1.7}$$	(5.22)
Round Hoods, Flanged		
DALLAVALLE $$0 < \frac{x}{D} < \infty$$	$$\frac{V}{V_f} = \frac{1}{9.6(x/D)^2 + 0.75}$$	(5.23)

APPENDIX: (Continued)

Applicable Range	Equation	Equation Number
FLYNN (APPROXIMATE) $$0 < \frac{x}{D} < \infty$$	$$\frac{V}{V_f} = \frac{\sqrt{3}\,\varepsilon^2}{2(3-2\varepsilon^2)^{1/2}}$$ where $\varepsilon^2 = \dfrac{1}{4(x/D)^2+1}$	(5.24)
FLYNN (EXACT) $$0 < \frac{x}{D} < \infty$$	$$\frac{V}{V_f} = \frac{1}{8(x/D)^2+1}$$	(5.25)
GARRISON $$0 < \frac{x}{D} < 0.5$$	$$\frac{V}{V_f} = 1.1\,(0.07)^{x/D}$$	(5.26)
$$0.5 \leqslant \frac{x}{D} \leqslant 1.5$$	$$\frac{V}{V_f} = 0.10\left(\frac{x}{D}\right)^{-1.6}$$	(5.27)

Square Hoods, Unflanged (side = a)

Applicable Range	Equation	Equation Number
DALLAVALLE $$0 < \frac{x}{a} < \infty$$	$$\frac{V}{V_f} = \frac{1}{10(x/a)^2+1}$$	(5.28)
FLETCHER $$0 < \frac{x}{a} < \infty$$	$$\frac{V}{V_f} = \frac{1}{8.6(x/a)^2+0.93}$$	(5.29)
GARRISON $$0 < \frac{x}{a} < 0.5$$	$$\frac{V}{V_f} = 1.07\,(0.09)^{x/\alpha}$$	(5.30)
$$0.5 \leqslant \frac{x}{a} \leqslant 1.5$$	$$\frac{V}{V_f} = 0.10\left(\frac{x}{a}\right)^{-1.7}$$	(5.31)

Square Hoods, Flanged

Applicable Range	Equation	Equation Number
DALLAVALLE $$0 < \frac{x}{a} < \infty$$	$$\frac{V}{V_f} = \frac{1}{7.5(x/a)^2+0.75}$$	(5.32)
GARRISON $$0 < \frac{x}{a} < 0.5$$	$$\frac{V}{V_f} = 1.07\,(0.11)^{x/a}$$	(5.33)
$$0.5 \leqslant \frac{x}{a} \leqslant 1.5$$	$$\frac{V}{V_f} = 0.12\left(\frac{x}{a}\right)^{-1.6}$$	(5.34)
FLYNN (EXACT) $$0 < \frac{x}{D} < \infty$$	$$\frac{V}{V_f} = \frac{1}{6.3(x/a)^2+0.25}$$	(5.35)

Rectangular Hoods, Unflanged (0.2 < w/L < 1)

Applicable Range	Equation	Equation Number
DALLAVALLE $$0 < \frac{x}{w} < \infty$$	$$\frac{V}{V_f} = \frac{1}{10\,(x^2/wL)+1}$$	(5.36)

APPENDIX: (Continued)

Applicable Range	Equation	Equation Number

FLETCHER

$0 < \dfrac{x}{w} < \infty$

$$\frac{V}{V_f} = \frac{1}{8.58\alpha^2 + 0.93}$$

$$\alpha = \frac{x}{(wL)^{1/2}} \left(\frac{w}{L}\right)^{-\beta}$$

$$\beta = 0.2\left[\frac{x}{(wL)^{1/2}}\right]^{1/3}$$

(5.37)

GARRISON [*]

At $w/L = 0.5$:

$0 < \dfrac{x}{w} < 0.5$ $\dfrac{V}{V_f} = 1.07(0.14)^{x/w}$ (5.38)

$0.5 \leqslant \dfrac{x}{w} < 1.0$ $\dfrac{V}{V_f} = 0.18\left(\dfrac{x}{w}\right)^{-1.2}$ (5.39)

$1.0 \leqslant \dfrac{x}{w} \leqslant 2.0$ $\dfrac{V}{V_f} = 0.18\left(\dfrac{x}{w}\right)^{-1.7}$ (5.40)

At $w/L = 0.25$:

$0 < \dfrac{x}{w} < 0.5$ $\dfrac{V}{V_f} = 1.07(0.18)^{x/w}$ (5.41)

$0.5 \leqslant \dfrac{x}{w} < 1.0$ $\dfrac{V}{V_f} = 0.23\left(\dfrac{x}{w}\right)^{-1.0}$ (5.42)

$1.0 \leqslant \dfrac{x}{w} \leqslant 2.5$ $\dfrac{V}{V_f} = 0.23\left(\dfrac{x}{w}\right)^{-1.5}$ (5.43)

Rectangular Hoods, Flanged

DALLAVALLE

$0 < \dfrac{x}{w} < \infty$ $\dfrac{V}{V_f} = \dfrac{1}{7.5\, x^2/wL}$ (5.44)

CONROY (EXACT)

$0 < \dfrac{x}{w} < \infty$ $\dfrac{V}{V_f} = \dfrac{1}{2\pi[0.25 + (x/w)^2]^{1/2}[0.25 + (x/L)^2]^{1/2}}$ (5.45)

GARRISON

For $w/L = 0.5$:

$0 < \dfrac{x}{w} < 0.5$ $\dfrac{V}{V_f} = 1.07\,(0.17)^{x/w}$ (5.46)

$0.5 \leqslant \dfrac{x}{w} < 1.0$ $\dfrac{V}{V_f} = 0.21\left(\dfrac{x}{w}\right)^{-1.1}$ (5.47)

$1.0 \leqslant \dfrac{x}{w} \leqslant 2.0$ $\dfrac{V}{V_f} = 0.21\left(\dfrac{x}{w}\right)^{-1.6}$ (5.48)

For $w/L = 0.25$:

$0 < \dfrac{x}{w} < 0.5$ $\dfrac{V}{V_f} = 1.07(0.22)^{x/w}$ (5.49)

$0.5 \leqslant \dfrac{x}{w} < 1.0$ $\dfrac{V}{V_f} = 0.27\left(\dfrac{x}{w}\right)^{-0.9}$ (5.50)

APPENDIX: (*Continued*)

Applicable Range	Equation	Equation Number
$1.0 \leqslant \dfrac{x}{w} \leqslant 3.0$	$\dfrac{V}{V_f} = 0.27\left(\dfrac{x}{w}\right)^{-1.4}$	(5.51)

Slot Hoods, Unflanged

SILVERMAN

$0 < \dfrac{x}{w} < \infty$	$\dfrac{V}{V_s} = 0.27\left(\dfrac{x}{w}\right)^{-1.0}$	(5.52)

GARRISON

$0 < \dfrac{x}{w} < 0.5$	$\dfrac{V}{V_s} = 1.07(0.19)^{x/w}$	(5.53)
$0.5 \leqslant \dfrac{x}{w} < 1.0$	$\dfrac{V}{V_s} = 0.24\left(\dfrac{x}{w}\right)^{-1.0}$	(5.54)
$1.0 \leqslant \dfrac{x}{w} \leqslant 3.5$	$\dfrac{V}{V_f} = 0.24\left(\dfrac{x}{w}\right)^{-1.2}$	(5.55)

FLETCHER

$0 < \dfrac{x}{w} < \infty$	$\dfrac{V}{V_s} = \dfrac{1}{8.58\alpha^2 + 0.93}$	(5.56)

$$\alpha = \frac{x}{(wL)^{1/2}}\left(\frac{w}{L}\right)^{-\beta}$$

$$\beta = 0.2\left[\frac{x}{(wL)^{1/2}}\right]^{1/3}$$

Slot Hoods, Flanged

SILVERMAN

$0 < \dfrac{x}{w} < \infty$	$\dfrac{V}{V_s} = 0.36\left(\dfrac{x}{w}\right)^{-1.0}$	(5.57)

GARRISON

$0 < \dfrac{x}{w} < 0.5$	$\dfrac{V}{V_s} = 1.07(0.22)^{x/w}$	(5.58)
$0.5 \leqslant \dfrac{x}{w} < 1.0$	$\dfrac{V}{V_s} = 0.29\left(\dfrac{x}{w}\right)^{-0.8}$	(5.59)
$1.0 \leqslant \dfrac{x}{w} \leqslant 4.0$	$\dfrac{V}{V_s} = 0.29\left(\dfrac{x}{w}\right)^{-1.1}$	(5.60)

CONROY (EXACT)

$0 < \dfrac{x}{w} < \infty$	$\dfrac{V}{V_s} = \dfrac{1}{2\pi[0.25+(x/w)^2]^{1/2}[0.25+(x/L)^2]^{1/2}}$	(5.61)

*Empirical coefficients were calculated only at $w/L = 0.5$ and $w/L = 0.25$.

6

Hood Designs for Specific Applications

The hood designs in Chapter 10 of the *Ventilation Manual* cover a range of industrial operations and include recommendations on hood geometry, airflow or critical control velocity, duct velocity, and entry loss. Caution is necessary in applying these designs to a specific plant operation since there may be plant-to-plant differences in industrial processes such that an exhaust system which is effective at one facility may not work at another. The ventilation designer must use knowledge of the specific materials in use at the plant, production rate, work practices, and special production requirements to modify the *Ventilation Manual* hood designs where appropriate.

In this chapter we cover the application of hood design data on selected industrial operations chosen because they are commonly encountered in general industry and may present potential occupational health risk to workers if air contaminants released from the process are not controlled effectively. In the case of electroplating, effective design data are available but their applications are frequently poor due to difficulty in interpreting the design data in Chapter 10 of the *Ventilation Manual* or in modifying the data to fit a related process. Painting and welding are included in this discussion due to their widespread application in industry. Material transfer operations are covered in the *Ventilation Manual*, but the designer should be aware of an alternative design technique, the induced airflow method, which complements the *Ventilation Manual* design approach. Chemical processing operations are given little attention in the *Ventilation Manual* and semiconductor gas cabinet design is not covered in any source book.

6.1 ELECTROPLATING

There are numerous metal treatment and coating operations associated with the electroplating industry. Normally, these processes are conducted in open surface tanks and the toxic air contaminants released by the process must be controlled by local exhaust ventilation techniques. The local exhaust design data on open surface tanks presented in Chapter 10 of the *Ventilation Manual* define hood geometry and airflow required for effective contaminant control and are the culmination of over five decades of research and field experience. In the early twentieth century, open-surface tanks were frequently placed on an outside plant wall and exhausted by a wall fan. Early ventilation designs also favored open canopy hoods; however, the deficiencies of these hoods limit their application in modern plants. Two hood types, enclosing hoods and lateral hoods, offer effective control of air contaminants released from open-surface tanks. The design of both hood types is discussed below.

6.1.1 Hood Design

As noted in Chapter 10 of the *Ventilation Manual*, enclosing hoods are preferred for many industrial operations since they permit efficient containment of toxic air contaminants with minimum airflow. The principal difficulty with this design is that the hood may restrict access to the operation. An enclosing hood with a conveyor access slot and open ends (Fig. 6.1a) finds wide application in automated plating operations; in a job shop with manual transfer of parts from tank to tank an enclosing hood with an open-front face may be acceptable (Fig. 6.1b).

The open-face area of an enclosing hood must be defined accurately in the design phase so that airflow can be calculated. A commitment must be obtained from production management on the planned degree of enclosure which will minimize airflow yet will not interfere with operations. Frequently, enclosing hood systems are properly designed and installed but are rendered useless when operating personnel remove a portion of the enclosure to gain access to the tank for the transfer of parts.

The three exhaust hoods in Figs. 6.2 and 6.3 are characterized as lateral exhaust hoods since they are designed to provide a horizontal sweep of air across the surface of the tank to convey the toxic air contaminants to the hood. This is the most popular open-surface tank hood in electroplating job shops; the operator has access to the tank contents from three or four sides, thus permitting easy movement of parts from one tank to another.

An important feature of the hoods with an upward plenum (Fig. 6.2) is that the fishtail taper from hood to duct provides a baffle to shield the tank from room drafts which can interfere with hood capture efficiency. The lip or perimeter slot exhaust hood (Fig. 6.3) requires that external baffles be installed or the tank must be positioned against a wall to provide equivalent shielding from interfering drafts.

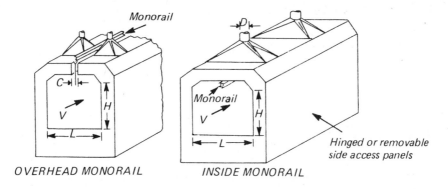

OVERHEAD MONORAIL   INSIDE MONORAIL

Q = *exhaust volume, cfm, equals A V*
H = *height of rectangular opening, feet*
L = *length of opening, feet*
A = *area of openings in enclosure*
   = *2 HL + area of monorail slot*
V = *minimum face velocity, fpm*
Duct velocity = 1500 fpm minimum
Entry loss = entry loss factor for tapered hood x duct VP
Make C, L, and H only large enough to pass work and hanger

Figure 6.1 (*a*) Enclosing hood in an electroplating shop with two open ends for material transport by conveyor.

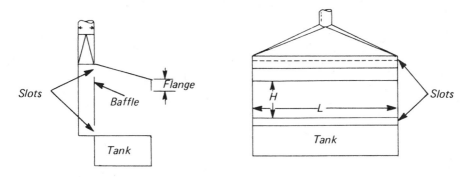

Q = *exhaust volume, cfm, equals A V*
H = *height of open area, feet*
L = *length of open area, feet*
A = *area of openings in enclosure, sq ft, equals L H*
V = *minimum face velocity, fpm*
Duct velocity = 1500 fpm minimum
Slot velocity = 2000 fpm minimum
Entry loss = 1.78 slot VP plus entry loss factor for tapered
   hood x duct VP

Figure 6.1 (*b*) Enclosing hood with an open front face which is acceptable for manual transfer of parts.

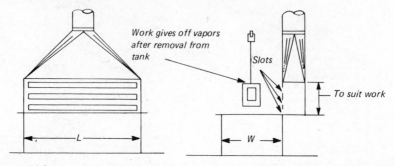

Figure 6.2 (*a*) So-called "pickling hood" with a slotted plenum. The parts removed from the tank can be positioned in front of this hood for collection of toxic air contaminants.

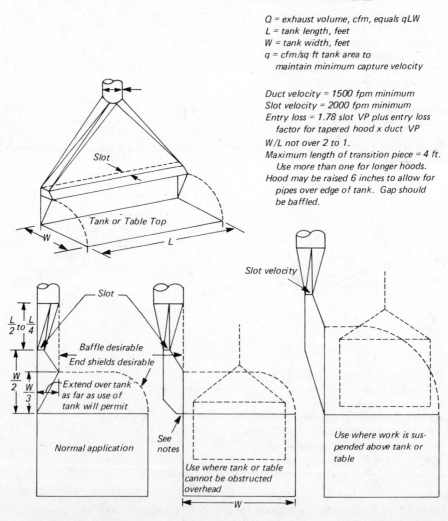

Q = exhaust volume, cfm, equals qLW
L = tank length, feet
W = tank width, feet
q = cfm/sq ft tank area to
   maintain minimum capture velocity

Duct velocity = 1500 fpm minimum
Slot velocity = 2000 fpm minimum
Entry loss = 1.78 slot VP plus entry loss
   factor for tapered hood x duct VP
W/L not over 2 to 1.
Maximum length of transition piece = 4 ft.
   Use more than one for longer hoods.
Hood may be raised 6 inches to allow for
   pipes over edge of tank. Gap should
   be baffled.

Figure 6.2 (*b*) Common fishtail hood on tanks with a width-to-length ratio of less than 0.5.

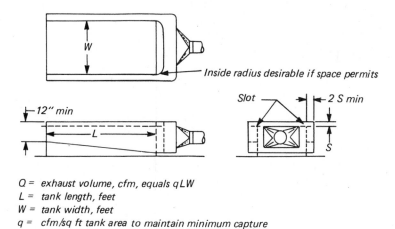

Q = exhaust volume, cfm, equals qLW
L = tank length, feet
W = tank width, feet
q = cfm/sq ft tank area to maintain minimum capture
      velocity
Duct velocity = 1500 fpm minimum
Slot velocity = 2000 fpm minimum
Entry loss = 1.78 slot VP plus entry loss factor for tapered hood x duct VP

Figure 6.3   Lateral exhaust hood with an end takeoff.

6.1.2 Airflow

The preliminary layout for an electroplating shop presented to the ventilation designer will include tank size, approximate location, tank contents, operating temperature, and materials of construction. As noted above, if the facility is highly automated, semi-enclosed hoods (Fig. 6.1) may be specified by the designer to provide efficient control of toxic air contaminants at minimal airflow. Job shop operations requiring direct tank access may use one of the hoods shown in Fig. 6.2 or 6.3. Tank width-to-length ratio has a critical impact on the minimum airflow necessary to control toxic air contaminants effectively. If the tank width is less than 3 ft, a lateral exhaust hood exhausting from one side of the tank may be adequate. A wider tank may require lateral slot exhaust from both sides of the tank, a centerline exhaust (*Ventilation Manual*, VS-503C), or a push-pull system (*Ventilation Manual*, VS-504).

The choice of hood type will have a significant influence on the airflow necessary to provide control of toxic air contaminants released from the tank. A rational approach to calculating the required airflow on open-surface tank operations has evolved during the past 50 years. Through the 1940s, the exhaust airflow for electroplating tanks was based on a fixed airflow per unit tank area (cfm/ft^2). The criterion varied with the agency or authority. During this period the New York Department of Industrial Hygiene specified 120 cfm/ft^2 for electroplating tanks handling toxic chemicals and operating at temperatures below 160°F.

The first step toward a more rational design approach occurred with the publication of American National Standards Institute (ANSI) Standard Z9.1-

1957. This standard proposed that the airflow per unit tank area be based on a two-part code that defines the hazard potential of the bath contents and the likelihood of release of the contents to the work environment. The first part of the code is the hazard potential classification (A through D); it is a rating based on the toxicity and the fire and explosion characteristics of the bath contents (Table 6.1). Classifications are chosen based on the ACGIH *threshold limit value* for the most toxic bath component and the flash point for the bath. The higher of the two hazard potential classes is used in the design.

The second part of the code (class 1 through 4) describes the potential for release of toxic air contaminants to the workplace based on tank operating temperature, evaporation rate, and gassing rate (Table 6.2). Gassing rate applies to electrolytic baths such as those encountered in plating and alkaline cleaning. In an electrolytic bath not all energy goes into the transfer of metal from the bath to the part. A portion of the energy, depending on bath efficiency, causes dissociation of water with hydrogen gas released at the cathode and oxygen at the anode. The gas bubbles produced in this fashion rise and burst at the surface of the bath, generating mist of the bath contents, which may be inhaled by the worker. Volatile bath contents may also be released when the bath is heated.

TABLE 6.1   Determination of Hazard Potential Class

Hazard Potential	TLV		Flash point
	Gas or vapor (ppm)	Mist (mg/m³)	
A	0–10	0.00–0.1	—
B	11–100	0.11–1.0	Under 100°F
C	101–500	1.1–10	100–200°F
D	Over 500	Over 10	Over 200°F

TABLE 6.2   Determination of Classification for Rate of Vapor, Gas, and Mist Evolution

Rate	Liquid Temperature (°F)	Degrees Below Boiling Point	Relative Evaporation[a]	Gassing
1	Over 200	0–20	Fast	High
2	150–200	21–50	Medium	Medium
3	94–149	51–100	Slow	Low
4	Under 94	Over 100	Nil	Nil

[a]Relative evaporation rate is determined according to the method described by A. K. Doolittle in *Ind. Eng. Chem.*, **27**: 1169 (1939), where time for 100% evaporation is as follows: fast, 0–3 hours; medium, 3–12 hours; slow, 12–50 hours; nil, more than 50 hours. For classification of specific contaminants, see Occupational Safety and Health Administration (1985), 29 CFR 1910. 1000 Washington, D.C.

Finally, air agitation used to improve product quality may generate a coarse mist. Using the technique shown in Table 6.2, the bath is assigned a numerical class based on the ability of the tank contents to be released to the workroom.

The combined hazard potential–contaminant evolution rate classification is used to choose the appropriate design velocity for a given operation and tank geometry (Table 6.3). The exhaust airflow is then calculated based on the required design velocity, the hood type, and the tank width-to-length ratio (Table 6.4), where width is the distance the hood must pull the air.

TABLE 6.3   Minimum Capture Velocity for Tank

Class	Enclosing Hood Face Velocity (fpm) One Open Side	Two Open Sides	Lateral Exhaust Capture Velocity (fpm)
A-1, and A-2	100	150	150
A-3, B-1, B-2, and C-1	75	100	100
B-3, C-2, and D-1[a]	65	90	75
A-4 C-3, and D-2[a]	50	75	50
B-4, C-4, D-3,[a] and D-4	General room ventilation required		

[a]Where complete control of hot water is desired, design as next highest class.

TABLE 6.4   Minimum Exhaust Airflow for Lateral Exhaust Hoods

Required Minimum Capture Velocity (fpm) (from Table 6.3)	Cubic Feet per Minute per Square Foot to Maintain Required Minimum Capture Velocities at Following Tank Width (W)/Tank Length (L) Ratios				
	0.0–0.09	0.1–0.24	0.25–0.49	0.5–0.99	1.0–2.0
Hood along one side or two parallel sides of tank when one hood is against a wall or baffle; also for a manifold along tank centerline					
50	50	60	75	90	100
75	75	90	110	130	150
100	100	125	150	175	200
150	150	190	225	260	300
Hood along one side or two parallel sides of free standing tank not against wall or baffle					
50	75	90	100	110	125
75	110	130	150	170	190
100	150	175	200	225	150
150	225	260	300	340	375

The ANSI approach described above results in a 16-step tank classification scale ranging from A1 (highest capture velocity) to D4 (lowest capture velocity). Although the procedure was a step forward, it did require familiarity with the open-surface tank processes, and for this reason many designers continued to use the earlier simple criteria of airflow per unit tank area.

In 1953 the New York State Division of Industrial Hygiene published a series of tables that encouraged the use of the ANSI rating approach. These tables covered the common electroplating processes and included information on the component of the bath released, physical form of the contaminant, rate of gassing, and tank operating temperature. The designer could take these data and use Tables 6.1 and 6.2 to assign a hazard potential–release rate classification for each individual open-surface tank and from that classification, choose an appropriate airflow. The New York tables were published in the *Ventilation Manual* in 1958. Recently, the tables were simplified by assigning the specific classification for each of the listed open-surface tank processes using the procedures described above (*Ventilation Manual*, Tables 10.5–6, 10.5–7, and 10.5–8).

Once the proper hazard potential–release rate class is assigned for a given tank process, the designer specifies the critical design velocity. For enclosing hoods (Fig. 6.1) the design velocity is the face velocity at the hood opening. The design velocity for lateral exhaust hoods (Figs. 6.2 and 6.3) is the capture velocity obtained from Table 6.3. For consistency in nomenclature it is important that the design velocity on lateral exhaust hoods be identified as a capture velocity since it is that velocity which must be established at the point of the tank furthermost from the hood to overcome air currents and capture the contaminated air, causing it to flow into the hood.

The information on minimum tank capture velocity (Table 6.3) is taken from *Ventilation Manual*, Table 10.5–3, with the elimination of the data for canopy hoods. It is strongly recommended that canopy hoods *not* be used for ventilation control of open-surface tanks. The *Ventilation Manual* (Table 10.5–3) states that canopy hoods should not be used for tanks with a hazard potential rating of A. *Ventilation Manual* design plate VS-903, however, *states* that canopy hoods are not to be used where toxic materials are present and the worker must bend over the tank or process. A conservative approach is recommended; since tanks are frequently converted from one use to another, it is prudent not to utilize canopy hoods on any open-surface tank, hence their omission from Table 6.3.

Once the critical design velocity is identified, Table 6.4 is used to choose the appropriate minimum exhaust airflow (cfm/ft^2 of tank surface). The minimum exhaust rate depends on the type of hood selected for the process and the width-to-length ratio of the tank. A review of Table 6.4 shows that airflow can be minimized by utilizing as small a tank as possible, keeping the width-to-length ratio low and using hoods with integral baffles or installing baffles or shields to prevent drafts.

A ventilation system for a plating shop is designed in Chapter 9. The procedure for calculating the exhaust rate on the tanks in this sample problem using the *Ventilation Manual* and the ANSI Z9.1-1971 procedure is shown in Example 6.1.

6.2 SPRAY PAINTING

Spray painting results in the exposure of workers to vapors of the volatile components of the vehicle and catalyst and to paint mist consisting of binder, fillers, and pigments. Engineering control is usually accomplished with ventilated spray booths which are classed as simple partial enclosures. These booths are frequently used for control of other industrial processes, including weighing and mixing of chemicals, solvent cleaning, and coating operations. Due to their widespread use, the designer should be acquainted with the types of paint spray hoods and their application.

6.2.1 Hood Design

More than two dozen companies in the United States manufacture paint spray hoods ranging from small benchtop units widely used in the jewelry industry to room-sized booths for painting automobiles and locomotives. Each vendor has a standard line of hoods with numerous design options. The hoods are designed to provide a lateral flow of clean air past the breathing zone of the operator with a uniform velocity distribution across the hood face, thereby minimizing worker exposure to toxic air contaminants. Downdraft hoods are installed in heavy industry for painting large, irregular-shaped objects such as castings; since these hoods are special designs used in limited applications, they will not be discussed in this section.

The conventional spray booth is a heavy-gage sheet metal enclosure with one open face (Fig. 6.4). The side panels and roof may have openings to allow a conveyor to bring the parts directly to the spray area. This feature is especially important in automated painting operations. The choice of hood size is obviously dependent on the size of the parts to be painted. It is important that the face area of the hood be large enough so that the part being painted does not block airflow through the hood face, which might result in poor capture efficiency. Geometry of the parts to be painted frequently requires a holding fixture or a turntable for complete spray coverage; the booth must be sized to accommodate this equipment.

Booth depth is critical to good performance since it helps to ensure a flat velocity profile at the hood face with resulting effective capture of toxic air contaminants (Chapter 5). Rebound from flat surfaces sprayed in a shallow booth may escape capture; a deep booth will assist in the recovery of such contaminants (Figure 6.5).

Air cleaning on spray booths is usually limited to removal of paint mist with little attention given to solvents. Sheet metal baffles hung at the rear of the hood are the simplest approach to air cleaning and improvement in air distribution

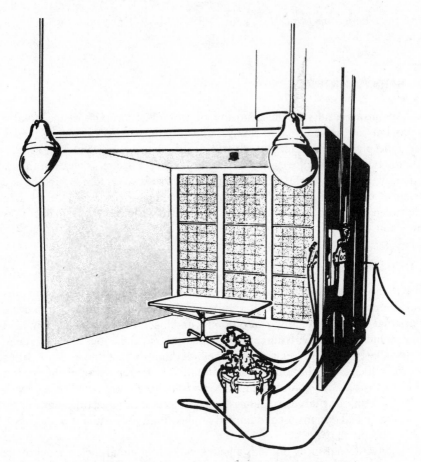

Figure 6.4   Main features of a conventional spray booth.

(Fig. 6.6). Baffles are simple to install and clean, and they present minimal resistance to airflow. DallaValle (1952) recommends the number of baffles to be used and their spacing as a function of cross section of the individual booth (Table 6.5). In addition to providing a flat velocity profile at the face of the hood, baffles remove a portion of the coarse paint spray mist.

A bank of disposable particle filters mounted in the rear of the hood provides a rear plenum and helps to establish a uniform velocity at the hood face. The filter medium is usually a coarse fibrous medium with a collection efficiency of less than 50% against overspray from conventional air-atomized paint. As these filters are coated with paint, their resistance will increase and airflow will drop off. Performance should be monitored continuously and filters should be changed periodically. Filter life may be extended by providing baffles in front of the filters to provide minimal collection of coarse paint mist.

The most effective air-cleaning approach for removal of paint mist while providing a good velocity profile in the hood is the water-wash booth (Fig. 6.7).

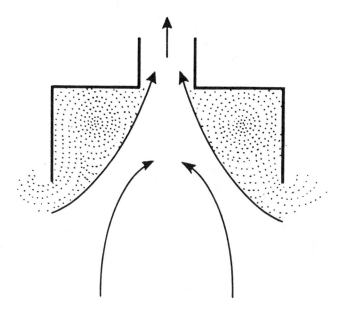

Figure 6.5 Hood depth is a critical design element. With a shallow booth leakage occurs from eddies as shown in this plan view. [From Harvey (1984).].

These booth designs vary but usually include a water wash curtain, counter-current sprays, and impingement plates for removal of water mist. It is stated that these systems are relatively effective against paint mist; however, no data are available on their performance. These simple water-wash booths are not effective for removal of the organic vapors encountered in painting.

The efficiency of a short-throat, concurrent-flow wet scrubber used with a downdraft paint spray booth has been evaluated by Chan et al. (1986) against a high-solids base coat and a clear-coat enamel paint mist generated by air-atomization spray guns and rotary atomizers. The base coat paint mist generated by air atomization had a mass median aerodynamic diameter (MMAD) ranging from 4.7 to 6.6 μm depending on gun pressure; the MMAD of the clear-coat paint generated by a rotary atomizer varied from 20 to 35 μm as the unit was operated at various rotational speeds. The collection efficiency of this one-stage wet scrubber was greater than 90% for particles with a MMAD greater than 2 μm. The authors stressed the importance of optimizing the spray operating conditions to minimize the generation of mist less than 2 μm in size.

6.2.2 Airflow

The choice of paint spray booth face velocity and therefore airflow varies with authority; a comparison of the *Ventilation Manual* and ANSI/OSHA requirements is shown in Table 6.6. The designer is cautioned that these criteria may not

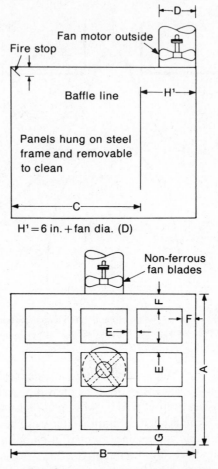

Figure 6.6 Hood baffles for air distribution. [From DallaValle (1952).]

be adequate to protect against paint containing heavy metals such as lead and cadmium or catalyst systems used in epoxy or urethane paints. Once the appropriate face velocity is chosen, the open area consisting of the booth face and any openings for conveyor access must be calculated.

Installation and proper operating practice are important in obtaining good performance from a paint spray booth. The following practices are important in controlling exposure:

- Minimize the distance between the paint spray nozzle and the part.
- Select spray conditions to generate particles with a diameter greater than 12 μm to minimize overspray.
- Use extension arms on the spray gun to provide access to voids in workpiece.

TABLE 6.5 Spray Booth Baffle Design
(Refer to Fig. 6.6)

Booth Cross-Sectional Dimensions, A × B[a] (ft)	No. of Panels, D	Spacing (in.)		
		Between Panels, E	Top and Side, F	Bottom, G
3×3	1			
$6\frac{1}{2} \times 4$	3	4	6	12
$6\frac{1}{2} \times 6$	4	4	6	12
$6\frac{1}{2} \times 8$	6	4	10	12
$6\frac{1}{2} \times 10$	9	6	10	18
9×12	9	8	12	18
12×14	12	8	18	24

[a]For booths of dimensions not shown, use the nearest size in the table for number and spacing of baffles.

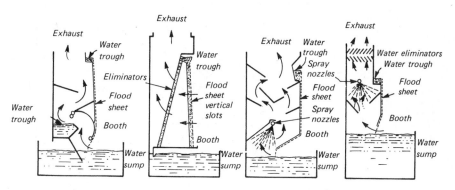

Figure 6.7 Typical water-wash spray booth designs. [From ANSI Z9.1–1971.]

- Apply robotic methods to the operation where possible to eliminate the worker at the booth.
- Stand to the side of the piece being sprayed.

Replacement air must be introduced in sufficient quantities to satisfy the exhaust requirements of the spray booth. Frequently, the air must be provided with endpoint filtration to minimize surface imperfections in the paint. The location and type of diffuser for replacement air are important considerations to booth performance and control of product quality. Although it is generally good practice to have replacement air introduced at some distance from the hood (see Chapter 12), this is not so in the case of paint spray booths. Travel distance will increase the potential for particle pickup and contamination of the painted surface. Proper placement of the booth is necessary. Hood placement in corners frequently results in nonsymmetrical flow and a poor velocity profile at the face.

TABLE 6.6   Ventilation Requirements for Spray Booths (cfm/ft^2 cross section)

	ACGIH		ANSI OSHA	
	Air Spray	Airless	Air Spray	Airless/ Electrostatic
Bench-type spray booth	150^a–200^b	100^a–125^b	150^c–200^d	100
Large spray booth				
Walk-in	100^e	60	100^c–150^d	—
Operator outside	100–150	60–100	100^c–150^d	—
Auto spray paint booth	100	60	—	—

^{a}Booth cross section less than 4 ft^2.
^{b}Booth cross section more than 4 ft^2.
^{c}Cross-drafts up to 50 fpm.
^{d}Cross-drafts up to 100 fpm.
^{e}75 cfm/ft^2 for very large, deep booth (operator may require an approved respirator).

Spray booths should not be placed near receiving or shipping doors, where disruptive drafts may exist.

It is important to select the correct fan, as described in Chapter 10. A fan is frequently supplied as a part of the paint spray booth package. Such fans are usually chosen based on a short duct run, no elbows or transitions, and direct discharge to outdoors. Since this is rarely the case in practice, the fan-booth system may not be matched and proper airflow will not be obtained when the installation is completed. A fan delivering 5000 cfm at $\frac{1}{8}$ in. H$_2$O may deliver only 3500 cfm at 1.0 in. H$_2$O when the usual additional resistance is encountered. It is good practice to size and purchase a fan separately based on total system loss calculations.

6.3   PROCESSING AND TRANSFER OF GRANULAR MATERIAL

The mining and recovery of minerals and the production and dispensing of granular chemicals present a major potential for the release of toxic air contaminants to the work environment from the numerous handling and transfer operations. The initial storage and feed operations for Portland cement (Fig. 6.8) illustrates the complexity of the problem with dozens of potential dust release points at each step of the process. The *Ventilation Manual* has a series of design plates covering many of these operations. In the authors' experience the recommended airflows based on these design plates may be insufficient for effective dust control. The induced air concept is introduced in this section to complement the *Ventilation Manual* design data. This concept provides a fundamental understanding of the dust generation and release mechanism in material-handling operations and introduces a rational local exhaust design base for such installations.

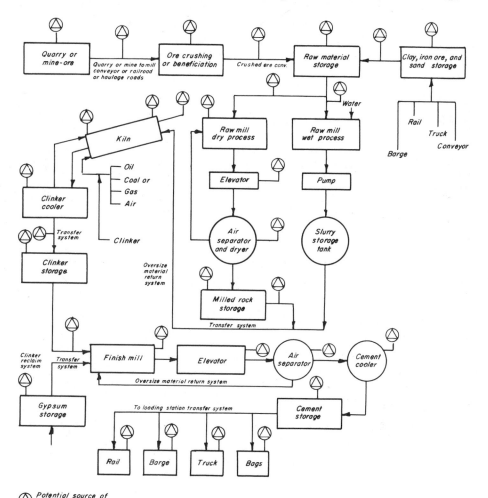

Figure 6.8 Multiple granular transfer points in the initial storage and feed operations in the manufacture of Portland cement.

Initial attempts at dust control at granular transfer points were based on the use of enclosing hoods with a selected face velocity at all open areas. This approach was typified by early New York State codes, which required 150 to 200 cfm/ft^2 at all hood openings. In many installations these recommendations did not provide adequate control. Vendors of packaged systems proposed designs based on 400 to 500 cfm/ft^2 at all hood openings. Frequently, this approach did result in improved control at the difficult transfer points, but in many cases it resulted in overdesign at other points. The second general design approach as cited in the *Ventilation Manual* (VS-303 to 307) specifies a given cfm/ft^2 through all hood openings, with the total airflow not less than a second design criterion such as 350 to 500 cfm/ft of conveyor belt width, 0.5 cfm/ft^3 of bin volume, or 50 fpm/ft^2 of screen area. Although this approach is the accepted design practice in the United States, it has not been thoroughly validated in the field.

A number of investigators have pointed out that installed systems based on the face velocity approach frequently do not result in effective control because this design procedure does not anticipate the quantity of air induced into the system. Pring (1949) was one of the first investigators to demonstrate that falling granular material draws air from the top of a system enclosure and forces it out the bottom, carrying dust with it. If the system component such as a bin, is tight, the induced air will reverse its path and carry the entrained dust back through the upstream opening as shown in Fig. 6.9. Hemeon (1963) also studied this problem and found that the maximum air volume induced into such a system is proportional to:

- The cube root of the mass rate of solids flowing
- The cube root of the specific gravity of solids
- The cube root of the particle diameter
- The two-thirds power of the solids cross-sectional area
- The two-thirds power of the falling distance

Although this information is of value in identifying those system parameters influencing the magnitude of the induced air, it did not provide a useful design

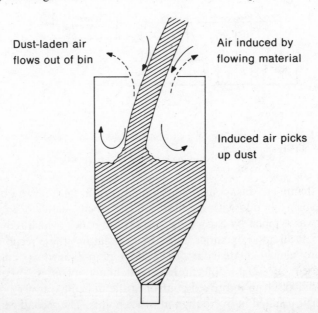

Figure 6.9 Induced-air concept showing how air induced into a bin from flowing granular material results in contamination of the workplace.

approach. A laboratory investigation conducted by Dennis (1954) of the airflow induced by falling granular material in material-handling operations led to the following empirical equation of induced airflow by Anderson (1964):

$$Q_i = 10.0 A_u \left(\frac{RS^2}{D} \right)^{1/3} \tag{6.1}$$

where Q_i = induced airflow, cfm

A_u = open area of the enclosure upstream of the dump location, ft^2

R = material flow rate, tons/h

S = height of fall of material, ft

D = average particle diameter, ft (D greater than 1/8 ft)

   The use of this design formula for a series of material transfer points is shown in Figs. 6.10 through 6.14. Anderson (1964) has provided examples of the differences between exhaust volumes calculated by the conventional face velocity and this induced air method. The total system airflow derived from the induced

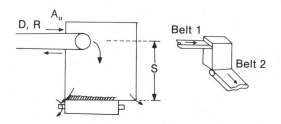

Figure 6.10   Induced air, belt to belt, chute to chute.

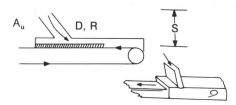

Figure 6.11   Induced air, chute to belt.

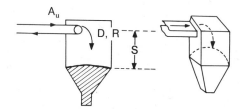

Figure 6.12   Induced air, belt to bin.

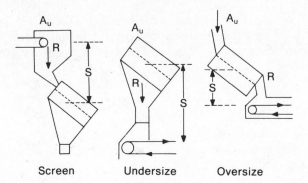

Screen   Undersize   Oversize

Figure 6.13   Induced air, screens.

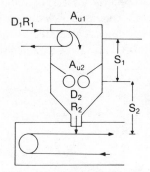

Figure 6.14   Induced air, crusher.

air method may not differ greatly from that specified by the *Ventilation Manual* face velocity approach. However, the induced-air approach has the advantage that it can assist in identifying the locations that warrant control and the exhaust rate necessary to minimize the release of dust. Due to limited design information, it is prudent to calculate the required exhaust for granular transfer points by both the *Ventilation Manual* and the induced-air methods and select the larger value for the design.

6.4 WELDING, SOLDERING, AND BRAZING

The required ventilation controls for welding operations vary greatly depending on the welding process utilized, the composition of the base metal, the coating on the base metal, and the degree of confinement existing in the welding area. Prior to World War II the majority of welding was done by the shielded metal arc welding (SMA) process (Fig. 6.15). Using this technique the welder is exposed to respirable metal particles in the form of the oxides of iron and other alloy metals present in the welding plume, concentrations of various contaminants released

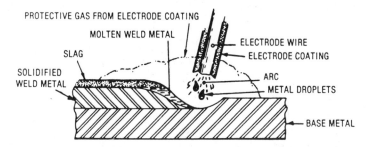

Figure 6.15 Shielded metal arc welding (SMA). The consumable electrode used in this process is coated with a flux that has a variety of functions, including forming a gas to shield the molten weld metal. The arc is established manually by the welder, who feeds the electrode into the work area. This process, commonly used on mild steels, may generate large quantities of metal fume from the electrode and base metal and gases from the vaporization of the welding rod coating.

from the rod and its coating, and minor quantities of ozone and nitrogen dioxide fixed by the arc. Low-hydrogen electrodes are used on selected alloys; the coatings on these electrodes contain soluble fluorides which release hydrogen fluoride and particles of soluble fluoride salts which are respiratory irritants and cause systemic effects.

Technical innovation in welding engineering in the 1940s and 1950s to improve welding quality resulted in the development of techniques to blanket the arc with an inert gas, thereby preventing defects in the weldment. Two methods, tungsten inert gas shielded arc welding (TIG) (Fig. 6.16) and metal inert gas shielded arc welding (MIG) (Fig. 6.17), frequently result in worker exposures to metal fume and toxic gases, such as ozone, nitrogen dioxide, and carbon

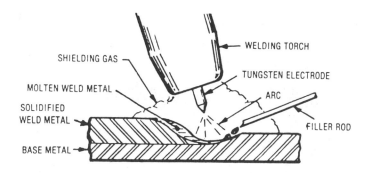

Figure 6.16 Tungsten inert-gas shielded arc welding (TIG). In this process the arc is established between the workpiece and a nonconsumable tungsten electrode. A filler rod is fed into the weld separately by the welder. Shielding gas such as argon or helium blankets the welding area. This process is used on special alloys and on more critical work on lighter sections than SMA. Airborne contaminants include relatively low metal fume concentrations from the variety of alloys, steels, and nonferrous materials welded with this technique; significant ozone concentrations may occur in poorly ventilated workstations.

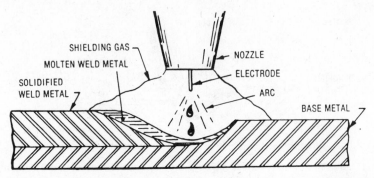

Figure 6.17 Metal inert gas-shielded arc welding (MIG). The arc is established with a consumable electrode and the inert shielding gas flowing from an annular space in the gun blankets the weldment. This process is frequently used at high current densities for a variety of metals, including steels and aluminum. High ozone concentrations are generated, especially when welding aluminum with argon as the shielding gas.

monoxide. In a fourth technique, submerged arc welding (Fig. 6.18), the arc is shielded by granular flux which presents minimal air contamination and normally does not warrant local exhaust ventilation. Plasma welding (Fig. 6.19) operates at still higher current densities than TIG and MIG techniques, resulting in concurrent increases in the generation of toxic air contaminants. Associated arc processes such as cutting, gouging, and flux-cored welding also require greater attention to the control of welding exposures.

Processes similar to welding such as soldering and brazing may require ventilation to control contaminants such as volatilization products from filler rod or solder, thermal degradation products of the flux, and carbon monoxide and oxides of nitrogen generated by gas torch operations.

Familiarity with each welding process is necessary to design the most effective ventilation system. An overview of these processes and the principal airborne contaminants which must be controlled is presented in Table 6.7.

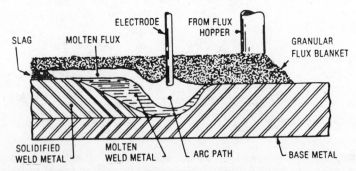

Figure 6.18 Submerged arc welding (SAW). This process does not normally require local exhaust ventilation. SAW is used to weld heavy sections of steel plate such as in shipbuilding; the consumable electrode establishes an arc under a granular flux which provides the shield. The flux does degrade, and minimal concentrations of soluble fluoride particles and hydrogen fluoride gas may be generated.

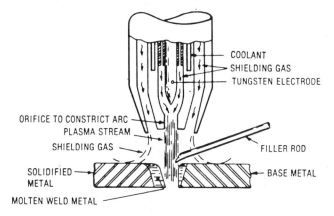

COOLANT
SHIELDING GAS
TUNGSTEN ELECTRODE

ORIFICE TO CONSTRICT ARC
PLASMA STREAM
SHIELDING GAS
FILLER ROD

SOLIDIFIED
METAL
BASE METAL

MOLTEN WELD METAL

Figure 6.19 Plasma arc welding (PAW). In this welding equipment the gun has both an inerting stream of argon and a separate orifice stream of argon. The orifice gas is placed under high voltage and results in a highly ionized gas stream or plasma with arc temperatures in excess of 60,000°F. This process can be used for a variety of metals, including titanium, stainless steels, aluminum, and refractory metals. Due to the ozone, nitrogen dioxide, and metal fume exposures, this process requires local exhaust ventilation.

In a comprehensive review of ventilation control on welding operations by Hagopian and Bastress (1976), air contaminants generated during welding operations were grouped by toxicity into three classes (Table 6.8). The suitability of various ventilation techniques and the minimum exhaust ventilation rates for each approach were assigned based on the contaminant class as shown in Table 6.9.

In this summary, general exhaust ventilation is not included as one of the candidate methods of control. The application of general exhaust for welding can be criticized in many ways. Since design criteria have not been validated, the installations rarely provide suitable control and empirical upgrading frequently is required. In addition, the airflows required for effective general exhaust designs make for very high operating costs. These limitations have been outlined in Chapter 4 and are reinforced by Hagopian and Bastress (1976). One of the local exhaust ventilation methods listed in Table 6.9 usually is feasible for control.

Of the four local exhaust ventilation methods listed in Table 6.9, the technique most frequently applied to welding is the freely suspended open hood (Fig. 6.20). In this system an exterior hood attached to a flexible duct is provided with a positioning system to permit the welder to move the hood close to the welding point. As noted in Fig. 6.20, the required airflow is modest if the hood can be positioned within 6 in. of the weldment. Obviously, this approach requires the cooperation of the welder to position the hood and achieve good control. Multiple workstations can utilize either a central system or a series of individual systems. Since 1975, a number of manufacturers have provided packaged systems consisting of a hood, flexible duct, positioner, fan, and air cleaner. The

TABLE 6.7 Potential Hazards from Welding[a]

Hazard	Welding Process						
	SMA[b]	Low Hydrogen	GTA[c]	GMA[d]	Submerged	Plasma	Gas
Metal fumes	M–H	M–H	L–M	M–H	L	H	L–M
Fluorides	L	H	L	L	M	L	L
Ozone	L	L	M	H	L	H	H
Nitrogen dioxide	L	L	L	M	L	H	L
Carbon monoxide	L	L	M	L, H if CO_2	L	L	M–H
Decomposition of chlorinated HC	L	L	M–H	M–H	L	H	L
Radiant energy	M	M	M–H	M–H	L	H	L
Noise	L	L	L	L	L	H	L

[a]Hazard codes: L, low; M, medium; H, high.
[b]Shielded metal arc welding.
[c]Gas tungsten arc.
[d]Gas metal arc.

TABLE 6.8 Contaminant Classification for Welding

Contaminant Class	Classification Criteria	Example Materials
I	Dusts, fumes, and mists with exposure limits of 5 mg/m^3 and above; gases and vapors with exposure limits of 100 ppm and above	Iron oxide
II	Dusts, fumes, and mists with exposure limits of 0.1 mg/m^3 and above (up to 5 mg/m^3); gases and vapors with exposure limits of 1 ppm and above (up to 100 ppm)	Fluorides; carbon monoxide; nitrogen dioxide
III	Dusts, fumes, and mists with exposure limits below 0.1 mg/m^3; gases and vapors with exposure limits below 1 ppm	Beryllium; mercury; ozone; cadmium

Sources: After Hagopian and Bastress (1976).

air cleaner is usually a high-efficiency particle filter or an electrostatic precipitator (Chapter 11). If aluminum is welded, the electrostatic precipitator may not be suitable since the electrodes can become coated with an insulating coat of aluminum oxide. Depending on the application, these packaged systems may be used in a recirculation mode. The freely suspended hood approach is commonly used in maintenance and repair shops where it can reach both a welding bench area and a portion of the shop floor if large equipment must be welded.

Production and maintenance welding operations on small parts can be accomplished at a small workstation provided with rear-slot ventilation (Fig. 6.21). This is an attractive approach since the welder need not position a hood for effective control. As discussed in Chapter 5, the hood should be provided with side baffles to minimize the effects of cross-drafts and improve collection of contaminants. Such stations frequently are used to weld long sections of pipe, strap, and angle, and welders often object to the interference presented by side baffles. This problem can be eliminated by providing slotted baffles, hinged side baffles, or heavy fabric side drapes. This rear-slot welding station is also an effective design for soldering and brazing operations.

A third local exhaust technique, a low-volume shill high-velocity hood, applicable to both TIG and MIG processes is shown in Fig. 6.22. This technique, providing effective capture at a minimal airflow, employs an exhausted annular ring around the welding head. The exhaust airflow varies depending on the manufacturer and the specific operation, but the usual range of 50 to 100 cfm is approximately 10% that of the freely suspended open hood. There is some reluctance on the part of welders to use this technique since the exhaust may strip the shielding gas from the welding point unless care is taken in setting the exhaust rate.

TABLE 6.9 Ventilation Requirements for Various Welding Techniques and Contaminant Class

Operation	Freely Suspended Open Hood	Rear Slot Table	Low-Volume/ High-Velocity Gun-mounted Hood	Enclosing Hood
Gas welding	Classes I and II—100-fpm capture velocity Class III—not recommended	Classes I and II—100-fpm capture velocity Class III—not recommended	Not applicable	Classes I and II—100-fpm face velocity Class III—150-fpm face velocity
Torch brazing and soldering	Classes I and II—100-fpm capture velocity Class III—not recommended	Classes I and II—100-fpm capture velocity Class III—not recommended	Not applicable	Classes I and II—100-fpm face velocity Class III—150-fpm face velocity
Shielded metal arc welding	Classes I and II—100-fpm capture velocity Class III—not recommended	Classes I and II—100-fpm capture velocity Class III—not recommended	Not applicable	Classes I and II—100-fpm face velocity Class III—150-fpm face velocity
Plasma arc welding and submerged arc welding	Classes I and II—100-fpm capture velocity Class III—not recommended	Classes I and II—100-fpm capture velocity Class III—not recommended	Not applicable	Classes I and II—100-fpm face velocity Class III—150-fpm face velocity

Gas shielded-arc welding and flux-cored arc welding	Classes I and II—100-fpm capture velocity Class III—not recommended	Classes I and II—100-fpm capture velocity Class III—not recommended	Classes I and II—exhaust rate to be determined by manufacturer or user Class III—not recommended	Classes I and II—100-fpr, face velocity Class III—150-fpm face velocity
Oxygen cutting, arc cutting, and gouging	Not recommended	Not recommended	Not applicable	Classes I and II—100-fpm face velocity Class III—150-fpm face velocity
Thermal spraying	Class I—200-fpm face velocity with operation at face of hood Classes II and III—not recommended	Not recommended	Not applicable	Class I—125-fpm face velocity Classes II and III—200-fpm face velocity

Source: After Hagopian and Bastress (1976).

171

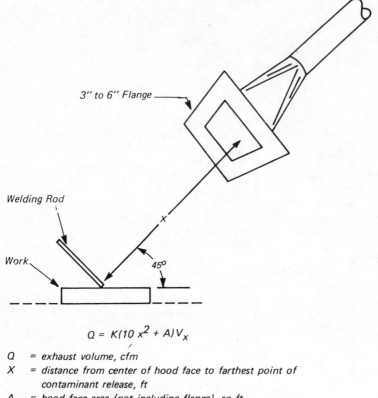

$$Q = K(10\,x^2 + A)V_x$$

Q = exhaust volume, cfm
X = distance from center of hood face to farthest point of contaminant release, ft
A = hood face area (not including flange), sq ft
V_x = minimum capture velocity, fpm
K = 1.0 for unflanged hood; 0.75 for flanged hood
Entry loss = entry loss factor for tapered hood X duct VP
Duct velocity = 2000 fpm minimum

Figure 6.20   Movable hood to be positioned by welder. [From Hagopian and Bastress (1976).]

If extremely toxic metals are processed, the best approach is an enclosing hood. In practice, this may be a welding bench placed in a modified paint spray booth. The downdraft table (Fig. 6.23) is a local exhaust hood designed principally for gas and arc cutting processes. The metal slab to be cut is placed on the table and the metal fume is pulled downward. Large molten droplets fall to the plenum and the fume is exhausted from the box plenum. The choice of a canopy hood for a welding station is a common mistake in industry. These hoods are not acceptable for control of toxic air contaminants for the reasons discussed in Chapter 5.

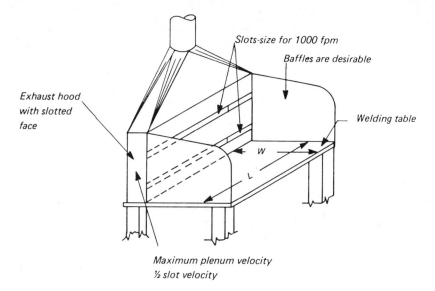

$$Q = KLWV_x$$

Q	= exhaust volume, cfm	K	= 2.4 with baffles; 2.8 without baffles
W	= table width, ft (not to exceed 4 ft)		Entry loss = 1.78 slot VP plus entry loss factor
L	= table length, ft		Duct velocity = 2000 fpm minimum
V_x	= minimum capture velocity, fpm		

Figure 6.21 Rear-slot bench exhaust suitable for welding and brazing of small parts. [From Hagopian and Bastress (1976).]

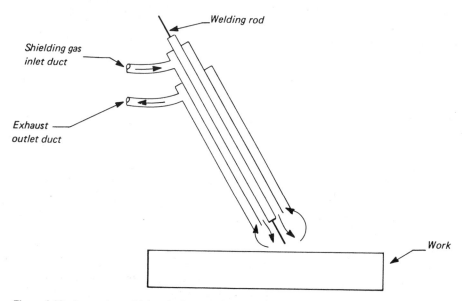

Figure 6.22 Low-volume/high-velocity exhaust for inert-gas shielded process. [From Hagopian and Bastress (1976).] Exhaust flow requirements must be determined for each welding operation and welding gun configuration by experimental testing with air contaminant sampling and analysis.

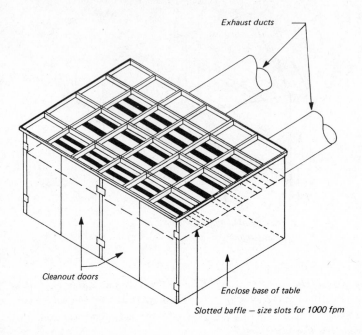

$$Q = A\,V_f$$

Q = exhaust volume, cfm

A = table top area, sq ft

V_f = minimum face velocity, fpm

Entry loss = 1.78 slot VP plus entry loss factor
                for tapered hood x duct VP

Duct velocity ≈ 2000 fpm minimum

Figure 6.23   Downdraft cutting table for gas cutting or plasma operations. [From Hagopian and Bastress (1976).]

6.5   CHEMICAL PROCESSING

Although no other industrial setting presents the range of contaminant generation points and the variety of air contaminants encountered in the chemical processing industry, standard ventilation designs are available for only a limited number of operations. In this section common operations in a batch chemical processing plant are reviewed and the possible approaches to ventilation control are discussed.

6.5.1 Chemical Processing Operations

A batch processing plant may synthesize several dozen chemicals in production scales ranging from tens to hundreds of thousands of kilograms per year (Fig. 6.24). Starting materials include granular chemicals, chemicals in liquid form such as solvents, and reactive gases. Granular materials utilized in small quantities are usually received in drums and may require preweighing of batches. This material frequently can be ordered in prill or flake form to minimize dusting. Regardless of material form, the weighing station normally warrants local exhaust hooding to control dust (Fig. 6.25). Granular chemicals used in large quantities are delivered to the plant by truck, railroad bulk cars, or transporters and may be transferred by pneumatic conveying directly to the mix or reaction vessel or to storage bins for later transfer to the vessel. This bulk transfer is usually performed in enclosed systems with minimal release of dust to the environment.

Liquid reagents may also be transported in bulk for tank storage at the plant, or, if used in small quantities, by drums held in special storage areas. Bulk solvents are frequently received at a loading dock equipped with a piping manifold for flexible hose transfer from truck or train to tank storage. Investigators have demonstrated that significant exposure to volatile liquids may occur during making up and draining of transfer lines. Even though this work is usually done outdoors, ventilation control may be necessary.

The first major in-plant processing equipment requiring ventilation control is the reaction vessel. As an example, consider the batch process for the synthesis of a dye, illustrated in Fig. 6.24. The various chemicals are added in a carefully planned manner to the reactor. Liquid is added by closing all vessel ports and valves and placing the reactor under vacuum. The liquid is then transferred from a storage tank to the reactor in an enclosed piping system with negligible exposure to workers.

Granular material is frequently added to the vessel. In large production operations the granular chemical may be pneumatically conveyed directly from storage bin to the vessel. In the more common batch process shown in Fig. 6.24, it is added manually through a vessel charging port or manway. In this case, two exposures can occur. Dumping the granular material into a reactor already charged with liquid solvent can induce airflow into the vessel, with the resulting displacement of particles and solvent-laden air into the workplace. This is an important exposure point and must be controlled by local exhaust ventilation. The usual approach is an exterior hood positioned at the charge port whose geometry and airflow permit effective capture of both solvent vapors and respirable dust (Fig. 6.26). Collection efficiency of well-designed hoods of this type is greater than 95% as measured by tracer studies.

Small quantities of solvent may be added to the reactor from a drum using a drum pump which discharges to the reactor through an open access port.

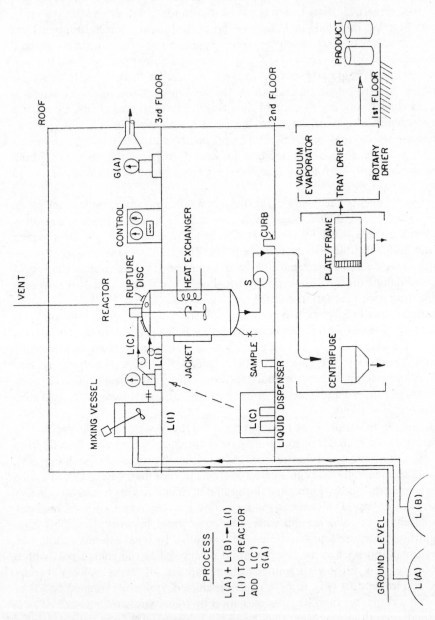

Figure 6.24 Chemical processing plant for batch processing. L, liquid; G, granular; S, suspension.

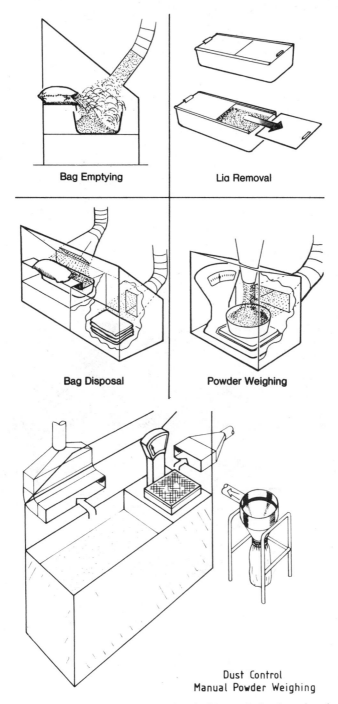

Figure 6.25 Weigh station dust control. [Reproduced with permission from *Annals of Occupational Hygiene* (1980).]

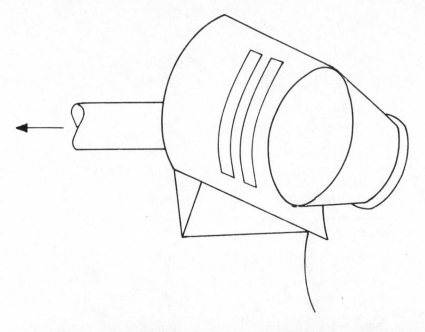

Figure 6.26 Reactor charge port ventilation showing exterior hood positioned to exhaust at top of funnel inserted in charge port. (Q varies from 1000 to 2000 cfm, depending on size, $C_e = 0.60$.)

178

Exposure to chemical vapors can occur both during preparatory mounting of the pump in the drum and as the wetted wand is removed from the drum when pumping is completed. There are two approaches to ventilation control during such operations. A bung hole exhaust hood may be positioned on the drum (Fig. 6.27) or a partial enclosure for the drum can be positioned adjacent to the reactor. The latter approach usually is more effective.

Periodically, samples must be taken from the reactor to check on the status of the reaction or as a final product quality check. This is frequently done by opening an access port on the top of the reactor and using a dipstick with a sampling cup. Although there is potential exposure from the headspace contaminants in the reactor and the vapor from the sample, exposures are usually negligible when the reactor is equipped with charge port ventilation as described above. Headspace exposure can also be controlled by placing the reactor under vacuum, causing an inward flow of air through the charge port. A second sampling technique involves the use of a special sampling line positioned near the bottom of the reactor. This sampling location should be equipped with a small exhausted enclosure (Fig. 6.28), as described in VS-211.

The final product, in our example a suspended particle, is now ready to be recovered from the reaction fluid. As shown in Fig. 6.29, three common techniques available to perform this function are the Neutsche filter, the centrifuge, and the plate and frame filter. Each of these operations results in potential exposure due to the residual solvents; the latter two require special attention.

If a centrifuge is used, the suspension in the reactor is pumped to the centrifuge, where it flows into a spinning basket lined with a filter pad (Fig. 6.30). The particles are trapped on the filter as the solvent flows through the filter and is pumped to a holding tank. The cake on the inner surface of the basket is washed serially with fresh solvent sprayed through the spray header. If the solvent presents a fire and explosion hazard, the centrifuge cavity is flushed with nitrogen or carbon dioxide. When the cake is completely washed a plow scrapes it from the basket surface and the recovered product drops to a container. Worker exposure to solvent occurs when the access doors are opened to scrape the basket manually or to inspect the condition of the basket filter. A secondary exposure occurs at the product collection drum or bin.

It is difficult to achieve control at the access port on the top of the centrifuge. One possibility is to pull sweep air through the centrifuge using the line to the holding tank. If this is not possible, an exterior hood must be installed at the access port as shown in Fig. 6.31. Solvent vapor control at the product collection drum can be accomplished by a barrel exhaust station.

The second technique in common use for recovery of the product is a plate and frame filter. Available in a variety of designs, the conventional system utilizes a number of filter pads placed in parallel to arrest the particles. The suspension in the reactor is pumped to the plate and frame filter, the particles are collected on the filter, and the stripped solvent is directed to a holding tank. The plate and

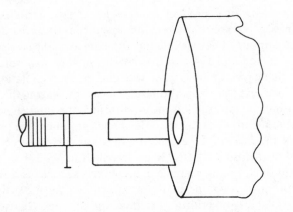

Figure 6.27 Drum bung hole exhaust. ($Q = 150$ cfm, $C_e = 0.50$.)

Figure 6.28 Sampling hood.

frame process also permits the washing of the collected product with clean solvent or water. When the washing is completed the plate clamping pressure is released and the plates are separated. The product cake, still wet with solvent, is manually scraped from the plates and drops to a recovery cart (Fig. 6.32). During this process the solvent evaporated from the wet cake results in significant air concentrations in the workplace, which require ventilation control.

Effective control of solvent vapors from plate and frame filters is extremely difficult. One approach utilizes a top exhausted plenum with a perimeter plastic curtain or horizontal sliding sash, which must be positioned by the operator for effective control (Fig. 6.33). If adequate overhead space is not available for the plenum, two fishtail hoods can be positioned at the ends of the press (Fig. 6.34).

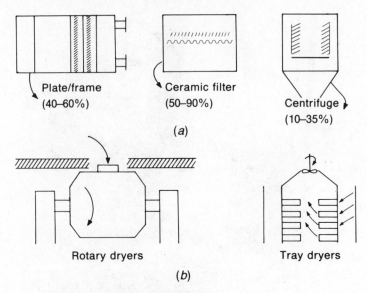

Plate/frame
(40–60%)

Ceramic filter
(50–90%)

Centrifuge
(10–35%)

(a)

Rotary dryers

Tray dryers

(b)

Figure 6.29 (a) Filtration and (b) drying techniques. The percent residual solvent in the wet cake is shown for each of the three filtration methods.

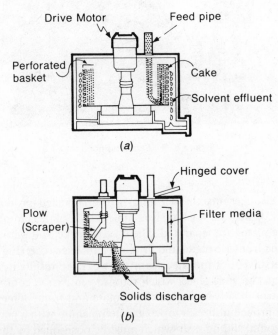

Drive Motor

Feed pipe

Perforated basket

Cake

Solvent effluent

(a)

Hinged cover

Plow (Scraper)

Filter media

Solids discharge

(b)

Figure 6.30 Details of basket centrifuge during two major operational steps: (a) active filtration; (b) removal of product. The product is also frequently removed manually with a small spade working through the hinged cover.

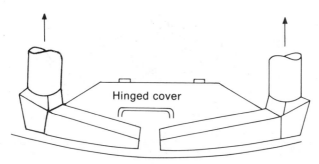

Two hinged hood elements
Slots on inner face

Figure 6.31 Exhaust hood for centrifuge hinged cover at access door. ($Q=250$ cfm per leg, C_e $=0$, 60.)

Figure 6.32   Dropping product from plate and frame filter.

Figure 6.33   Ventilation of plate and frame filter enclosure by top plenum.

Figure 6.34   Ventilation of plate and frame filter enclosure by fishtail hoods.

Again, the effectiveness of control depends on proper positioning of the curtains by the operators to ensure proper airflow past the operator breathing zone.

Drying of the product cake in a batch operation is normally accomplished by a tray dryer, a conical dryer, or a fluidized-bed dryer. Loading of each operation involves a solvent exposure; unloading may result in an exposure to respirable dust during packaging and transport.

6.6   SEMICONDUCTOR GAS CABINETS

A variety of highly toxic gases, including arsine and phosphine, are used in semiconductor manufacturing operations. The gases are normally dispensed from pressurized cylinders through a pressure-regulating manifold. The principal control to minimize the potential for worker exposure to these gases in the event of a leak during storage and dispensing operations is an exhausted enclosure. These enclosures, known in the semiconductor industry as gas cabinets, fulfill the definition of an enclosing-type hood and should provide a high level of control with minimum airflow. It should be noted that gas cabinets are designed for control of fugitive emissions and not violent release of total gas cylinder contents.

Manufacturers make standard gas cabinets to house up to four gas cylinders. The gas cylinder is positioned in the cabinet and connected to a panel-mounted manifold which permits accurate flow metering of the gas to the various fabrication operations. The major structural variables in the design of the gas cabinet are the air inlet and outlet configurations. In the conventional gas cabinet design, air sweeps through an opening in the bottom of the door and exhausts through a straight-duct takeoff at the top of the cabinet. The inlet is usually a series of slots in the door and may include mounting for a "roughing" particle filter, which acts as a diffuser.

Although these hoods are widely used in industry, there are no published data on their performance, nor has a method of defining optimum design been published. Furthermore, the air flows specified by various manufacturers may be quite different for similar equipment. Recognizing this deficiency, the authors recently used a tracer gas to evaluate the performance of gas cabinets (Forster and Burgess, 1986). The study evaluated optimum gas cabinet geometry and exhaust airflow for these cabinets.

A two-cylinder test gas cabinet equipped with a variety of inlet and outlet configurations and accommodations for a range of cylinder sizes was made available for this study from Matheson Gas Products, Inc. (Fig. 6.35). The cabinet is provided with a full-length front door used to change cylinders; pressure regulation and other adjustments are done through an access port to minimize the potential for personnel exposure to the toxic gas.

The test cabinet was provided with three interchangeable inlet air configurations: (1) a plain slotted inlet in the door, (2) a perforated plate inlet in the base of the cabinet, and (3) a diffuser inlet in the door. The perforated bottom plate inlet was included to test the theory that a "piston" flow of air through a perforated bottom plate to a takeoff at the top would provide excellent clearance of any fugitive gas leakage.

The conventional gas cabinet for semiconductor gases has a flat exhaust duct takeoff, a geometry that may encourage corner eddies and possibly increase clearance time. To minimize these effects, conical exhausts have been considered in cabinet design. Shallow and deep exhaust cones representing more conventional aerodynamic design were tested in addition to the flat exhaust takeoff (Fig. 6.36).

6.6.1 Entry Loss

The coefficients of entry (C_e) calculated at various flow rates for each of the different inlet and exhaust configurations are shown in Table 6.10. The air inlet geometry was found to have a major effect on the coefficient of entry, while differences in outlet geometry had little effect on entry loss. Specifically, C_e for the slot inlet ranged from 0.44 to 0.47, while the best values of 0.68 to 0.75 were obtained with the diffuser and perforated base inlets.

Figure 6.35 Two-cylinder test gas cabinet showing front door with access port, diffuser air inlet at base of door, shallow cone exhaust, and clear side panels for flow-visualization studies.

6.6.2 Optimum Exhaust Rate

In a series of tests to define the minimum airflow exhaust rate to maintain control, a tracer gas was released at a rate of 1.9 lpm from a critical leak location inside the cabinet. The downstream concentration of the tracer gas was

Figure 6.36 Three air exhaust outlets for mounting on top of test cabinet. Left front is flat, left rear is shallow cone, and right is deep cone.

TABLE 6.10 Coefficient of Entry C_e for Gas Cabinet Inlet and Outlet Configurations

Inlet	Outlet	Coefficient of Entry, C_e
Slot	Flat	0.44
	Shallow cone	0.47
	Deep cone	0.47
Diffuser	Flat	0.70
	Shallow cone	0.75
	Deep cone	0.72
Perforated base	Flat	0.68
	Shallow cone	0.72
	Deep cone	0.72

Source: Forster and Burgess (1986).

measured continuously at the sampling port in the exhaust duct. When the system had stabilized and a steady downstream concentration had been established, the leak was stopped and the time noted on the chart recorder. The decay of the downstream concentration was then used to define the performance of the system. For these tests, the time of decay to 5% of the original concentration was used to evaluate performance.

The clearance time data for the various test configurations without cylinders in place are summarized in Fig. 6.37. These data demonstrate several points.

- The slotted inlet consistently takes longer to clear a leak than do the other two inlets.
- The clearance time for all three inlets is reduced significantly by increasing the airflow from 150 to 250 cfm for this two-cylinder cabinet.
- For the perforated plate and the diffuser inlet, little improvement in performance is gained by increasing the airflow at 350 cfm.
- There is little difference in the performance of the perforated base plate and the diffuser front at 250 and 350 cfm.
- The exhaust configuration has little bearing on clearance time.

The clearance time for the gas cabinet was also evaluated with either two 60-cm or two 130-cm cylinders in place. The results of this testing with the flat plate and shallow cone exhausts are presented in Figs. 6.38 and 6.39. The deep cone was not evaluated since it did not demonstrate an advantage over the shallow cone in the previous tests. These results demonstrate the same trends as the data generated with no cylinders in place, thereby validating the earlier findings. Therefore, a diffuser inlet in the bottom of the cabinet door with an exhaust rate of 250 cfm for a two-cylinder cabinet is recommended.

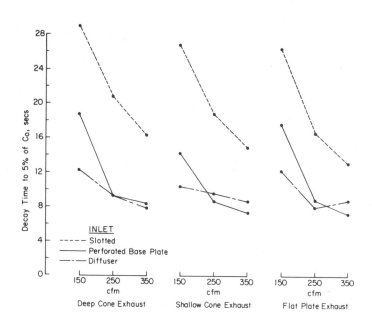

Figure 6.37 Clearance times of three test inlets for each test configuration without cylinders.

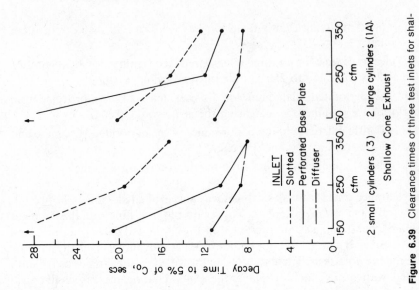

Figure 6.39 Clearance times of three test inlets for shallow cone exhaust with small and large cylinders.

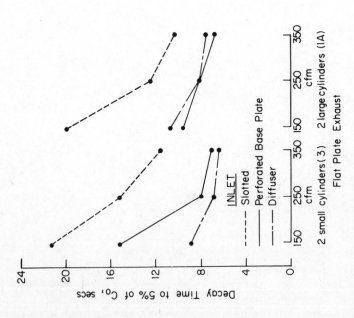

Figure 6.38 Clearance times of three test inlets for flat plate exhaust with small and large cylinders.

6.7 LOW-VOLUME/HIGH-VELOCITY SYSTEMS FOR PORTABLE TOOLS

The low-volume/high-velocity capture system uses an extractor hood designed as an integral part of the tool or positioned very close to the tool operating point. The hood is empirically designed to provide high capture velocities, frequently greater than 10,000 fpm, at the contaminant release point. This velocity is usually achieved at airflows less than 50 cfm by employing a small hood face area.

This approach was applied in the 1950s in the United States for the control of highly toxic materials such as beryllium released from machining operations (Chamberlin et al., 1958). Although application on machine tools has continued to this date, the major impetus in the acceptance of low-volume/high-velocity systems has resulted from its application to portable tools pioneered in the United Kingdom and the Scandinavian countries since 1965 (Fig. 6.40). It is now widely used on asbestos fabrication at job sites, buffing and grinding operations at widely dispersed plant operations, and pneumatic drilling in mining and quarrying.

There are a number of advantages to this ventilation control technique. The collection system is an integral part of the tool and does not rely on the operator to position an exhaust hood close to the work for proper control each time the work position is changed. In addition, the exhaust airflows required to achieve

(*a*)

Figure 6.40 Application of low-volume/high-velocity hood on portable tools: (*a*) disk sander.

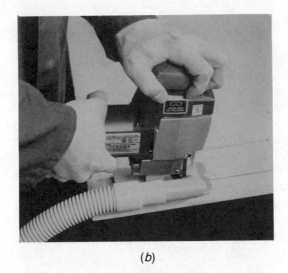

(b)

(c)

Figure 6.40 (*Continued*) Application of low-volume/high-velocity hood on portable tools: (b) saber saw; (c) oscillating saw. (Courtesy of Nilfisk of America, Inc.)

the high collection efficiency are frequently 10% of that required with conventional exterior hoods. Due to the expense of conditioning replacement air in northern latitudes, this has great cost-reduction potential in a plant containing a large number of workstations. Finally, this may be the only feasible approach for the control of highly toxic air contaminants in field operations such as construction.

It might appear on first glance that an additional advantage to the system is reduced air-mover operating cost. This turns out not to be the case since the energy cost to operate an air mover is proportional not only to airflow but also to total resistance (Chapter 10). In low-volume/high-velocity systems the airflow is, as noted above, frequently reduced by a factor of 10, but the system resistance may be increased tenfold; hence the air horsepower is approximately the same as that for a conventional system and the cost of operating the air mover will be approximately the same.

The principal disadvantages of this system are the high noise levels caused by the air moving through the restricted area of the nozzle at high speed and the propensity for high-vacuum hoods to pick up small tools and material placed close to the inlet.

The design of a low-volume/high-velocity system is largely empirical and is based on a general understanding of the release mechanism of the air contaminant. The extractor hood is connected to the exhaust system with a length of high-resistance flexible duct which permits its use on a portable tool and allows latitude for movement of movable work beds on fixed machine tools. The flexible duct is connected to the main duct, which is a heavy-walled, smooth-bore tubing with no inner surface discontinuity. Because conventional pipe or tubing would quickly erode on a particle collection application due to the abrasive action of debris carried by the high-velocity airstream, the main tubing must be equipped with large sweep elbows and low-angle entries. Due to static charge accumulation, the flexible hose at the tool must be conductive and static proof.

In the United Kingdom the extractor hoods are designed so that a hood static pressure of 5 in. Hg will provide the necessary airflows (U.K. Department of Employment, 1974). Typical system losses downstream of the hood due to the flexible hose, smooth duct, and air cleaner are 5 to 7 in. Hg, for a total system static pressure requirement of 10 to 12 in. Hg. A single tool can, therefore, be operated by a suitable vacuum cleaner; a multihood system requires the greater power of a multistage turbine fan or a positive-displacement blower.

The design of the typical low-volume/high-velocity systems in the United States is also empirical; the recommended minimum extraction volumes in Table 6.11 are approximately three to four times those recommended in the United Kingdom (Stephenson and Nixon, 1987). The extractor hood static pressure is claimed to be 6 in. Hg by one U.S. manufacturer and the *Ventilation Manual* cites pressures in the range 7 to 14 in. Hg. The design procedure used by one manufacturer of these systems is noted in Example 6.2 and Fig. 6.41 (Stephenson and Nixon, 1987). The nozzle entry loss is obtained from a chart and combined with the losses due to the flexible hose, the air cleaner, and the system tubing running to the multistage turbine. Since the system operates at nearly one-half an atmosphere, air volumes must be corrected to standard conditions.

The performance of low-volume/high-velocity systems has been evaluated by American (Burgess and Murrow, 1976; Garrison and Byers, 1980a,b,c), British (Dalrymple, 1986), and French (Regnier et al., 1986) investigators. Recent studies

TABLE 6.11   Minimum Volumes and Hose Sizes[a]

Hand Tool	CFM	Plastic Hose Inside Diameter(in.)
Disc sander		
3 in.—10,000 rpm	125	1
7 in.—5,000 rpm	200	2
9 in.—2,500 rpm	200	2
Vibratory pad sander—4 × 9	150	$1\frac{1}{2}$
Router		
$\frac{1}{8}$ in.—50,000 rpm	125	$1\frac{1}{2}$
$\frac{1}{4}$ in.—30,000 rpm	150	$1\frac{1}{2}$
1 in. Cutter—20,000 rpm	150	$1\frac{1}{2}$
Belt sander 3 in.—4000 rpm	125	$1\frac{1}{2}$
Pneumatic chisel	125	$1\frac{1}{2}$
Radial wheel grinder	150	$1\frac{1}{2}$
Surface die grinder, $\frac{1}{4}$ in.	125	$1\frac{1}{2}$
Cone wheel grinder, 2 in.	150	$1\frac{1}{2}$
Cup stone grinder, 4 in.	200	2
Cup-type brush, 6 in.	250	2
Radial wire brush, 6 in.	175	$1\frac{1}{2}$
Hand wire brush—3 × 7	125	$1\frac{1}{2}$
Rip-out knife	175	$1\frac{1}{2}$
Rip-out cast cutter	150	$1\frac{1}{2}$
Saber saw	150	$1\frac{1}{2}$
Swing frame grinder—2 × 18	600	3
Saw abrasive, 3 in.	150	$1\frac{1}{2}$

[a]Based on low production requirement with ideal pickup hoods and branch static at 6 in. Hg (200 mm Hg).
Source: Stephenson and Nixon (1987).

indicate high collection efficiencies on a variety of particulate contaminants using air volumes in the same range as those cited in the *Ventilation Manual*.

In addition to their use on particle control for portable tools, these systems have been applied to other air contaminants, such as soldering and welding fumes (as noted previously in this chapter). Since the metal fume contaminant is released at low velocity, excellent control can be achieved with very low airflows. Dalrymple (1986) describes a soldering application where pyrolysis products of solder fluxes can be captured with high efficiency at flows of 1 to 2 cfm.

Recently, there has been a movement to design machine tools to take advantage of the low-volume/high-velocity concept. One application on a milling machine utilizes a hollow milling cutter exhausted at 50 cfm with a hood static pressure of 7.5 in. Hg. In this system the milling cutter is mounted in a hollow spindle drive to which is attached the vacuum source and filter. This system is now widely used in the United States for milling "proof" board made of Styrofoam and machining structural plastic materials.

Example 6.1 Calculation of Exhaust Rate for Open-Surface Tanks in Example 9.2

(1) Tank No.	(2) Solution	(3) Temp.	(4) Class	(5) Tank Dimensions				(6) V capture (fpm)	(7) Min. Rate (cfm/ft²)	(8) Q (cfm)
				L	W	W/L	A			
1	Hot Water	200	D-2/1	3	2	0.67	6	25	130	780
2	Cr Strip	130	C-3	2⅔	2	0.86	4⅔	50	90	420
3	Acid Cu	110	B-3	6	3½	0.58	21	25	130	2730
4	Cr Plate	115	A-1	4	3	0.75	12	150	250	3000
5	Ni Plate	90	B-2	4	3	0.25	12	100	175	2100
6	Alk. Cleaner	200	C-1	5	2	0.40	10	.100	150	1500

Column 1 gives tank identification numbers to be used in Example 9.2.

Column 2 shows contents of the plating tanks and *column 3* shows operating temperature specified by the customer. Tank 1, hot water; 2, chromium strip (alkaline); 3, bright acid, copper plate; 4, chromium plate (acid); 5, nickel plate (sulfate); 6, alkaline cleaner.

Column 4 gives the hazard potential-evolution rate class, obtained from various tables in the *Ventilation Manual* as shown below.

Column 5 has tank dimensions, given to the designer by customer.
Column 6 is obtained from Table 6.3.
Column 7 is obtained from Table 6.4.
Column 8 equals column 7 × column 5A.

Tank	Table	Special Conditions	Class
1	10-5-5	Complete collection of steam	D-1
2	10-5-8		C-3
3	10-5-7	Higher temperature range	B-3
4	10-5-7		A-1
5	10-5-7		B-2
6	10-5-6	Higher temperature range	C-1

195

Example 6.2   Calculation for Low-Volume/High-Velocity Exhaust System.
Determine the Inlet-cfm (ICFM) and suction in inches of mercury existing at the exhauster inlet, where:

1. The distance from the most remote inlet valve to the inlet of the exhauster pipe-wise with due allowance for fittings is 250 ft.

2. The piping loss in the system is assumed at 1.1 in. Hg per 100 ft of pipe.

3. There are 10 hoses in simultaneous operation on the entire system.

4. Each hose is $1\frac{1}{2}$ in. in diameter, 8 ft long, with $1\frac{1}{4}$-in.-ID nozzle.

5. Each nozzle will draw in 200 cfm.

6. A separator (bag filter) loss of 1 in. Hg is assumed.

7. The installation is at 400 ft above sea level.

Using Fig. 6.41a, locate 200 cfm along bottom scale of center graph, rise to curve representing $1\frac{1}{2}$-in.-diameter 8-ft-long hose and move horizontally to the far left scale and read out 4.34 in. Hg and mark as point 1. Locate 1.10 in. Hg loss per 100 ft of pipe value along the bottom of the right-hand graph and rise vertically to the sloping line representing 250 ft run; move horizontally to the right and read out 2.75; mark as point 2. Connect points 1 and 2 with a straight line and read out 7.09 (point 3).

196

Using Fig. 6.41b, locate a point representing 200 cfm along the bottom of the scale and rise vertically, stopping first at the sloping line representing a $1\frac{1}{2}$-in. nozzle. Move horizontally to the right scale and read out 0.94 in. Hg, continue vertically to the sloping line representing a $1\frac{1}{4}$-in-nozzle and move horizontally to the right scale and read out 2.85 in. Hg.

System loss with $1\frac{1}{2}$-in.-diameter nozzle	7.09
less $1\frac{1}{2}$ in. nozzle loss	-0.94
	6.15
plus $1\frac{1}{4}$ in. nozzle loss	2.85
Total loss in nozzle, hose, and piping	9.00 in. Hg

To this value add 1 in. Hg separator loss, making a total suction S of 10 in. Hg at the exhauster inlet.

$$ICFM = \frac{N(SCFM/hose)B}{B-S} \quad \text{where } N = \text{number of hose}$$

$$B = \text{barometric pressure, in. Hg}$$
$$S = \text{static pressure, in. Hg.}$$

$$= \frac{10 \times 200 \times 29.92}{29.92 - 10.00} = 3004$$

$$= 3004 \text{ cfm at } 10.0 \text{ in. Hg suction at exhauster inlet}$$

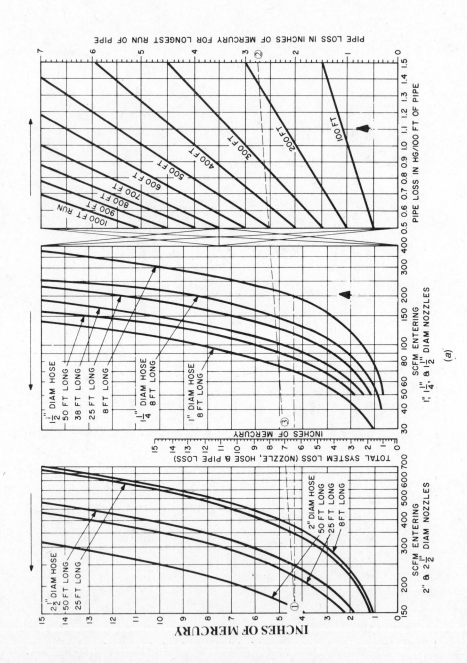

(a)

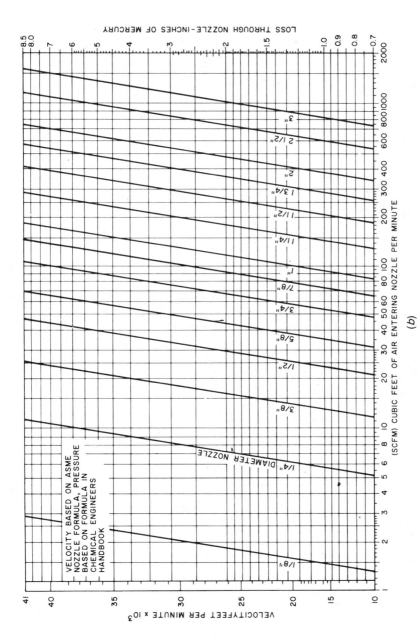

Figure 6.41 Design procedure for low-volume/high-velocity system. (Courtesy of Hoffman Air and Filtration Systems.)

LIST OF SYMBOLS

A_u	open area of the enclosure upstream at dump location
C_e	coefficient of entry
D	average particle diameter
L	tank length
Q	airflow
Q_i	airflow induced by falling granules
R	material flow rate
S	height of fall
W	tank width

REFERENCES

American National Standards Institute, (1964), "American Standard Safety Code for the Design, Construction, and Ventilation of Spray Finishing Operations," ANSI Z9.3–1964 (R 1971), ANSI, New York.

American National Standards Institute, (1971), "American National Standard Practices for Ventilation and Operation of Open-Surface Tanks," ANSI Z9.1-1971, ANSI, New York.

Anderson, D. M. (1964), "Dust Control by Air Induction Technique," *Ind. Med. Surg.* **34**: 168.

Burgess, W. A., and J. Murrow (1976), "Evaluation of Hoods for Low Volume-High Velocity Exhaust Systems," *Am. Ind. Hyg. Assoc. J.*, **37**: 546.

Chamberlin, R. (1959), *Arch. Ind. Health*, "The Control of Beryllium Machining Operations," **19**(2): February :231.

Chan, T. L., J. B. D'Arcy, and R. M. Schreck (1986) "High-Solids Paint Overspray Aerosols in a Spray Painting Booth: Particle Size Analysis and Scrubber Efficiency," *Am. Ind. Hyg. Assoc. J.*, **47**: 411.

DallaValle, J. M. (1952) *Exhaust Hoods*, Industrial Press, New York.

Dalrymple, H. L. (1986), "Development and Use of a Local Exhaust Ventilation System to Control Fume from Hand Held Soldering Irons," in *Proceedings of the 1st International Symposium on Ventilation for Contaminant Control*, October 1–3, 1985, Toronto, Canada, H. D. Goodfellow, ed, Elsevier, Amsterdam.

Dennis, R. (1954), *personal communication*.

Forster, F., and W. A. Burgess (1986), in *Proceedings of the 1st International Symposium on Ventilation for Contaminant Control*, October 1–3, 1985, Toronto, Canada, H. D. Goodfellow, ed., Elsevier, Amsterdam.

Garrison, R. P., and D. H. Byers (1980a), "Noise Characteristics of Circular Nozzles for High Velocity/Low Volume Exhaust Ventilation," *Am. Ind. Hyg. Assoc. J.*, **41**: 713.

Garrison, R. P., and D. H. Byers (1980b), "Static Pressure and Velocity Characteristics of Circular Nozzles for High Velocity/Low Volume Exhaust Ventilation," *Am. Ind. Hyg. Assoc. J.*, **41**: 803.

Garrison, R. P., and D. H. Byers (1980c), "Static Pressure, Velocity, and Noise Characteristics of Rectangular Nozzles for High Velocity/Low Volume Exhaust Ventilation" *Am. Ind. Hyg. Assoc. J.*, **41**: 855.

Hagopian, J. H., and E. K. Bastress (1976), "Recommended Industrial Ventilation Guidelines," U.S. Department of Health, Education, and Welfare, National Institute for Occupational Safety and Health, Publication No. 76–162, Washington, DC.

Hemeon, W. C. L. (1963), *Plant and Process Ventilation*, 2nd ed., Industrial Press, New York.

Nations Fire Protection Association (1969), "Standard for Spray Finishing Using Flammable and Combustible Materials," No. 33-1969, Sect. 201–206, NFPA Boston.

Occupational Safety and Health Administration (1985), OSHA Standard 29, *CFR* 1910.94, 1910.107, Washington, DC.

Pring, R. T., J. F. Knudsen, and R. Dennis (1949), "Design of Exhaust Ventilation for Solid Materials Handling, " *Ind. Eng. Chem.*, **41**: 2442.

Regnier, R., R. Braconnier, and G. Aubertin (1986), "Study of Capture Devices Integrated into Portable Machine-Tools," in *Proceedings of the 1st International Symposium on Ventilation for Contaminant Control*, October 1–3, 1985, Toronto, Canada, H. D. Goodfellow, ed, Elsevier, Amsterdam.

Stephenson, R. L., and H. E. Nixon (1987), "Design of Industrial Vacuum Cleaning Systems and High Velocity, Low Volume Dust Control," Hoffman Air & Filtration Systems, Inc., New York.

U.K. Department of Employment, Her Majesty's Factory Inspectorate (1974), "Dust Control, The Low-Volume High-Velocity System," Technical Data Note 1 (2nd rev.), H.M. Factory Inspectorate, London.

ADDITIONAL READINGS

Goodfellow, H. D., (1986), in *Proceedings of the 1st International Symposium on Ventilation for Contaminant Control*, October 1–3, 1985, Toronto, Canada, Elsevier, Amsterdam.

Harvey, B. (ed.) (1983). *Handbook of Occupational Hygiene*, Kulwer, Middlesex, England.

Peterson, Jack (1985), "Selection and Arrangement of Process Equipment," in L. V. Galley and L. J. Galley, eds., *Industrial Hygiene Aspects of Plant Operations*, Vol. 3, Wiley, New York.

Piney, M. et al. *Controlling Airborne Contaminants in the Workplace*, BOHS Technical Guide No. 7, Science Reviews Ltd., Leeds, England.

7

Chemical Laboratory Ventilation

Early attempts to exhaust noxious air contaminants from chemical laboratory operations were crude. Chemists conducted their experiments near a window or door to disperse air contaminants. Later a storage cupboard was converted to a work area and equipped with an exhaust stack to provide natural ventilation. Some improvement in control was gained by using burners in the hood to achieve convective air movement. These work stations were first provided with mechanical ventilation in the late nineteenth century and the first proprietary laboratory hoods were manufactured shortly after the turn of the century. Principal design innovations to improve collection efficiency have occurred since 1950. At this time the laboratory hood remains the primary ventilation control device in teaching, material control, and research laboratories. In this chapter attention will be given to the laboratory hood, the integration of the hood into the total ventilation scheme, and the effective utilization of the hood by laboratory personnel.

7.1 DESIGN OF CHEMICAL LABORATORY HOODS

The design evolution of the modern "bench-type" laboratory hood will be traced. The engineer must understand the reason for the principal features of the standard laboratory hood to design properly an effective chemical laboratory ventilation system.

7.1.1 Vertical Sliding Sash Hoods

The first laboratory hoods took the form of the simple exhausted storage cabinet shown in Fig. 7.1 and are classified as partial enclosures using the definitions presented in Chapter 5. The size and geometry of these early laboratory hoods varied greatly; in most cases, however, the exhaust stack takeoff was at the top of the hood. Due to this geometry the velocity profile across the face of the hood was poor (Fig. 7.1) and the efficiency of collecting air contaminants released in the hood varied greatly depending on the release point.

The first major improvement to this simple hood was the addition of a vertical sliding sash (Fig. 7.2). The tempered or laminated safety glass sash slides out of the way while the apparatus is set up and during manipulation. When access is not required the sash is partially or totally closed to improve contaminant capture efficiency and to protect the operator from chemical splashes or projectiles from minor explosions or exothermic reactions.

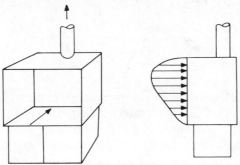

Figure 7.1 Simple exhausted box or enclosure with a top "takeoff." This laboratory hood provides a dedicated work space but rather poor ventilation control. Since most of the air enters the top half of the hood, the velocity profile is not uniform. A rear takeoff would improve the velocity profile, but that geometry requires more space and is rarely used in laboratory hoods.

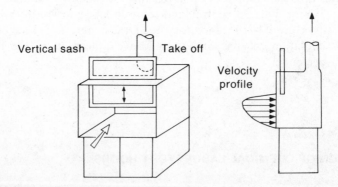

Figure 7.2 The addition of a vertical sliding sash permits variable face area and velocity. In early designs the sash projected through a slot in the top of the hood and leakage could occur under high contaminant generation rates. In later designs the sash slides into a pocket or the hood bonnet and this leakage path is eliminated.

With the sash in the upper position the velocity can vary greatly across the face of the hood, and hood capture efficiency may be poor. A simple modification to flatten the velocity profile is the introduction of a rear baffle with slots (Fig. 7.3). In initial designs a single slot was located at the work surface; although this provided a flat horizontal velocity profile at the work surface, it did not improve the vertical velocity profile. Improvement in the vertical profile was obtained by adding a second adjustable slot in the upper bonnet of the hood. The flattening of the velocity profile with this slot geometry permits the face velocity at all locations to be maintained within $\pm 20\%$ of the average face velocity under ideal test conditions. The number of slots required to achieve a reasonable velocity profile depends on the hood geometry and slot placement (Fig. 7.4). The impact on performance of additional slots has not been studied thoroughly; it is possible that poorly designed intermediate slots may contribute to contaminant "roll", with resulting leakage at the face.

In general, the sash and the rear baffle have improved hood performance. However, when the sash is in a lowered position the air inside the hood may have a characteristic rollback toward the face of the hood, which can be visualized with smoke (Fig. 7.5). In a poorly designed hood, contaminant may escape from the face of the hood by this mechanism.

With the introduction of the vertical sliding sash an additional problem was presented to the hood designer. As the sash is pulled down the face area is decreased and the face velocity is increased. This increased velocity may be

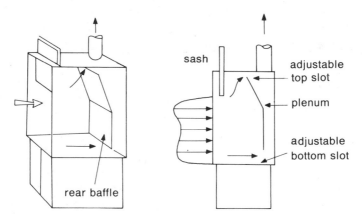

Figure 7.3 The plenum-slot arrangement introduced in the 1930s provided a flat velocity profile across the face of the hood. Slots sized for 1000 to 2000 fpm presented a resistance element which resulted in even distribution of airflow along the length of the slot (see Chapter 5). Normally, the bottom and top slots are adjustable and the intermediate slots are fixed. The slots must be set at the optimum position when installed and maintained in that position unless conditions change. A perforated plate defining the plenum is an alternative to the slots, although manufacturers hesitate to provide this option since it would plug with dust and require periodic cleaning. With proper slot dimensions a velocity profile at the hood face is obtained with variations around the mean velocity of less than 20%. Such a flat velocity profile will ensure a consistent collection efficiency across the face of the hood.

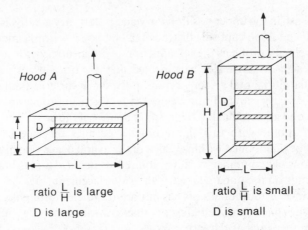

ratio $\frac{L}{H}$ is large
D is large

ratio $\frac{L}{H}$ is small
D is small

Figure 7.4 Hood A is a deep hood with minimum height and may require only one or two slots for proper airflow distribution across the hood face. A large shallow walk-in hood design such as hood B requires multiple slots to produce a uniform face velocity.

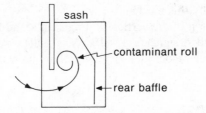

Figure 7.5 The characteristic roll noted behind the sash in a conventional hood may result in loss of contaminant to the workplace.

disruptive to open flames, sensitive balances, filter aids, and certain chemical operations. Frequently, a screen must be placed over the slots to prevent materials from being swept from the work surfaces to the duct. To resolve this problem the bypass hood (Fig. 7.6) was introduced in the 1950s. In its upper position the sash covers the bypass opening and the total exhaust passes through the face opening at the design velocity. As the sash is pulled down the bypass grille is uncovered and a percentage of the exhaust is pulled through the bypass, thereby maintaining a relatively constant face velocity.

The evolution of the contemporary hood was completed by modifying the entry to improve the airflow pattern at the hood face. Studies by Schulte et al. (1954) at the Los Alamos Scientific Laboratory revealed that the square vertical posts and front edge of the work surface generated a slight vena contracta at the hood face. The resulting turbulent eddies caused flow reversal and impaired the capture efficiency of the hood. To minimize this problem the corner posts were "softened" by an air-foil section, as shown in Fig. 7.7. Although this change had a negligible effect on the coefficient of entry, it did reduce contaminant loss by

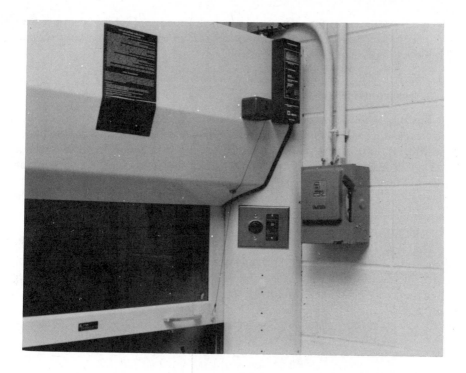

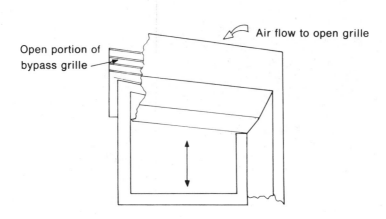

Open portion of
bypass grille

Air flow to open grille

Figure 7.6 A simple bypass or bleeder hood ensures that a critical face velocity limit of 300 fpm is not exceeded when the sash is down. As the sash is lowered there is an increase in face velocity until the upper edge of the sash uncovers the bypass grille permitting air to enter the hood. The face velocity then remains relatively constant until the bottom sash stop is reached. Manufacturers have various designs to provide this bypass feature. In all cases the bypass grille should provide line-of-sight protection against debris generated by explosions in the hood.

boundary eddies. The turbulent losses at the front edge of the work surface were reduced with a displaced air-foil section (Fig. 7.7). A well-designed hood with all of these design features is shown in Fig. 7.8.

7.1.2 Horizontal Sliding Sash Hoods

In the basic laboratory hood described above a variable face area is achieved by a vertical sliding sash. This goal can also be achieved by horizontal sliding panels positioned on tracks at the hood face (Fig. 7.9). Although available for several decades, these hoods have only recently gained popularity since they provide an effective means of minimizing face area, with a resulting reduction in airflow and operating cost. In addition to minimizing face area, an advantage claimed for the horizontal sliding sash hood is the availability of the panel as a protective shield for laboratory personnel.

Detractors of horizontal sliding sash hoods claim that the chemist frequently rejects this design since it requires different work practices than those associated with the traditional vertical sash hood. If not accepted, the sash may be removed by laboratory personnel, resulting in an unsafe hood.

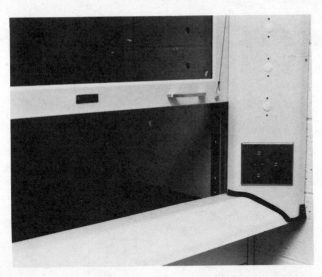

Figure 7.7 The vena contracta, which causes turbulent eddies at the hood corner posts, can be minimized by modifying the hood inlet geometry. An ideal corner post would have the form of a bellmouth. In practice this is frequently approximated with a simple angied detail which is inexpensive to fabricate and install but effectively minimizes the entrance eddies and improves contaminant control at the entrance to the hood. Although the same simple geometry used on the corner posts could be used to improve the entrance geometry at the front lip of the work surface, a more effective approach is the displaced air foil. This element breaks turbulent eddies, provides a well-defined airflow path to sweep the hood work surface when the sash is down, and may be used to define the minimum working distance from the face of the hood. The space between the work surface and the air foil should be large enough so that the operator can pass through tubing and electrical cables.

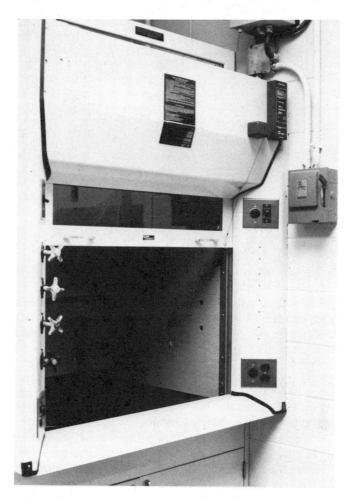

Figure 7.8 Modern vertical sliding sash hood which incorporates the various design features, including air-foil elements, bypass feature, and multiple rear slots.

A vertical sash hood has one sharp edge, the bottom of the sash, which will cause a rolling eddy in the back of the sash. In the horizontal sliding sash hood, air may pass two vertical sharp-edged entrance elements. In the vertical sliding sash hood the edge turbulence may occur with the sash up, away from the point of contaminant generation, and the effect of the eddy may be minimized. In the horizontal sliding sash system the vertical turbulence is generated along the full height of the sash and, therefore, through the zone where the contaminant is generated. This may result in significant loss of contaminant from the hood. The eddies formed at the edges of both vertical and horizontal sashes can be minimized by an air-foil element similar to that shown in Fig 7.7, although this is rarely included in a conventional hood.

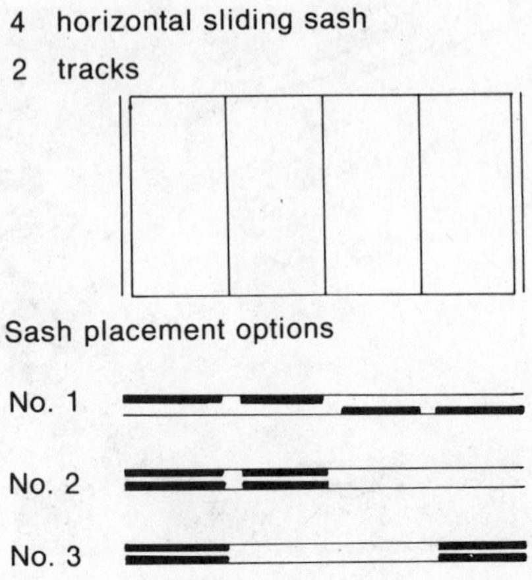

4 horizontal sliding sash

2 tracks

Sash placement options

No. 1

No. 2

No. 3

Figure 7.9 Horizontal sliding sash on an 8-ft hood which reduces the required airflow by approximately 50%.

The horizontal sliding sash designs vary, with the principal differences being the width and number of panels and the number of tracks (Table 7.1). A minimum panel width of 1.5 ft is recommended to minimize this edge effect.

7.1.3 Auxiliary-Air-Supply Hoods

The auxiliary-air-supply hood, also known as a supply-air or compensating hood, is designed to minimize the quantity of conditioned replacement air to be supplied to the laboratory space during the cooling season. The system, shown in Fig. 7.10, may be purchased as a complete hood system or as a canopy system for retrofitting conventional hoods. It is designed for application on a vertical sliding sash hood, although it can be used on a horizontal sliding sash system. When used with a horizontal sliding sash hood it does not realize its full energy conservation potential and may cause additional turbulence around the hood face. Rather than taking 100% of the total hood exhaust from the room, a minimum of 30% is removed from the laboratory space and up to 70% of the

TABLE 7.1 Horizontal Sliding Sash

Number of Panels	Number of Tracks	Maximum Open Area Fraction of total
2	2	$\frac{1}{2}$
3	2	$\frac{1}{3}$
3	3	$\frac{2}{3}$
4	3	$\frac{1}{2}$
4	4	$\frac{3}{4}$

Figure 7.10 Well-designed auxiliary-air-supply hood. The outside air is introduced through a canopy designed to provide a laminar flow pattern which blankets the hood face and blends with room air as it enters the hood. The hood is designed to provide air at room temperature during the heating season; unconditioned air is supplied during cooling periods, thereby saving air-conditioning costs for the air supplied through the canopy. An additional advantage is less well known. When a person stands in front of a laboratory hood a slight negative pressure is created between his or her body and the hood, causing turbulence and loss of contaminant from the hood. The auxiliary air supply fills this space, eliminates turbulence, and improves collection efficiency. (Courtesy of Kewaunee Scientific Corp.)

exhaust is outside air delivered directly to the face of the hood through a canopy. The outside air supply may be unconditioned air in the summer, but in the winter it must be tempered to room temperature to eliminate operator complaints and to minimize condensation problems in the hood.

If convenient, the canopy supply can be secondary air, that is, air that has been conditioned and must be exhausted for general ventilation needs. It is important that the supply be delivered to the outside face of the hood through a well-designed canopy. Typically, the canopy is provided with an eggcrate grille to more closely approach laminar flow across the hood face. Designs that provide air directly to the hood cavity are disruptive, pressurize the hood, and will cause loss of contaminant to the room.

A well-designed auxiliary-air-supplied hood reduces air-conditioning costs and improves hood capture efficiency (Chamberlin and Leahy, 1978). The disadvantages of this hood include higher initial cost and maintenance due to the additional replacement air system, the possible introduction of airborne dust into the laboratory, discomfort to operators during winter conditions unless the air is tempered, disturbance of the exhaust velocity profile at the hood face with poor canopy designs, and the possible hazard that exists in a poorly designed system if the exhaust fan fails and the auxiliary air fan continues to operate (BOHS, 1975). Although these objections can be resolved through appropriate system design, they do demonstrate the complexity of the auxiliary-air-supply system. One facility requirement that frequently cannot be met is the overhead space required for the air supply duct and the auxiliary air plenum.

7.2 FACE VELOCITY FOR LABORATORY HOODS

A discussion of face velocity for a laboratory hood may seem unnecessary since extensive advice is available on face velocity in the *Ventilation Manual* for similar hoods such as paint spray hoods, (VS-603, 604). Experience has shown that this is not the case. Special attention must be given to face velocity criteria for laboratory hoods due to their unique application, the range of toxic chemicals handled, and the siting of the hood in the laboratory.

By definition, the laboratory hood is an enclosure with an adjustable front-access panel. Face velocity (*Ventilation Manual*, Fig. 3–1) is the critical design velocity for a laboratory hood. Capture velocity, the velocity established outside the physical boundary of a hood to capture the contaminated air, is often incorrectly used to identify face velocity. Face area denotes the working opening at which the face velocity shall be achieved. Other terms, such as "sash opening" or "door area," are used interchangeably with face area.

The design airflow is based on the average face velocity and the maximum face area that can be achieved with the vertical or sliding sash configuration. If the sash is constrained to a restricted position by mechanical means that cannot be bypassed, that restricted face area may be used in defining the minimum airflow.

The standard opening height for a bench-mounted, vertical sliding sash hood is 30 in., although a range of 26 to 36 in. may be encountered. Individual laboratories interested in minimizing energy costs have found that an opening as low as 20 in. may be acceptable for normal laboratory work. In this configuration, the full standard opening height of 30 in. is available during equipment setup; the reduced height is used during the chemical procedure.

A major investigation of laboratory hood face velocity was completed by Schulte et al. (1954). Subsequent studies have been summarized by Hughes (1980) with over 20 literature citations for the period 1950 to 1980. Recommendations for face velocity for chemical laboratory hoods may be proposed in three specific forms. The first approach, best characterized by Peterson (1959), provides a rational format to calculate the face velocity for a specific chemical used in a given laboratory environment. This calculation requires the user to consider a range of critical variables influencing capture efficiency.

In the second approach graded classes are established and a minimum face velocity is specified for each class. The classifications may be based (1) on the application realm with some inferred level of risk (Coleman, 1951); (2) solely on the toxicity of the chemicals handled in the hood (Brief et al., 1963); or (3) on a classification scheme that includes consideration of the toxicity of the chemical and the laboratory geometry [Scientific Apparatus Manufacturers Association (SAMA), 1980].

The third approach utilizes a single recommended face velocity for all hoods in the belief that the minimum face velocity necessary to overcome background air movement in the laboratory will control any chemical air contaminant in a well-designed hood (Schulte et al., 1954).

Although the response to these approaches, as introduced over several decades, has varied, general trends have emerged. The comprehensive chemical-by-chemical procedure provides extensive insight into hood performance design criteria; however, it has not been widely accepted, probably due to its complexity. The broad classification schemes have been used by many laboratories, but critics of this approach state that it is difficult to anticipate the chemicals and procedures to be used in laboratory hoods even on a short-term basis. This procedure also limits hood use to selected chemicals, thus complicating laboratory operation and supervision. By far the most common procedure for setting minimum face velocities is the third approach, the use of a single face velocity that will permit application of the hood for a range of chemical activities and laboratory environments.

The recommended face velocities have ranged widely. The minimum velocity needed to overcome diffusion of gases and vapors from the hood is 20 fpm (Ettinger et al., 1968). During the 1950s when the classification schemes were popular, velocities of 50 to 150 fpm were widely cited. A velocity of 100 fpm emerged during the 1960s under the single-value approach and was widely supported until 1980. It can be shown that 100 fpm provides control for most chemicals handled in industrial laboratories using the rational design procedure proposed by Peterson (1959). This velocity also has been demonstrated to overcome the disruptive air movement generated by a chemist working at a hood (Schulte et al., 1954). For these reasons 100 fpm is thought to constitute a good general design value.

The energy constraints instituted in the mid-1970s provided an impetus to investigate methods to minimize hood operating cost by lowering the total quantity of exhaust air. One such approach is to reduce face velocity. Investigators have demonstrated that face velocities as low as 60 fpm are adequate if an aerodynamic hood is used in a nearly ideal setting and the operator follows rigorous work practices. This philosophy is clearly stated in the design guidelines for new laboratories presented in the *Ventilation Manual* (VS-204.1). The working distance behind the sash must be kept to a maximum and replacement air must be introduced through perforated ceiling panels or by grilles or diffusers (the "throw" velocity from the diffuser at the hood face must be less than 30 to 50 fpm). Furthermore, air movement from doors, windows, and traffic must be minimized, as should the amount of apparatus in the hood. Critics of this approach agree that reasonable control may be achieved under ideal conditions with such low face velocities, but contend that it is difficult to provide such operating conditions in a "real-world" laboratory.

Although there is disagreement on the minimum face velocity needed for laboratory hoods, most investigators agree that increasing the face velocity above 120 fpm will not improve capture efficiency. At some higher face velocity, depending on the hood design, the hood performance may actually be diminished, due to turbulence at the hood face.

For effective containment there must be a uniform velocity across the laboratory hood face. A hood tested under ideal conditions should exhibit the same velocity at all points across face; an acceptable variation between locations is ± 10 to 20% of the mean. The following recommendations for variation in face velocity have been proposed for actual conditions in the laboratory with equipment in the hood:

- ASHRAE (1982). The minimum velocity at any point should not be less than 80% of the mean value.
- BOHS (1975). The mean value at specific points should be within 15% of the overall mean in a well-designed system.

7.3 SPECIAL LABORATORY HOODS

The laboratory hoods described above are valuable general–purpose control devices in the modern chemical laboratory. The laboratory hood provides a physical barrier to contain spills, splashes, and mild overpressure accidents. A modest face velocity will contain air contaminants released from chemical operations. Although it has these impressive features, the laboratory hood should not be considered the only ventilation control available in chemistry laboratories. Standard procedures such as evaporation, distillation, and digestion are routinely conducted in laboratories and a range of local exhaust hoods are available for these operations. Examples of these special-purpose hoods, which have the advantages of low initial cost, low airflow, and excellent capture efficiency, are described below.

Specimen digestion using strong oxidizers such as perchloric acid is frequently encountered in settings ranging from biological to metallurgical laboratories. This procedure requires excellent containment and air cleaning to ensure that violent reactions do not occur. Organic materials cannot be used in the hood or duct construction, and spray nozzles must be installed to wash down the duct work to prevent the buildup of unstable perchlorates (*Ventilation Manual*, VS-205.1). The hood used for perchloric acid digestion must be committed to this activity exclusively. Silverman and First (1962) have designed a scrubber for perchloric acid digestion which is placed directly in a standard laboratory hood. After cleaning, the scrubber exhaust stream is discharged to the laboratory hood exhaust stream. An integral scrubber effective against perchloric acid has been described by Renton and Duffield (1986).

A second common procedure in chemical laboratories is paper chromatography. In this process a paper sheet onto which the liquid sample has been placed is put in a large jar in a solvent-saturated environment for the development of the chromatograph. The changing of samples results in release of the solvent vapors to the air. Again, this work could be done in a conventional laboratory hood, but it can be controlled more efficiently by utilizing a mobile exterior hood which is placed directly behind the jar during transfers (Brief et al., 1963). This design provides excellent control using modest airflows of 250 to 300 cfm. The savings in installation and operating cost utilizing this special hood over the conventional laboratory hood are impressive.

A common procedure in material control laboratories is the evaporation of a large number of samples. Rather than conducting this procedure in a chemical laboratory hood it can be controlled more efficiently by spot local exhaust (VS-206). A more difficult chemical laboratory procedure to control is distillation or reaction equipment mounted on a large rack with the potential for release of air contaminants at multiple locations. A "walk-in" hood can be provided for control, but the penalty of limited accessibility and large exhaust volumes must be accepted. An alternative approach is to mount the equipment in the open and provide one or more flexible "drops" which can be positioned at critical release points. Normally a 3-in.-diameter flexible duct with an exhaust volume of 200 cfm is adequate to control all but major accidental releases.

7.4 LABORATORY EXHAUST SYSTEM FEATURES

Whether to use a separate fan for each hood or to manifold a number of hoods together to be served by a single fan is one of the first major decisions to be made in the design of a laboratory ventilation system. Each installation must be considered separately.

The "one-hood/one-fan" system might at first glance seem desirable. This approach provides ultimate flexibility for the individual operator and the manager of the laboratory. The chemist can command a total system deciding when to turn it on or off. Furthermore, it is available for use during off-shift periods. If a classification scheme has been used to choose the hood face velocity, the hood can be upgraded to a higher classification by changing fan speed or installing a new fan to achieve a higher face velocity. If air cleaning is necessary on the total hood exhaust stream the additional resistance to airflow presented by the air cleaner may be handled by a change in sheave size or air mover. If a need arises for a new specialty hood such as a perchloric acid digestion hood, a one-hood system can be modified and committed to this service without compromising other hood applications. The design of such systems is simple and a modular design can be used throughout the facility. If maintenance is necessary, the system can be shut down and the activities of only one operator will be disrupted.

The disadvantages of a one-hood/one-fan system include higher initial cost and additional space requirement for ducts and fans. If there are several hoods in one laboratory and one hood is shut down, lack of replacement air may cause backflow to the laboratory through the duct of the inoperative hood. At best, this could cause temperature extremes in the vicinity of the hood; at worst, if the duct terminated at a location where contaminated air is present, it could introduce toxic air contaminants into the laboratory. The installation of backflow dampers to prevent this should be discouraged since these devices will frequently fail in the closed position.

A central system that utilizes one fan serving a number of hoods can be used for hoods within the same laboratory as long as the air contaminants released from the various hoods are compatible. There have been few documented cases of serious fire, explosion, or health hazards resulting from the mixing of two chemistry laboratory hood exhaust streams. This favorable history is probably due to the low concentrations of contaminants commonly present. There are other more serious disadvantages to a central system. If the fan serving multiple hoods fails, all hoods served by the system are out of commission and backflow may occur from one laboratory to another. The initial cost of a central system may be lower than that of a family of single-hood systems, although balanced design of such a system takes more engineering skill. If a blastgate design is utilized, the system must be properly balanced. Since all hoods in a central system operate continuously the replacement-air system can be a fixed-flow design to respond to the total system needs.

If the single-unit design approach is chosen, the hoods should not be purchased with an integral fan since this configuration places the ductwork under positive pressure and leakage of contaminated air may occur. Additionally, integral fans provided with laboratory hoods are usually low-static, direct-drive fans and cannot handle the resistance to airflow presented by long duct runs.

In any system the fan should be positioned so that the duct run inside the building is under negative pressure. This normally means a roof-mounted fan for a multistory building or a roof or ground-level fan for a single-story building. If a short portion of the duct run inside a building must be under positive pressure, the duct should have welded seams, flanged joints with gaskets, or joints sealed with an impervious material.

Fans for large laboratories using single-hood systems are frequently placed in a roof penthouse dedicated to mechanical services. Such placement keeps the fans out of the weather and encourages maintenance. However, this practice may present a risk to maintenance personnel in case of duct failure within this penthouse downstream of the fan. Such equipment space should be provided with dilution ventilation and all joints on duct runs under positive pressure inside the penthouse should be sealed.

Materials of construction for laboratory systems are discussed in detail elsewhere (DiBerardinis et al., 1987). The current practice is to use 304 stainless steel, or if the duct chase is provided with sprinklers, polyvinyl chloride (PVC) or fiberglass-reinforced polyester (FRP). The initial duct run from the hood should be designed with an elbow close to the hood to prevent material falling back into the hood. If the hood is to handle condensible vapor, the first few lengths of duct should be pitched to a drainage point and the drain trapped. All longitudinal seams on the horizontal duct runs should be positioned on the top of the duct.

Since particles are infrequently encountered in laboratory systems, high transport velocities are not required, although space constraints in service chases may occasionally require small duct sizes and corresponding high duct velocities. The duct should normally be sized for a velocity of 1000 to 2000 fpm (*Ventilation Manual* VS-203), which will minimize airflow resistance and air noise but will increase duct construction costs.

The choice of fan will depend on the type of system in use. Single-hood systems are frequently served by axial flow fans due to their low cost, straight-through geometry, and ease of installation. This choice precludes the many advantages of a centrifugal fan, including the flexibility of fan speed change. Systems serving multiple hoods are normally equipped with centrifugal blowers due to their ruggedness and high performance. Fan impellers can be fabricated of PVC, FRP, or steel with an impervious plastic coating. If flammable gases or vapors or explosive dusts are encountered, an aluminum, bronze, or FRP impeller may be required. Occasionally, a serious corrosion problem exists which cannot be handled by fan construction materials. In this case an ejector (see Chapter 10) may be the air mover of choice.

Reentry of odorous or toxic air contaminants from exhaust stacks is a chronic problem in many industrial and academic chemical laboratories. The design data provided in Chapter 16 will help minimize this problem, but it can be eliminated only by appropriate air and gas cleaning.

The conventional industrial approach to air cleaning is to direct the entire hood exhaust stream through the air cleaner. This approach is not effective for the general-purpose laboratory hood since the exhaust stream has very low concentrations of contaminants in a relatively large airflow. Such systems are not amenable to air cleaning. The most successful approach to this problem is to direct the low-volume release from the reaction under way in the hood to an air-cleaning unit in the hood before it is released to the main hood exhaust. Most general chemistry laboratories handle 0.10 to 1 molar quantities. The contaminant release rates, presuming a 20-min reaction time, are therefore in the range 0.1 to 1.0 1pm. Acidic and alkaline air contaminants can be scrubbed in simple packed columns, and other streams can be treated as described in Chapter 11.

The air-cleaning technology evolved over the past five decades in the respiratory protection field may also be applied to such small streams. A range of adsorbents and absorbents available as cartridges, chin canisters, and chest canisters may be effective against certain off-gas streams. The self-contained air-cleaning hoods with direct return to the laboratory space are not suitable in their present form for serious application in the chemistry laboratory.

7.5 FACTORS INFLUENCING HOOD PERFORMANCE

A number of factors that influence the field performance of the laboratory hood must receive the attention of the design engineer. One set of factors relates to the physical location of the hood and other elements in the laboratory. In addition, the performance of the hood depends on its proper use by the chemist. Although the latter issue is beyond the responsibility of the engineer, the impact of the chemist's work practices on hood performance must be considered.

7.5.1 Layout of Laboratory

Doorways. A pedestrian door positioned close to the face of a laboratory hood may seriously impair hood performance. A door opened in a normal fashion acts as an air mover. If swung open in a 90° arc in 2 seconds, a door 30 in. by 80 in. has an edge speed of approximately 200 fpm and displaces room air at a rate of 1200 cfm. This effective pumping rate is approximately equal to the airflow through a 5 ft hood with a sash height of 30 in. operating with a face velocity of 100 fpm. Positioned near a hood, such a door can therefore temporarily induce an air flow across the hood face equivalent to the hood exhaust, disrupting the air flow pattern and sweeping contaminant from the hood. It is important that hoods be positioned some distance from laboratory access doors. If it is absolutely necessary that a door be positioned near a hood, its disruptive effects can be minimized by using a door governor to ensure a low opening and closing speed. Another solution is to restrict access to the laboratory through the critical door. Door placement is also important in ensuring a safe means of egress. Since a hood may be the site of a fire, explosion, or toxic chemical spill, it is obvious that the egress route should not be past the hood.

Pedestrian Traffic. Schulte et al. (1954) demonstrated that pedestrian traffic past a hood can affect its performance, and this should be reflected in the laboratory layout. Walking while wearing a buttoned lab coat sweeps out a cross-sectional area of approximately 10 to 20 ft^2. At a walking speed of 200 fpm (2.3 mph), 2000 to 4000 cfm is displaced. This is equivalent to the exhaust of an 8-ft-long laboratory hood with a 30-in. sash opening operating at a 100-fpm face velocity. If the walking occurs in front of a laboratory hood, it can result in escape of contaminant from the hood. To minimize this problem, the British Standards Institution, in unpublished material, recommends that a free zone be established 1 m from the hood face, which should be entered only by the chemist.

Windows. Window placement is equally important. Most laboratories are air conditioned and windows, if available, will have permanently closed sashes. If it is possible to open a window in a laboratory, an outdoor wind gust of as little as 3 to 5 mph (300 to 450 fpm) across the hood face can be quite disruptive.

Equipment. The optimum performance of a laboratory hood occurs when the movement of air in front of the hood approaches laminar flow in a direction perpendicular to the hood face. Placement of laboratory benches, hoods, and partitions can interfere with this airflow. To minimize this interference, the British Standards Institution has recommended the minimum clearances shown in Fig. 7.11.

Another important feature of the laboratory which affects hood performance is the introduction of replacement air. If the laboratory is small and equipped with several hoods, it may be difficult to provide the required supply from the HVAC diffusers without disrupting air movement at the hood face. A basic rule is to introduce the air at the maximum possible distance from the hood and permit it to sweep through the laboratory to achieve maximum effective general ventilation. Another good rule is to maintain the discharge velocity from air supply grilles below 400 fpm, and ensure that the velocity caused by the grille or diffuser does not exceed 30 to 50 fpm at the hood face. If the laboratory is small, the coupling between diffuser and hood can be great. In this case the replacement air can be introduced through a perforated ceiling plenum at velocities of 70 to 100 fpm. Such a system ensures excellent mixing in the laboratory space and will minimize disruptive effects at the hood face. If introduced just in front of the hood, it will provide some of the advantages of the auxiliary air supply hood.

7.5.2 Work Practices

Even if the basic design is excellent and the hood is placed in a "sheltered" location in the laboratory, it may still present worker exposure unless properly used. Although the engineer will not have primary responsibility for proper hood use, the impact of poor work practices on hood performance should be understood. One common poor work practice is the use of the hood as a chemical storage cabinet. Typically, a large number of chemical bottles are placed deep in the hood against the lower exhaust slot. Blocking this slot causes the airflow balance between the lower and upper slots to be disrupted and the face velocity profile will suffer. Separate storage cabinets should be provided to

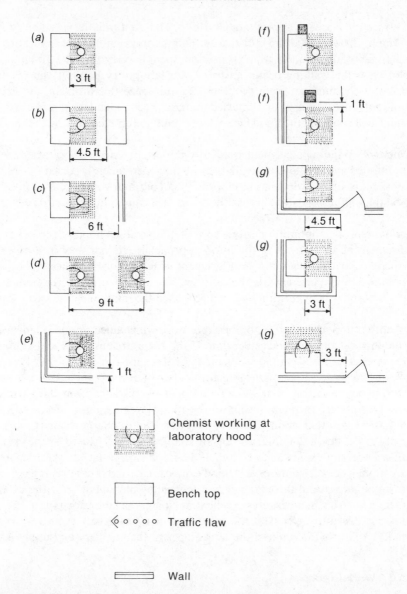

Figure 7.11 Minimum clearance of laboratory hoods from benches and structural details. (a) separation of undisturbed zone from traffic route; (b) spacing where same operator uses hood and bench top or where occasional traffic only is anticipated; (c,d,e) spacing determined by airflow requirements; (f-g) spacings which avoid undue disturbance of airflow. [From BSI (1979)].

prevent this practice. A flammable liquid storage cabinet requires an exhaust flow of only 50 cfm. Storage can also be provided in an exhausted base cabinet under the hood. A hood shelf can be installed above the bottom slot if it is absolutely necessary to keep large numbers of bottles in the hood.

Another disruptive work practice is the storage of miscellaneous boxes of supplies or equipment in the hood. If a large piece of equipment such as an oven, blender, or hot plate must be in the hood, it should be positioned to minimize disruption of the hood velocity profile. The equipment should be placed deep in the hood on legs so that the bottom slot will not be blocked. The equipment should be positioned so that the minimum cross-sectional area is normal to the direction of airflow.

Depth of the working area is an important hood performance variable. For example, in industrial ventilation systems the larger the paint spray booth or the deeper the work location in the hood, the lower the required face velocity (*Ventilation Manual*, VS-604). These same guidelines apply to laboratory hoods. If work is done directly at the face, a contaminant may escape the hood and not be recaptured. If the maneuver is repeated deep in the hood, the ability to capture the contaminant is improved. It may thus be possible to reduce face velocity in a chemical laboratory hood if a deep working depth is acceptable. The proper working depth can be designated by marking the desired position on the laboratory hood work surface with tape, placing a physical lip on the work surface, or using an air-foil section which extends into the hood and thus defines the minimum working depth. Proponents of reduced face velocities recommend a minimum working depth of 5 to 20 cm, depending on the chemicals (Table 7.2). This practice, of course, results in a restrictive work area and frequently is disregarded by the chemist.

Placing extending wings on the hood minimizes the effects of drafts and can improve capture efficiency. Manufacturers are limited in the construction of deeper hoods by poor operator acceptance and the fact that hood depth is governed by the dimensions of construction materials (typically Transite) used as the interior hood sheathing. In recent years a variety of alternative sheathing materials (e.g., other mineral-filled board and polyester, epoxy, and polypropylene panels have been introduced).

The convective head produced by a heat-generating apparatus placed in the hood can also impair performance. Schulte et al. (1954) found that three 1200-W electric hot plates were acceptable, but that four Bunsen burners in a 4-ft hood

TABLE 7.2   Hood Working Depth

Permissible Exposure Limit (ppm)	Minimum Distance between Source and Sash (cm)
1000	5
100	10
10	15
1	20
< 1	5 at 120 fpm

Source: Fuller and Etchells (1979).

resulted in air temperatures in the upper part of the hood 38 °C above room temperature. The resulting convective velocities were disruptive to effective contaminant collection; smoke released in the hood during this temperature excursion escaped to the room through openings in the bypass mechanism. Brief et al. (1963) states that the application of electric heaters should be restricted to 100 W/ft of hood length to minimize this effect. If there is a high heat load in the hood, the top slot should be adjusted to draw a greater share of the airflow.

Laboratory personnel should place all equipment and supplies in the hood before starting a procedure, since manipulation during the chemical procedure may result in contaminant escape. It is also good practice to close the sash as far as possible, thereby approaching a complete enclosure and minimizing loss of contaminant. In this position the sash is also a barrier that can provide protection from spray, splashes, and minor explosions.

7.6 ENERGY CONSERVATION

The rapidly escalating price of energy has called attention to the cost of laboratory ventilation. A standard 5-ft hood installed in an air-conditioned laboratory in the northeastern United States and operating continuously (8760 hr per year) at a face velocity of 100 fpm costs approximately $4000 to $5000 per year to operate at 1988 energy prices. If an industrial or academic laboratory complex has 100 such hoods, the annual operating cost to run the fans and heat and cool the replacement air is nearly $500,000. Four approaches to minimizing this cost are to reduce the operating time, limit the air quantity being exhausted from each hood, recover heat from exhaust air, and directly recirculate from an air cleaner.

7.6.1 Reduce Operating Time

There are several reasons for operating laboratory hoods 24 hours a day, seven days a week:

- Personnel might work in the laboratories any time of the day or night, on weekends, or on holidays.
- Hazardous chemicals are frequently stored inside the hood overnight when the hood is not being used.
- Some untended reactions are left ongoing in the hood for extended periods.
- The exhaust hood frequently provides the minimum general ventilation for the space and establishes the required pressure difference between spaces.

Even given these reasons, a number of major laboratories have found that hoods can be shut off at least 50% of the time with significant savings. In research and development laboratories this rigorous protocol can be relaxed by giving the hood user the option of turning on the hood when it is needed. Since the hoods are not operated continuously, alternative ventilated storage space must be provided for selected chemicals. The storage space may be a flammable-liquid storage cabinet meeting National Fire Protection Association requirements. Nonflammable hazardous liquids can be placed in a separately exhausted storage cabinet, an exhausted storage area under the laboratory hood, or in a specially designed chemical storage box (DiBerardinis et al, 1987).

It is important to note that in reducing the hood operating time the air balance within the laboratory and the building may be compromised. Buildings frequently have one supply system for a group of laboratories. It is difficult to maintain labs at negative pressure with respect to corridors when the exhaust rate varies as a result of hood utilization. Variable-air-volume supply systems (Chapter 15) can be used to handle this replacement-air problem.

7.6.2 Limit Airflow

By design, the required laboratory hood airflow is based on the largest possible opening, under the assumption that the operator will utilize the hood with the maximum opening some fraction of the time. On a vertical sash hood this opening is usually the width of the hood work surface, multiplied by the height of the hood opening with the sash fully raised. The maximum opening and thus the airflow can be reduced by limiting the vertical sash travel or by installing a horizontal sliding sash. Laboratory fume hood sashes commonly open to a height of 30 in., but if sash height is limited to 20 in., a height many chemists find acceptable, the amount of air being exhausted can be reduced by 33% while maintaining the required face velocity.

This restricted opening may present a problem to the chemist while installing equipment, in that it may be difficult to reach the upper part of the hood. This problem can be resolved by equipping the hood with an alarm system. If the sash is raised above 20 in. for short periods to allow setting up apparatus, an audio and visual alarm is activated to let the hood user know that the hood is not in a safe operating mode. If the sash must be positioned above 20 in. for long periods, the audible alarm can be temporarily disarmed. The alarm should have a memory so that it will reset if the sash is still raised after some time period. As soon as the sash is lowered below the proper height, the alarm is reset and is available to respond the next time the hood sash is raised above the proper height.

If a reaction is not running, the exhaust can be reduced to a minimum rate that will control any chemicals left in the hood while providing the minimum general exhaust ventilation of the laboratory space. As described previously, the exhaust airflow can be reduced with a horizontal sliding sash. In the typical case illustrated in Fig. 7.9, a maximum of one-half of the normal opening of the 8 ft hood is available; an energy savings of 50% is therefore realized. As mentioned earlier, a potential problem with this approach is the poor acceptance by hood users. To reach every part of the hood the individual sash elements must be moved horizontally, which may provoke the chemist to remove the sashes, thus creating an unsafe condition. Some method of monitoring is necessary to ensure that horizontal sashes are used correctly.

The use of special local exhaust hoods to reduce airflow has also been discussed previously. The major disadvantage of this approach is that such hoods are restricted to specific tasks and cannot be used for general laboratory purposes. In addition, special local exhaust hoods do not provide the protection of a chemical laboratory hood in terms of containment of spills or protection from small explosions or fires.

Auxiliary-air-supply hoods draw only 30 to 50% of the hood exhaust from the laboratory space. A slight economic advantage occurs during the heating season when the outside air entering the hood from the canopy needs only to be tempered to room temperature, not to the room supply state point. The major savings, however, are realized in the cooling season since the air delivered directly from the hood canopy does not have to be cooled.

7.6.3 Heat Recovery

Heat recovery systems employ a heat exchanger in the exhaust airstream. The application of this type of system must be based on a building-by-building evaluation. This may not be a particularly useful technique for laboratory spaces since the exhaust stream may contain corrosive or reactive contaminants which will damage the heat recovery system.

7.6.4 Recirculating Laboratory Hoods

Portable laboratory hoods with an integral enclosure, fan, and activated charcoal bed permitting direct return of air to the laboratory workplace were introduced in the 1970s. Manufacturers claim that these hoods provide good capture efficiency while conserving operating costs since replacement air is not needed. This concept is attractive to laboratory management since there is a chronic need for additional exhausted workstations in laboratories with space limitations and without adequate replacement air to handle a conventional hood. In addition, the initial cost of these recirculating laboratory hoods may be 10 to 20% the total cost of a small conventional laboratory hood, including the initial hood cost, hood installation, and the provision of replacement air.

Investigations of the performance of these hoods indicate that caution should be exercised before purchasing them for conventional laboratories (Abrams et al., 1986). Such hoods should only be considered for laboratories where the chemicals in use are well defined and for which the manufacturer's sorbent bed is known to be effective. Four major problems have been identified.

- The hood geometry is frequently a simplified version of a laboratory hood without the design features that minimize eddy formation and contaminant loss at the hood face.

- To extend the life of the activated charcoal bed the airflow through the bed must be minimized, and therefore, a low face velocity (60 to 70 fpm) typically is utilized, which may not be adequate in other than quiescent conditions.
- Placement of the portable hood near doors, pedestrian walkways, or near the discharge from an HVAC diffuser will result in poor performance and loss of contaminants from the hood.
- The most serious objection to the recirculating laboratory hood is that the operator does not know when the adsorption bed is depleted. For this reason, these hoods should not be used for toxic air contaminants which do not have adequate warning properties below their TLV concentration unless a suitable air monitor can be installed to monitor breakthrough of the bed. One manufacturer of recirculating laboratory hoods provides a system that releases an offensive odorant when the bed is depleted and breakthrough occurs.

As discussed, there are several alternatives that apply to energy conservation in the use of laboratory hoods. In most cases a combination of the alternatives described above may be required, particularly in new buildings. In the renovation of older buildings, it may be difficult to apply these methods, and each situation will have to be evaluated on a case-by-case basis. In any energy conservation approach the following issues must be considered:

- Acceptance by lab personnel of any restrictions that are placed on the use of the hood
- The effect of each alternative on building HVAC systems
- The effect of each alternative on safety in the laboratory
- Options for providing additional storage space

Any program to alter the current use of laboratory chemical hoods must include a detailed education and training program for laboratory personnel. This program should include information on how laboratory chemical fume hoods operate; what restrictions, if any, are being placed on their use; and how this may affect work in the area.

7.7 GENERAL LABORATORY VENTILATION

Laboratory hoods perform not only as local exhaust ventilation hoods but also provide general exhaust ventilation in the space. Such ventilation is necessary to establish comfort conditions and provide dilution ventilation for any minor contamination generated outside the hood. The design for general ventilation for a chemical laboratory is normally specified as 4 to 12 air changes in the space each hour (DiBerardinis et al., 1987). A single 4 ft hood with a sash opening of 30 in. and an average face velocity of 100 fpm will exhaust 1000 cfm; for a room with dimensions of $10 \times 20 \times 30$ ft this hood provides an air change every 6 min or 10 air changes per hour. The limitations in the use of the air-change-per-hour concept have been discussed in Chapter 4; however, the concept is widely used in laboratory ventilation specifications and for that reason is presented here.

The laboratory should be maintained under slight negative pressure relative to the rest of the building. This can be done by a judicious choice of the replacement-air supply, as described in Chapter 12.

REFERENCES

Abrams, D. S., P. C. Reist, and J. Dement (1986), "An Evaluation of the Effectiveness of a Recirculating Laboratory Hood," *Am. Ind. Hyg. Assoc. J.*, **47**(1): 22–26.

American Society of Heating, Refrigerating and Air Conditioning Engineers, (1982), "1982 Applications," ASHRAE, New York.

Brief, R. S., F. W. Church, and N. V. Hendricks, (1963), "Design and Selection of Laboratory Hoods," *Air Eng.*, **5**: 70.

British Occupational Hygiene Society (BOHS) Technology Committee (1975), "A Guide to the Design and Installation of Laboratory Fume Cupboards," *Ann. Occup. Hyg.*, **18**: 273.

Chamberlin R. I., and J. E. Leahy (1978), "Laboratory Fume Hood Standards, Recommended for the U.S. Environmental Protection Agency," Contract No. 68–01–4661.

Coleman, H. S. (1951), "Laboratory Design: National Research Council Report on Design, Construction and Equipment of Laboratories," Reinhold, New York.

Ettinger, H. J., M. W. First, and R. N. Mitchell (1968), "Industrial Hygiene Practices Guide: Laboratory Hood Ventilation," *Am. Ind. Hyg. Assoc. J.*, **29**: 611–617.

Fuller, F. H., and A. W. Etchells (1979a), "The Rating of Laboratory Hood Performance," *ASHRAE J. Trans.* **49**: October 49–53.

Fuller, F. H., and A. W. Etchells (1979b), "Safe Operation with the 0.3 m/s (60 fpm) Laboratory Hood," *ASHRAE J. Trans.* **49**: October 53.

Peterson, J. E. (1959), "An Approach to a Rational Method of Recommending Face Velocities for Laboratory Hoods," *Am. Ind. Hyg. Assoc J.*, **24**: 259–266.

Renton, R.,and G. Duffield (1986), "Preliminary Evaluation of an Integral Scrubber/ Fume Hood Design For Control of Noxious Vapors" in *Proceedings of the 1st International Symposium on Ventilation for Contaminant Control*, October 1–3, 1985, Tononto, Canada, H. D. Goodfellow, ed., Elsevier, Amsterdam.

Schulte, H. F., E. C. Hyatt, H. S. Jordan, and R. N. Mitchell, (1954), "Evaluation of Laboratory Fume Hoods," *Am. Ind. Hyg. Assoc. J.*, **15**: 195.

Scientific Apparatus Makers' Association (1980), "Standard for Laboratory Fume Hoods," Standard LF-1980, SAMA, Washington, DC.

Silverman, L., and M. W. First (1962), "Portable Laboratory Scrubber Unit for Perchloric Acid," *Am. Ind. Hyg. Assoc. J.*, **23**: 462–472.

ADDITIONAL READING

Caplan, K. J., and G. W. Knutson (1982), "A Performance Test for Laboratory Fume Hoods," *Am. Ind. Hyg. Assoc. J.*, **43**: 722–737.

Caplan, K. J., and G. W. Knutson (1982), "Influence of Room Air Supply on Laboratory Hoods," *Am. Ind. Hyg. Assoc. J.*, **43**: 738–746.

DiBerardinis, L., et al. (1987), *Guidelines for Laboratory Design: Health and Safety Considerations*, Wiley, New York.

Hughes, D. (1980), *A Literature Survey and Design Study of Fume Cupboards and Fume Dispersal Systems*, Occupational Health Monograph No. 4, Science Reviews Ltd., Leeds, England.

"Laboratory Fume Hoods," Data Sheet No. 1-687-80, *Nat. Saf. News*, May 1980.

8

Design of Single-Hood Systems

About 50,000 new or redesigned local exhaust ventilation (LEV) systems are installed annually in U.S. industry. It is estimated that the total capacity of these new systems approaches 50 million cfm with an installation cost of $250 million and an annual operating cost of $50 million. The majority of these systems are designed by personnel who have little training in this field and who do not consider rigorous engineering design to be necessary. As a result, a large fraction of existing LEV systems are not effective in protecting worker health. The availability of source information such as the ANSI Standard (ANSI, 1979), the ACGIH *Ventilation Manual* (ACGIH, 1988), a self-study companion to the manual (Burton, 1986), and computer design programs for both hand-held calculators and personal computers have improved this situation somewhat. However, a large group of plant engineering and health and safety professionals remain uninformed or believe that the design of LEV systems is not in their purview. Chapters 8 and 9 are intended to provide a basic understanding of the importance of engineering design and to introduce the available design techniques.

In this chapter local exhaust ventilation systems will be designed for single-hood systems using two widely accepted procedures, the *velocity pressure* and *equivalent foot* design methods. The velocity pressure method has been in widespread use for over 30 years; its popularity results from the ease of initial calculation and the ability to redesign system elements when necessary. The concept that all turbulent and frictional losses are directly proportional to the velocity pressure provides a rational basis to the method that can be readily understood by the novice designer. This method has been identified as the preferred one in the 20th edition of the *Ventilation Manual*. Prior to the introduction of the velocity pressure method the standard design procedure was the equivalent foot method, which is presented here as an alternative since it is still encountered in industrial practice.

8.1 DESIGN APPROACH

The standard design procedure is a five-step process that relies on the design information in Chapter 5 of the *Ventilation Manual.*

Step 1: Choose Hood Geometry. A great deal of hood design information for common industrial operations is contained in Chapter 10 of the *Ventilation Manual.* Processes unique to certain industries (e.g., semiconductor manufacturing) may not be covered adequately, however, and the designer must go to other sources. In some cases, the system designer must develop a hood design for a unique operation.

Step 2: Calculate the Required Airflow. The hood airflow specified in the design plate is an integral part of any hood design. The specification may be a given airflow or an airflow calculated from a specified face or capture velocity. The design data in the *Ventilation Manual* are usually supported by observed performance of the hood in an industrial setting.

Step 3: Specify the Minimum Duct Velocity. Duct transport velocity information is also included in the design plate specifications. If the hood is to be used for gases and vapors, a duct velocity of 1500 to 2000 fpm is usually specified; if particles are to be conveyed, the minimum transport velocity is 3500 to 4500 fpm.

Step 4: Choose a Duct Size. The size of a round duct is chosen to provide the required minimum duct velocity for the given airflow. The maximum duct cross-sectional area is obtained by dividing the airflow by the minimum velocity. Choosing the standard duct diameter with the next smaller area will assure that the minimum velocity is met. The actual duct velocity is then calculated based on the duct area and airflow.

Step 5: Calculate Energy Losses. The layout of the duct in the plant is chosen to minimize energy loss due to straight duct runs and elbows while observing the space constraints of both architectural and equipment obstructions. Frequently, several optional layouts must be considered. Once the duct layout is defined, the frictional and turbulent losses due to air flowing through the system can be determined from equations, charts, or tables. These losses may be calculated directly in in. H_2O, or as a convenience, the tabulation may be done with losses expressed as multiples of velocity pressure which are later converted to in. H_2O. Normally, airflow is specified to the nearest 10 cfm and pressures are rounded off to 0.01 in. H_2O.

8.2 DESIGN OF A SIMPLE ONE-HOOD SYSTEM (BANBURY MIXER HOOD)

The first system design example is a hood for a Banbury mixer (Fig. 8.1) used in the rubber and plastics industries to mix the elastomeric material with accelerators, antioxidants, antiozonants, pigments, plasticizers, and vulcanizing agents. The "feeding" or "charging" of these granular materials to the Banbury is a dusty operation and requires excellent local exhaust ventilation. The accepted technique as described in *Ventilation Manual* VS 901 is a partial enclosure over the charging port (Fig. 8.2). Two additional small hoods are usually provided to handle dust escaping from the dust rings at the ends of the trunnion. In this example, only the charge port will be provided with ventilation. The ventilation system to be designed consists of a hood, fan, and connecting ductwork. The goals of the design process are the determination of the minimum airflow required (cfm) and the total resistance to airflow (in. H_2O) for this system. These design data can then be used to select an appropriate fan using the techniques described in Chapter 10.

8.2.1 Velocity Pressure Design Procedure (Example 8.1)

Step 1: Choose Hood Geometry. The design data presented in Fig. 8.2 will be used to develop the system design for the hood on the Banbury mixer.

Figure 8.1   The Banbury mixer is used to mix rubber and plastic stock with various additives. This is a view of the charging platform on the upper level where the stock and additives are fed to the mixer. After feeding, the charging door closes, a ram seals the chamber, and the material is mixed. When mixing is completed, the bottom dump door opens and the blended material drops to a mill positioned directly under the Banbury on the lower level.

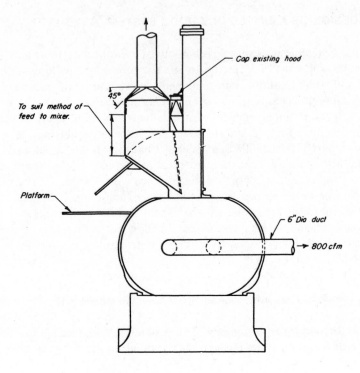

$Q = 200 - 300$ cfm/sq ft open face area.
  500 cfm/ft of belt width if belt feeder used.
Duct velocity = 3500 fpm minimum.
Entry loss = 0.25 VP at hood
            1.0 VP at trunnion

Figure 8.2 This design plate of a Banbury mixer provides the engineer with the four key design elements: (1) design and placement of hood, (2) minimum airflow to achieve control, (3) minimum duct velocity, and (4) hood entry loss (From *Ventilation Manual*, VS-901. Used by permission of Committee on Industrial Ventilation, ACGIH, Lansing, MI).

Step 2: Calculate the Required Air Flow. Upon inspection, the Banbury mixer is found to be fed manually, not by belt. In addition, there are no serious drafts from doors, windows, or traffic. Based on this information, a minimum exhaust of 200 cfm/ft^2 of open face area (equivalent to a face velocity of 200 fpm) is chosen. If the Banbury were located near a receiving door and serious drafts occurred or if drafts resulted from pedestal-type cooling fans, the higher value of 300 fpm from VS-901 would be chosen. A belt feeder would require 500 cfm/ft^2 of opening. As shown in Fig. 8.3, the hood is 5 ft wide and 4 ft high, so the airflow is calculated as

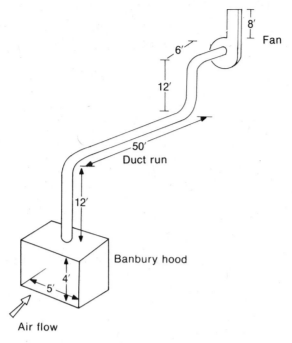

Figure 8.3 Layout of the Banbury, duct, and fan. The hood face is 5 ft wide and 4 ft high and the duct lengths are indicated. The sources of static pressure loss in this system include the hood entry loss, the inertial loss, the frictional loss of straight duct, and the losses due to elbows.

$$Q = VA \qquad\qquad (2.5)$$

$$= 200 \, \frac{\text{ft}}{\text{min}} \, (5 \text{ ft} \times 4 \text{ ft})$$

$$= 4000 \text{ ft}^3/\text{min}$$

Step 3: Specify the Minimum Duct Velocity. The minimum duct velocity recommended in VS-901 is 3500 fpm.

Step 4: Choose a Duct Size. The maximum allowable duct area is found by dividing the airflow by the minimum duct velocity:

$$A = \frac{Q}{V}$$

$$= \frac{4000 \text{ ft}^3/\text{min}}{3500 \text{ ft/min}} = 1.14 \text{ ft}^2$$

Any duct with an area smaller than 1.14 ft² and conveying 4000 cfm will satisfy the minimum duct velocity requirement of 3500 fpm. From Table 8.1 the round duct with a cross-sectional area just smaller than 1.14 ft² is a 14-in. duct, with an area of 1.069 ft².

The actual duct velocity and the associated velocity pressure must be calculated since they are needed in this design process. The velocity is

$$V = \frac{Q}{A}$$

$$= \frac{4000 \text{ ft}^3/\text{min}}{1.069 \text{ ft}^2} = 3740 \text{ ft/min}$$

The associated velocity pressure is

$$p_v = \left(\frac{V}{4,000}\right)^2 \tag{2.17}$$

$$= \left(\frac{3740}{4000}\right)^2$$

$$= 0.87 \text{ in. H}_2\text{O}$$

Step 5: Calculate Energy Losses. The energy losses occurring in simple systems such as this have been discussed in detail in Chapter 2. As noted in that discussion, both the turbulent losses due to changes in direction and velocity of airflow and the frictional losses due to the air flowing in the duct are directly proportional to the velocity pressure. For this reason, each loss can be expressed

Table 8.1 Choosing the correct duct diameter*

Diam.	Area		Circumference	
(in.)	in.²	ft²	in.	ft
12	113.1	0.7854	37.70	3.142
13	132.7	0.9218	40.84	3.403
14	153.9	1.069	43.98	3.665
15	176.7	1.227	47.12	3.927
16	201.0	1.396	50.26	4.189
17	226.9	1.576	53.41	4.451
18	254.4	1.767	56.55	4.712
19	283.5	1.969	59.69	4.974
20	314.1	2.182	62.83	5.236
21	346.3	2.405	65.97	5.498

*   See Ventilation Manual for complete data.

Table 8.2 Velocity pressure chart.*

p_v in. H_2O	V fpm
0.82	3622
0.83	3644
0.84	3666
0.85	3688
0.86	3709
0.87	3731
0.88	3752
0.89	3774
0.90	3795
0.91	3816

* See *Ventilation Manual*, for complete data.

as a loss factor times the velocity pressure in the duct. The losses in the Banbury hood design expressed in inches of water will be calculated in this fashion.

Hood Static Pressure in the Duct As described in Chapter 2, the hood static pressure as measured in the duct just downstream of the hood consists of the hood entry loss and the acceleration factor (one velocity pressure):

$$p_{s,\,h} = p_v + h_e \tag{2.41}$$

where $p_{s,h}$ = hood static pressure (in. H_2O)
p_v = velocity pressure (in. H_2O)
h_e = hood entry loss (in. H_2O)

The hood entry loss h_e is usually specified as the hood entry loss factor F_e multiplied by the velocity pressure:

$$h_e = F_e \times p_v \tag{2.42}$$

so that
$$p_{s,h} = p_v(1 + F_e) \tag{2.43}$$

The entry loss factor F_e for the Banbury hood is 0.25 as shown in Fig. 8.2. Equation 2.43 can now be used to calculate the hood static pressure.

$$p_{s,h} = p_v(1 + F_e)$$

$$= p_v(1 + 0.25)$$

$$= 1.25 p_v$$

This hood static pressure must be established in the system to provide an airflow of 4000 cfm through this hood.

Duct Friction   The frictional loss due to the 4000 cfm flowing through the 14-in.-diameter duct can also be expressed as a product of a loss factor and the velocity pressure. As shown in Fig. 8.3, the length of straight duct upstream of the fan is $12+50+12+6=80$ ft and there is an additional 8 ft downstream of the fan. Since air flowing through a duct exhibits the same frictional losses whether the duct is upstream or downstream of the fan, the total duct run of 88 ft can be considered as a unit.

The frictional loss of straight duct expressed as velocity pressure is obtained from Fig. 8.4. The intersection of the 4000 cfm airflow and the 14-in.-duct-diameter curves is found and the corresponding loss for each foot of duct is $0.0155 p_v$. Since the Banbury system has 88 ft of straight duct, the loss is

$$88 \text{ ft} \times 0.0155 \frac{p_v}{\text{ft}}$$

or

$$1.36 p_v$$

Alternatively, the loss per foot can be calculated using the equation in Fig. 6-13b of the *Ventilation Manual*:

$$p_{s,f} = \frac{0.4937 p_v}{Q^{0.079} \times D^{1.066}} \tag{8.1}$$

where D is the duct diameter (in.).
In this case

$$p_{s,f} = \frac{0.4937 p_v}{4000^{0.079} \times 14^{1.066}}$$

$$= 0.0155 p_v$$

Elbow Losses   The turbulent loss in elbows is also directly proportional to velocity pressure; typical loss factors are shown in Fig. 8.5. More recent data have been developed by Durr et al. (1987). The larger the radius of curvature, R, of an elbow, the lower the turbulent loss. However, large sweep elbows are expensive to fabricate and take up valuable space, so elbows in most local exhaust ventilation systems have a radius of curvature equal to 2 or 2.5 duct diameters (D). An elbow with $R/D=2.0$ is chosen for the Banbury; the 14-in.-diameter elbow thus has a sweep radius of 28 in. and the turbulent loss in one elbow is $0.27 p_v$ (Fig. 8.5). Since there are three 90° elbows, the total elbow loss is $0.81 p_v$.

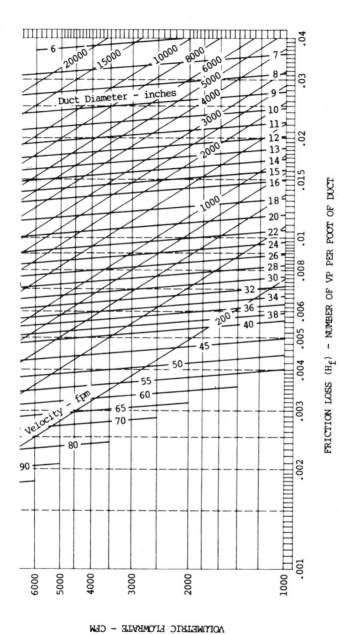

Figure 8.4 Frictional loss of straight duct, using the velocity pressure method. (From *Ventilation Manual*, Fig. 5–18b.)

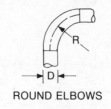

ROUND ELBOWS

R/D	Pressure Loss, Fraction of P_V
2.75	0.26
2.50	0.22
2.25	0.26
2.00	0.27
1.75	0.32
1.50	0.39
1.25	0.55

Figure 8.5 Geometry of a 90° elbow showing duct diameter (D) and the radius of curvature (R) and loss for given R/D values.

The total frictional and turbulent loss for the hood and duct elements in the Banbury mixer exhaust system is as follows:

Hood entry loss h_e	$0.25p_v$
Acceleration (inertial loss) p_v	$1.00p_v$
Straight duct frictional loss $p_{s,f}$	$1.36p_v$
Elbow loss	$0.81p_v$
Total static pressure loss (p_s)	$3.42p_v$

The velocity pressure was previously calculated to be 0.87 in. H_2O therefore the actual static pressure required to move 4000 cfm through this system is

$$p_s = (3.42)(0.87 \text{ in. } H_2O) = 2.98 \text{ in. } H_2O$$

Use of Calculation Sheet for Example 8.1.   The detailed procedure employed above is useful for demonstrating the individual steps necessary to calculate static pressure losses. However, it is much too laborious and inconvenient for routine calculation, especially for the complex multibranch systems to be considered in Chapter 9. The calculation sheet provided in the *Ventilation Manual* provides a convenient format to organize the design process. The Banbury mixer hood design is repeated using this modified calculation sheet (see Example 8.1).

In the design of a complex local exhaust system it is important to use a consistent method of identifying the various hoods and duct sections. The *Ventilation Manual* uses the convention shown in Fig. 8.6. Each hood is assigned a number and each duct junction or transition is assigned a letter. A hood-to-junction or junction-to-junction duct run represented by a two-character combination (e.g., 1*A*, *AB*) is used to label the duct section under design. Each column in the calculation sheet corresponds to one section of duct.

8.2.2 Equivalent-Foot Design Procedure (Example 8.2)

The equivalent-foot design method follows the same general procedure as the velocity pressure method. However, duct frictional losses and losses due to elbows and branch entries are all calculated in terms of equivalent frictional loss in straight duct and these individual losses are summed to provide a pseudo-length for the system. The actual static pressure loss is then determined from Fig. 5.21*a* of the *Ventilation Manual*, knowing the system element duct diameter and air flow. A special calculation sheet for this design technique is available in the *Ventilation Manual*. The Banbury hood system is recalculated in Example 8.2 using this sheet.

8.3 DESIGN OF A SLOT HOOD SYSTEM FOR A DEGREASING TANK

Because of the toxicity of chlorinated hydrocarbon solvents used in degreasers, it is common practice to install a lateral slot hood on degreasing tanks (Fig. 8.7). The value of such a slotted hood in providing effective exhaust distribution over the tank surface has been discussed in Chapter 5. The hood entry loss for such hoods is more complex than for simple hoods due to this slot, the plenum, and the duct transition.

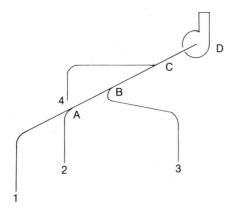

Figure 8.6 Nomenclature for major duct elements in a local exhaust ventilation system. Hoods are identified by a number and junctions by a letter.

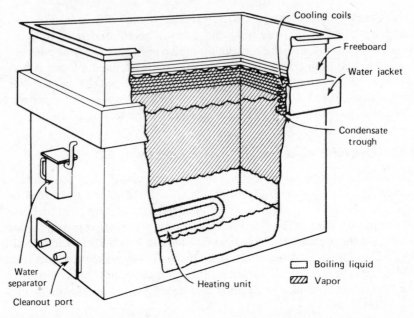

Figure 8.7 The solvent in a vapor-phase degreaser is heated with steam or electricity and a dense vapor is formed which fills the tank. The tank is equipped with a cold-water condenser jacket. The hot vapor condenses on the jacketed side of the tank and runs back into the sump. When parts are lowered into the vapor cloud, the hot solvent vapor condenses on the cold parts and strips off the oily residue, which drops to the tank bottom. Fugitive losses from the tank are commonly controlled with lateral slot exhaust hoods. [From J. A. Murphy, *Surface Preparation and Finishes for Metals*. Copyright 1971. Used with permission of McGraw-Hill Book Company.]

8.3.1 Loss Elements in a Complex Hood

The entry loss in a slotted hood depends on the relationship between the slot and duct velocity and the geometry of the plenum. A general view of the slot, plenum, and duct transition is shown in Fig. 8.8.

The hood entry loss for a slot hood has two components, the entry loss through the slot itself $(h_{e,s})$ and the entry loss in the transition from the plenum to the duct $(h_{e,d})$. Mathematically, we have

$$h_e = h_{e,s} + h_{e,d} \tag{8.2}$$

The air velocity and velocity pressure through a slot hood changes as it passes through the slot $p_{v,s}$, the plenum $p_{v,p}$, and finally into the duct $p_{v,d}$. For the general case in which the slot velocity is high, there is a residual plenum velocity $p_{v,p}$ (Fig. 8.8) which is reaccelerated to the duct velocity $p_{v,d}$. The required velocity pressure is

$$p_v = p_{v,s} + (p_{v,d} - p_{v,p}) \tag{8.3}$$

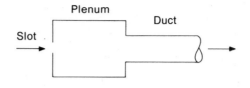

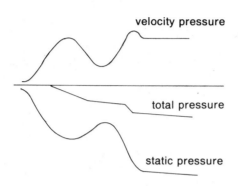

Figure 8.8   Pressures in slot hoods including velocity pressure at slot $p_{v,s}$, velocity pressure at plenum $p_{v,p}$, and velocity pressure at duct $p_{v,d}$.

The hood static pressure,

$$p_{s,h} = p_v + h_e \tag{2.41}$$

is obtained by combining Eq. 8.2 and 8.3:

$$p_{s,h} = p_{v,s} + (p_{v,d} - p_{v,p}) + h_{e,s} + h_{e,d} \tag{8.4}$$

If the duct velocity is greater than the slot velocity $(p_{v,d} > p_{v,s})$

- For a shallow plenum $(p_{v,p}$ is approximately equal to $p_{v,s})$

$$p_{s,h} = p_{v,d} + h_{e,s} + h_{e,d} \tag{8.5}$$

- For a deep plenum $(p_{v,p}$ is approximately zero)

$$p_{s,h} = p_{v,s} + p_{v,d} + h_{e,s} + h_{e,d} \tag{8.6}$$

If the duct velocity is less than the slot velocity $(p_{v,d} < p_{v,s})$

- For a shallow plenum $(p_{v,p}$ is approximately equal to $p_{v,s})$ and both $p_{v,p}$ and $p_{v,s}$ are greater than $p_{v,d}$

$$p_{s,h} = h_{e,s} + h_{e,d} + p_{v,s} \tag{8.7}$$

(no credit for static regain, $p_{v,d} - p_{v,s}$)

- For a deep plenum ($p_{v,p}$ is approximately zero)

$$p_{s,h} = p_{v,s} + p_{v,d} + h_{e,s} + h_{e,d} \tag{8.8}$$

For the most common case, a shallow plenum hood where the duct velocity is greater than the slot velocity, the hood static pressure is

$$p_{s,h} = h_{e,s} + h_{e,d} + p_{v,d} \tag{8.9}$$

If the slot is modeled as a sharp-edged orifice, the slot entry loss is

$$h_{e,s} = 1.78 \, p_{v,s} \tag{8.10}$$

The plenum-to-duct entry loss is

$$h_{e,d} = F_{e,d} \times p_{v,d} \tag{8.11}$$

where $F_{e,d}$, the plenum-to-duct entry loss factor, depends on the taper angle of the transition. Values are given in *Ventilation Manual*, Fig. 5–15.
As an example, if the included taper angle for a round duct is 120°,

$$h_{e,d} = 0.25 \, p_{v,d} \tag{8.12}$$

substituting Eqs. 8.10 and 8.12 into Eq. 8.2 gives the hood entry loss:

$$h_e = 1.78 \, p_{v,s} + 0.25 \, p_{v,d}$$

and the hood static pressure from Eq. 8.2 is thus

$$p_{s,h} = 1.78 \, p_{v,s} + 1.25 \, p_{v,d} \tag{8.13}$$

8.3.2   Degreaser Hood Design Using Velocity Pressure Calculation Sheet (Example 8.3)

The system design using the velocity pressure calculation sheets is shown in Fig. 8.9 and Example 8.3. The duct design for the degreasing tank is conducted in a fashion similar to the Banbury hood. The slot loss is calculated using lines 8 to 14 on the standard velocity pressure method calculation sheet.

The design process indicates that the system will provide 1500 cfm with a static pressure loss of 1.15 in. H_2O. The selection of an 11-in.-diameter duct in solution A provided the minimum duct velocity of 2000 fpm specified in Fig. 8.9 for this example. Maintaining a minimum duct velocity is important in a system handling particles since dust settling in the duct should be avoided. In a system conveying gases or vapors the duct velocity is less important and lower velocities

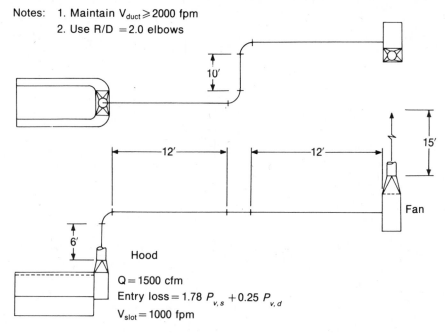

Notes: 1. Maintain $V_{duct} \geqslant 2000$ fpm
            2. Use R/D $= 2.0$ elbows

10′

12′   12′   15′

Fan

6′

Hood

$Q = 1500$ cfm

Entry loss $= 1.78\ P_{v,s} + 0.25\ P_{v,d}$

$V_{slot} = 1000$ fpm

Figure 8.9   Design plate for a vapor-phase degreaser.

can be chosen if space is available for the larger duct. The energy reduction in going to a lower duct velocity can easily be calculated. The use of the next larger duct (12. in.) in solution B reduces the duct velocity from 2270 fpm to 1910 fpm. This modest reduction in velocity will result in the pressure loss dropping from 1.15 in. H_2O to 0.85 in. H_2O.

8.4   PRESSURE PLOT FOR SINGLE-HOOD SYSTEM

The pressures existing in a duct run were defined in Chapter 2. Velocity, static, and total pressures for the degreasing tank system (using the 11-in. duct) can be plotted as shown in Fig. 8.10.

The most convenient approach to calculating the pressures at any point in the system is first to calculate the velocity pressure for each duct element. An 11-in.-diameter duct is used in the system, so the velocity pressure of 0.32 in. H_2O exists throughout the duct portion of the system. The velocity pressure at the slot can be calculated, but the velocity change in the plenum can only be approximated. The individual frictional and turbulent losses are calculated for each system element shown in Fig. 8.9 and added incrementally starting with a static pressure of 0 in. H_2O at the hood to obtain the static pressure at each benchmark. The total pressure at each benchmark is then calculated based on $p_t = p_v + p_s$. As illustrated, the total pressure line is parallel to the static pressure line and offset

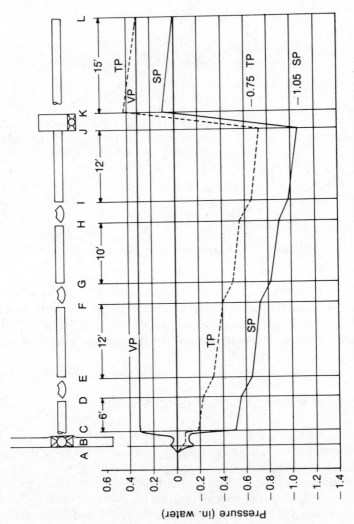

Figure 8.10 Plot of duct pressures for vapor-phase degreaser system using 11-in. duct.

by an amount equal to the velocity pressure. Connecting the fan inlet and outlet pressures (shown as *J-K*) is a graphic presentation of the function of the fan as a static pressure generator.

LIST OF SYMBOLS

A	area
D	diameter
F_e	hood entry loss factor
$F_{e,d}$	duct entry loss factor
h_e	hood entry loss
$h_{e,d}$	entry loss for transition plenum to duct
$h_{e,s}$	entry loss for hood slot
p_s	static pressure
$p_{s,h}$	hood static pressure
p_t	total pressure
p_v	velocity pressure
$p_{v,d}$	duct velocity pressure
$p_{v,p}$	plenum velocity pressure
$p_{v,s}$	slot velocity pressure
Q	volume flow
R	radius of curvature
V	average velocity

VELOCITY PRESSURE METHOD CALCULATION SHEET

(adapted from A.C.G.I.H. COMMITTEE ON INDUSTRIAL VENTILATION)

#			Description	Reference / Calculation	Units	1A
1			Duct Segment Identification			1A
2			Volumetric Flowrate		cfm	4000
3			Minimum Transport Velocity		fpm	3500
4			Duct Diameter		inches	14
5			Duct Area		square feet	1.069
6			Actual Duct Velocity		fpm	3740
7			Duct Velocity Pressure (p_v)		"H_2O	0.77
8	H		Slot Area		square feet	
9	O	S	Slot Velocity		fpm	
10	O	L	Slot Velocity Pressure		"H_2O	
11	D	O	Slot Loss Factor	Section 5 or Fig. 6-10*	0 or 1	
12		T	Acceleration Factor		0 or 1	
13		S	Plenum Loss per p_v	Items 11 + 12		
14		U	Plenum p_s	Items 10 x 13	"H_2O	
15		C	Hood Entry Loss Factor Section 5 or Fig. 6-10			0.25
16		T	Acceleration Factor	1 or 0		1.00
17		I	Entry Loss Factor p_v	Items 15 + 16		1.25
18		O	Entry Loss	Items 7 x 17	"H_2O	1.09
19		N	Other Loss		"H_2O	
20			Hood Static Pressure	Items 14 + 18 + 19	"H_2O	1.09
21			Straight Duct Length		feet	88
22			Friction Factor	Figure 6-13 or equation		0.0155
23			Friction Loss per p_v	Items 21 x 22		1.36
24			No. 90 degree Elbows			3
25			Elbow Loss per p_v	Item 24 x loss factor		0.91
26			No. Branches			1
27			Branch Loss per p_v	Item 26 x loss factor		1
28			Special Fittings Loss Factors			
29			Duct Loss per p_v	Items 23 + 25 + 27 + 28		2.17
30			Duct Loss	Items 7 x 29	"H_2O	1.89
31			Duct Static Pressure Loss	Items 20 + 30	"H_2O	2.98
32			Cumulative Static Pressure		"H_2O	
33			Governing Static Pressure		"H_2O	
34			Corrected Volumetric Flowrate		cfm	
35			Resultant Velocity Pressure		"H_2O	

Example 8.1 Banbury Mixer System Designed by the Velocity Pressure Method.

Example 8.1 Design Notes for Calculation Sheet

Line 2: The design airflow for each duct segment should be entered in this line. The entry for a single-hood system such as the Banbury mixer is the minimum design volume specified in the VS design plate. However, in the design of a multiple-hood system (Chapter 9) the velocity pressure design technique frequently results in corrected airflows for the individual hoods. The corrected airflow is entered in line 34.

Lines 3–6: The maximum duct area to achieve the minimum duct velocity is $A = Q/V$, where Q is the airflow from line 2 and V is the minimum duct velocity from the *VS* design plate. The next smaller duct is chosen from Table 8.1, and this diameter is entered in line 4 with the duct area noted on line 5. The actual velocity in the duct is entered on line 6 as

$$V = \frac{Q}{A} = \frac{\text{line 2}}{\text{line 5}}$$

Line 7: The velocity pressure resulting from the true duct velocity calculated in line 6 is obtained from Table 8.2 and entered in line 7.

Lines 8–14: Entries are required in these lines only if the system includes a complex hood, that is, one with slots. Since the Banbury hood is not a slotted hood, this part of the calculation form does not apply to this example.

Line 15: The hood entry loss is stated on the VS design plate (0.25 p_v from VS-901).

Line 16: The air entering the hood must be accelerated to duct velocity; therefore, 1.0 is entered on this line.

Lines 17, 18, 20: The loss through the hood and into the duct is expressed in terms of p_v and is the sum of lines 15 and 16. This loss factor multiplied by p_v is the hood entry loss in in. H_2O. Added to the plenum static pressure of the slotted hood, it represents the hood

static pressure. If there are no slots, as in this example, the hood entry loss is equivalent to the hood static pressure.

Line 21: The length of straight duct is determined from plan and elevation views of the system. It is practice to use centerline distances and round off to the nearest foot. The additional distance represented by elbows is not included.

Line 22: This straight duct frictional loss factor can be calculated or solved graphically. In the velocity pressure method of calculation the resistance is expressed as a function of the velocity pressure and the length of straight duct (*Ventilation Manual*, Fig. 5–18b).

Line 23: The loss is expressed as the product of the loss factor for straight duct and the length of duct.

Lines 24, 25: Elbow losses are expressed as loss factors and values are obtained from the data presented at the bottom of the calculation sheet. Summation of elbow losses as a single loss factor is obtained by multiplying the equivalent number of 90° elbows in the system by the loss factor per elbow.

Lines 26, 27: Caution! These lines are designed for branch entry losses and *not* hood entry losses. They are used in multiple-hood systems and are discussed in Chapter 9.

Line 28: Other loss factors from expansions, contractions, or other duct details are noted here.

Line 29: All loss factors included in lines 23, 25, 27, and 28 are summed and entered here.

Line 30: Duct velocity pressure from line 7 multiplied by the total loss factor in line 29 represents the static pressure in in. H_2O lost in moving the desired airflow through the duct. It does not include the loss due to the hood entry (line 20). Line 30 plus line 20 represents the total loss in the system.

Lines 32–35: These lines apply only to multiple-hood systems and are discussed in Chapter 9.

Plant name _____
Location _____
Department _____

Elevation _____
Temperature _____

Remarks
Example 8.2

1	2	3	4	5	6	Refer to 7	8	9	10	11	12	13	Factor 14	15	16	17	18	19
			Air volume CFM					Length of duct in feet		Friction loss, inches of water						Pressure, inches of water		
No of br or main	Dia duct in in	Area duct in sq ft	in branch	in main	Vel. in FPM	straight runs	Number of elbows/entries	equiv length (From Fig. 5–20)	total length (Col.7 plus Col.9)	per 100 (From Fig 5–21)	of run (Col.10×Col.11/100)	VP (From Col.6 Tab. 5–4)	one entry loss(VP) (From Fig. 5–15)	hood suct(VP) (100 plus Col.14)	hood suct (Col.13 times Col.15)	static press. (Col.12 plus Col.16)	gov. SP	corrected CFM
1	14	.069	4000	— 3740		PF	3	21	161	1.3	1.96	.87	1.25	1.15	1.09	3.05		

Remarks: At junction

Example 8.2   Banbury Mixer System Designed by the Equivalent Foot Method.

Example 8.2   Design Notes for Calculation Sheet

The duct size and actual duct velocity are calculated as in the velocity pressure method using the first major block (columns 1 to 6) in the calculation sheet.

The second block of the sheet (columns 7–10) is used to calculate all losses except hood entry losses as equivalent feet of straight duct (*Ventilation Manual* Fig. 5–20). Once this is done, Fig. 5–21 of the *Ventilation Manual* is used to calculate the static pressure losses in in. H_2O (columns 11 and 12). The velocity pressure in in. H_2O is entered in

column 13. The hood entry loss factor is then entered in column 14 and column 15 is the acceleration loss plus the hood entry loss factor, for a total of $-1.25\,p_v$. The hood static pressure is then calculated in column 16 and is added to the frictional and elbow losses (column 12) and entered in column 17. as the total system static pressure loss. This loss of 3.05 in. H_2O is comparable to the 2.98 in. H_2O determined by the velocity pressure method in Example 8.1.

#	Item	Units / Reference	Solution A	Solution B
1	Duct Segment Identification			
2	Volumetric Flowrate	cfm	1500	1500
3	Minimum Transport Velocity	fpm	2000	2000
4	Duct Diameter	inches	11	12
5	Duct Area	square feet	0.66	0.7854
6	Actual Duct Velocity	fpm	2270	1910
7	Duct Velocity Pressure (p_v)	"H2O	0.32	0.23
8	Slot Area	square feet	1.5	1.5
9	Slot Velocity	fpm	1000	1000
10	Slot Velocity Pressure	"H2O	0.06	0.06
11	Slot Loss Factor Section 5 or Fig. 6-10*		1.78	1.78
12	Acceleration Factor 0 or 1		0	0
13	Plenum Loss per p_v Items 11 + 12		1.78	1.78
14	Plenum p_s Items 10 x 13	"H2O	0.11	0.11
15	Hood Entry Loss Factor Section 5 or Fig. 6-10		0.25	0.25
16	Acceleration Factor 1 or 0		1.00	1.00
17	Entry Loss Factor p_v Items 15 + 16		1.25	1.25
18	Entry Loss Items 7 x 17	"H2O	0.40	0.35
19	Other Loss	"H2O	0	0
20	Hood Static Pressure Items 14 + 18 + 19	"H2O	0.51	0.46
21	Straight Duct Length	feet	55	55
22	Friction Factor Figure 6-13 or equation		0.022	0.0059
23	Friction Loss per p_v Items 21 x 22		1.21	0.87
24	No. 90 degree Elbows		3	3
25	Elbow Loss per p_v Item 24 x loss factor		0.81	0.91
26	No. Branches		0	0
27	Branch Loss per p_v Item 26 x loss factor		0	0
28	Special Fittings Loss Factors		0	0
29	Duct Loss per p_v Items 23 + 25 + 27 + 28		2.02	1.68
30	Duct Loss Items 7 x 29	"H2O	0.65	0.39
31	Duct Static Pressure Loss Items 20 + 30	"H2O	1.16	0.85
32	Cumulative Static Pressure	"H2O		
33	Governing Static Pressure	"H2O		
34	Corrected Volumetric Flowrate	cfm		
35	Resultant Velocity Pressure	"H2O		

Example 8.3   Degreaser System Designed by the Velocity Pressure Method.

Example 8.3 Design Notes for Calculation Sheet

Line 2: Required airflow from Fig. 8.9.

Line 3: From Fig. 8.9.

Line 4: $A = Q/V = (1500 \text{ ft}^3/\text{min})/(2000 \text{ ft/min}) = 0.75 \text{ ft}^2$. Next smaller duct is 11 in.

Line 5: From *Ventilation Manual*, Table 5–5.

Line 6: $V = Q/A$ (1500 ft^3/min)/(0.66 ft^2) = 2270 ft/min.

Line 7: From *Ventilation Manual*, Fig. 5–4B.

Line 8: Maximum area, $A = Q/V = (1500 \text{ ft}^3/\text{min})/(1000 \text{ ft/min}) = 1.5 \text{ ft}^2$.

Line 9: VS-501 and Fig. 8.9.

Line 10: *Ventilation Manual*, Fig. 5–4B.

Line 11: Figure 8.9.

Line 12: Since $V_s < V_d$ and shallow plenum assign entire acceleration factor to line 16.

Line 21: Figure 8.9, length is $6 + 12 + 10 + 12 + 15 = 55$ ft.

Line 23: $55 \times 0.022 = 1.21$.

Line 25: $3 \times 0.27 = 0.81$ pv.

Lines 26, 27: Applies only to multiple-hood systems.

Line 28: No expansions or contractions.

REFERENCES

American Conference of Governmental Industrial Hygienists, Committee on Industrial Ventilation (1988), *Industrial Ventilation*, 20th. ed., ACGIH, Lansing, MI.

American National Standards Institute (1979) "Fundamentals Governing the Design and Operation of Local Exhaust Systems," ANSI Standard Z9.2-1979, ANSI, New York.

ADDITIONAL READINGS

Burton, D. J. (1986), *Industrial Ventilation—A Self Study Companion to the ACGIH Ventilation Manual*, 3rd ed., IVE, Inc., Los Alamos, NM.

Durr, D., N, Esmen, C. Stanley, and D. Weyel (1987), "Pressure Drop in Elbows," *Appl. Ind. Hyg.*, **2**:57.

Hemeon, W. C. L. (1963), *Plant and Process Ventilation*, 2nd ed., Industrial Press, New York.

9

Design of Multiple-Hood Systems

The design technique for a single-hood system is straightforward and a collection of single-hood systems provides the ultimate in flexibility for the user. However, single-hood systems do have many drawbacks and it is usually more economical to group a series of hoods into a system with a single fan. Such multiple-hood systems can be designed in a variety of ways, depending on the application.

A key goal in the design of multiple-hood systems is establishment of the proper airflow at each hood. This is attained by a combination of two factors. The correct *total* flow through the system is obtained by selection of the proper fan; this process is discussed in Chapter 10. The proper *distribution* of the flow among the various hoods is obtained by balancing the airflow resistance of the hoods and their associated duct work. The process by which such balance is obtained is the subject of this chapter.

9.1 APPLICATIONS OF MULTIPLE-HOOD SYSTEMS

The most common application for a multiple-hood system is a manufacturing facility with a number of workstations that release the same airborne contaminant. Buffing and polishing lines are examples of this application. All workstations are provided with identical hoods exhausted at the same airflow and ducts are designed for a minimum transport velocity.

Another example is a foundry cleaning room, where the air contaminants released from the process include fused silica sand and metal particles. Since the individual operations and equipment differ from station to station, the hood geometry and exhaust rate will vary. It is common practice to group these hoods in a multiple-hood system since the same contaminants are generated on each operation and one air cleaner can be used. Such a system will be designed in this chapter.

Grouping exhaust hoods into one system for operations that produce different airborne contaminants can present design and operational difficulties. As an example, assume that a series of hoods in a metal finishing shop handles noncorrosive gases and vapors. However, one hood exhausts a stripping tank releasing a corrosive acid mist. If the latter hood is included in a multiple-hood system, the entire system must be constructed of corrosion-resistant material. In this case, the solution may be to group all hoods handling noncorrosive air contaminants in one system and install a single-hood system constructed of materials appropriate for the corrosive bath. The grouping of hoods handling high concentrations of noncompatible or reactive materials is also poor practice, due to the potential for violent reaction.

In the foregoing examples of multiple-hood systems it was assumed all hoods are exhausted continuously. In many cases this may not be necessary. Each process may operate for a small fraction of the total shift and the operating times may be scattered throughout several shifts. Operating all hoods continuously will waste energy. One obvious solution is to install a separate hood-fan system for each process. Another approach is to utilize a multiple-hood system but equip each hood with a blast gate or damper which is opened only when the operation is under way. The required system exhaust capacity can be determined by a time study of all operations. Such an approach works when gases and vapors are to be conveyed; it frequently fails when particles are handled since variable airflows make it difficult to maintain proper transport velocities. A plenum exhaust system (*Ventilation Manual*, Chapter 5) is another technique available for this application. This system maintains a minimum transport velocity in the branches, but the low-velocity main or plenum acts as a settling chamber requiring periodic cleanout.

Several other cautions should be advanced. The installation of multiple-hood systems to provide exhaust for operations located in different departments under different supervision should be avoided if possible. Just as it is difficult to share production tooling, it is difficult to share local exhaust ventilation capacity. Another precaution should be observed during the system design phase. If a particular hood design is tentative because the hood shape and exhaust rate have not been validated, the hood should be a candidate for a separate fan. The reason is that extensive retrofitting of that hood might be necessary, possibly causing disruption of the rest of the system.

The permanence of the plant operations to be exhausted is critical to the successful design and operation of a large multiple-hood system. If frequent changes in process and equipment occur, large multiple-hood systems should be avoided. Modifications of existing multiple-hood systems to add hoods to the system without engineering review is one of the most flagrant problems encountered in local exhaust ventilation systems and frequently results in malfunctioning of the total system.

9.2 BALANCED DESIGN APPROACH

The process of obtaining the proper airflow distribution among hoods is termed *balancing*. Balancing is achieved either by the static pressure balance method or by the use of blast gates. In both approaches the desired flow through each branch and submain is obtained by providing the necessary static pressure and airflow resistance at each junction.

In the static pressure balance method (Fig. 9.1) the required static pressure at each junction is obtained by the choice of hood entry loss, duct size, and other resistance elements. If the design is executed properly and the appropriate fan is installed, the static pressures established in the design will be achieved and the required airflow will occur at each hood. This design approach is the preferred one and will be followed in this chapter.

In the blast gate balancing process, static pressures equal to or in excess of that required to maintain the desired flows are selected at each junction in the system. Blast gates are installed on each branch and are adjusted in the field to provide the branch resistance associated with the desired airflow (Fig. 9.2). An excellent critique of the two balancing methods by Caplan (1983) indicates the serious limitations in the blast gate method and describes why the static pressure balance method is preferred. A brief comparison of the two balancing methods is provided in Table 9.1.

A third system design procedure (Fig. 9.3) can be used on a multiple-hood system when the individual hoods are used intermittently. In this procedure the

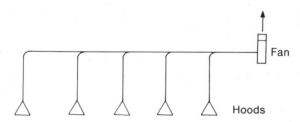

Figure 9.1 Multiple-hood system on similar operations with balance obtained by choice of duct size and airflow.

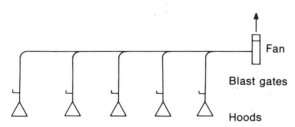

Figure 9.2 Multiple-hood system with adjustable blast gates used to obtain balance and correct airflow through each hood.

Table 9.1 Comparison of Balancing Methods

Balance without Blast Gates	Balance with Blast Gates
Air volumes cannot easily be changed by workers	Air volumes may be changed relatively easily.
Little flexibility for future equipment changes or additions.	Greater degree of flexibility for future changes or additions.
Choice of exhaust volumes for a new operation may be incorrect.	Improperly estimated exhaust volumes can be corrected.
No unusual duct erosion or deposits in duct.	Partially closed blast gates may cause erosion to slides changing air flow and may cause dust accumulations. Ductwork may plug if the blast gate adjustments are tampered with.
Design calculation is more time-consuming than blast gate method.	Design calculations are relatively brief.
Total air volumes are slightly greater than design air volume due to corrected airflows.	Balance may be achieved with design air volume.
Poor choice of "branch of greatest resistance" will show up in design calculations.	Poor choice of "branch of greatest resistance" may remain undiscovered. In such a case the branch of greater resistance will be "starved."
Layout of system must be in complete detail and installations must exactly follow layout.	Allowance for variations in duct location to avoid obstructions or interferences not known at time of layout.

Source: Adapted from ACGIH *Industrial Ventilation Manual.*

blast gates are not used for balancing but are merely on-off valves. This process requires that the designer define the utilization time of the various processes and the system is designed to provide the necessary airflow through a limited number of hoods. It is obvious that if the blast gates on more than the specified hoods are open at one time the system will malfunction. The best way to prevent this is to interlock the operation with the hood dampers and use controls to restrict the number of operations that can be conducted at one time. It may be possible to operate such systems without elaborate controls if the users are informed of the

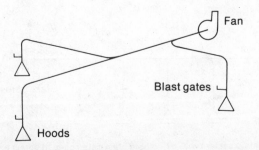

Figure 9.3 Multiple-hood system with on-off blast gates to activate hood only when needed.

operating instructions and are diligent about closing dampers on hoods not in use.

A conventional multiple-hood system has a main duct run with individual branches joining the main at several locations. The design of multiple-hood systems is conducted in the same manner as the single-hood systems (Chapter 8). The individual branches can be viewed as single-hood systems and the main as a network of these simple systems. The designer defines the frictional and turbulent energy losses incurred by air flowing through each branch at the desired flow up to the junction at the submain or the main. An additional energy loss not encountered in the one-hood system is incurred as the branch enters the main duct. The goal in the static pressure balance design approach is to match the static pressure required by each branch with that available at the main where the branch enters. If these static pressures are matched and a fan of suitable capacity is provided the design airflow will be drawn through the hood.

To optimize the design it is prudent to evaluate the product of airflow and total resistance for a number of different layouts of duct and fan. In each case the design should be started with the branch of highest resistance, since this tends to simplify the design procedure. The choice of this branch is frequently obvious, but the following guidelines may be helpful in choosing the design starting point.

- If hoods and exhaust airflows are all similar, the branch farthest from the fan is probably the one with the highest resistance.
- A hood with a combination of high entry loss and high airflow is a good candidate for the branch of highest resistance.
- A branch with high duct velocity may require the greatest static pressure.
- The use of long runs of flexible duct for hood mobility will frequently result in the branch of maximum resistance.

In any case, the static pressure balancing method is forgiving in this respect. If the branch of highest resistance is not chosen a priori, it will identify itself as the system is designed. This is not the case with the blast gate balancing method.

Once the branch of maximum resistance is chosen, the frictional and turbulent losses are calculated starting at the hood and proceeding to the first junction at the submain or main. As discussed in Chapter 8, two design approaches can be utilized: the velocity pressure method and the equivalent-foot method. Only the velocity pressure method is used in the examples presented in this chapter.

The basic procedure is illustrated in Fig. 9.4. At the part of the system most distant from the fan, two branches join to form one end of the main (point A). The required static pressure in the first branch ($1A$) is calculated followed by the required static pressure in the second branch ($2A$). The two static pressures are then compared; if they differ, one of the branches will not have the proper airflow when the system is operated. This is true because only one static pressure can exist at point A. There are several design options available to balance these static pressures and obtain the desired airflows in both branches.

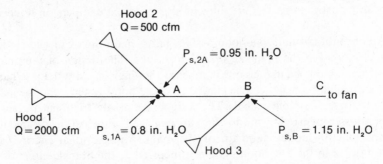

Figure 9.4   Branches 1*A* and 2*A* join at junction *A*. The design airflow and the total losses for the two runs are presented to permit the design of a balanced system.

In the example, hood 1 requires an airflow of 2000 cfm and a static pressure at junction *A* of 0.8 in. H_2O. Hood 2 has a static pressure requirement of 0.95 in. H_2O at this junction to supply the required airflow of 500 cfm. In operation, the fan will supply only one value of static pressure to point *A*. If 0.8 in. H_2O is selected, hood 1 will have the proper airflow, but the design airflow of 500 cfm through hood 2 will not be attained. Since the duct velocity and airflow vary as the square root of static pressure, the flow through hood 2 will be (500) $(0.8/0.95)^{0.5} = 460$ cfm. Alternatively, the fan can be designed to provide 0.95 in. H_2O at point *A*. Now hood 2 will have the proper airflow, but hood 1 will have a new airflow of (2000) $(0.95/0.8)^{0.5}$ or 2180 cfm.

The three available options are to reduce the resistance in branch 2*A*, increase the resistance in branch 1*A*, or allow the increased flow in branch 1*A*. First, the design of hood 2 and its ductwork should be reviewed to determine if the resistance to airflow can be reduced. If the hood has a high entry loss, it might be replaced with a hood with a higher coefficient of entry. It might be possible to eliminate elbows. If the duct run is long and the duct velocity is high, it might be possible to increase the duct diameter, thereby reducing the duct velocity and resulting friction losses, but only if a minimum transport velocity is not required. In this example, hood 2 cannot be redesigned and the transport velocity in that branch must be maintained. Under this condition, redesign of branch 2*A* is not the solution.

The design of the hood 1 branch should now be reviewed. Since hood 2 cannot be changed, the static pressure in branch 1*A* must be increased to 0.95 in. H_2O to match that required by branch 2*A*. Boosting the required static pressure of the hood 1 branch can be accomplished in one of several ways. If the 2000 cfm airflow is to remain unchanged, the resistance to airflow must be increased. A high-loss hood could replace the original or a resistance element such as a blast gate could be installed in the duct. The duct size can also be reduced to increase duct velocity and frictional loss. A final alternative is to allow the increased airflow of 2180 as calculated above. This option is selected here. The increased airflow might improve hood performance but has the disadvantage of increased operating cost. If an airflow is increased during this phase of the design process, it is designated as the corrected airflow.

The static pressure of 0.95 in. H_2O is defined as the governing static pressure or design point at junction A. In calculating the frictional loss from air flowing through the main (AB), the corrected flow of 2180 cfm from hood 1 is added to the 500 cfm through hood 2 to obtain the total airflow of 2680 cfm. The duct size and frictional loss factor are obtained in the conventional way, and the static pressure loss for this segment is calculated as 0.2 in. H_2O. This static pressure is added to the governing static pressure at A (0.95 in. H_2O) to provide the required static pressure (1.15 in H_2O) at junction B. The required pressure in branch $3B$ is then calculated and compared to the value at junction B. The design approach described above is then repeated for branch $3B$ and the final duct elements (BC) until the fan and outlet are reached.

In general, static pressures will not match perfectly at junctions. What is an acceptable balance at a junction, and what is the preferred design approach to correct this difference? The convention followed by most designers is to calculate the ratio of the two static pressures and follow the recommendations given in Table 9.2.

9.3 STATIC PRESSURE BALANCE METHOD

9.3.1 Foundry Cleaning Room System (Example 9.1)

A small foundry cleaning room will be used as an introductory multiple-hood design problem. This hypothetical foundry casts water-meter components from a brass alloy containing lead. After casting, the parts are taken to a shakeout operation, where the molding sand and core debris are removed. The parts are then cleaned in an abrasive blasting unit. Finally, the parts are "rough dressed" using a swing grinder, a belt sander, and a pedestal grinder. Since the alloy

Table 9.2 Benchmarks for Balancing Multiple Hood Systems

If the Ratio of the Higher to Lower Static Pressure Is	Action
Less than 1.05	Acceptable; do not change flow in calculations; use higher static pressure as governing value.
Between 1.05 and 1.20	Allow increased flow in branch of lowest static pressure. Recalculate flow as $$Q_{corrected} = Q_{design}\left(\frac{p_s \text{ of duct with higher loss}}{p_s \text{ of duct with lower loss}}\right)^{0.5}$$
Greater than 1.20	Resize duct with lower static pressure.

contains lead, the dust generated from these operations must be controlled using local exhaust ventilation.

A local exhaust ventilation system will be designed using the static pressure balancing method. This example requires balancing only by correcting airflow; duct resizing will be introduced as a balancing technique later in this chapter (Example 9.2).

Comments on the design:

- It is not necessary to use the *Ventilation Manual* VS design plates in this example, since the necessary design data are presented directly on Fig. 9.5.
- Because the system is designed for particle collection a minimum velocity of 3500 fpm is chosen in this example.
- The resistance to airflow of the air cleaner normally is obtained from the manufacturer. In this example the static pressure loss incorporates all losses associated with the air cleaner, including turbulent losses at the entry to the filter, the resistance to airflow of the filter medium, and the transition from the air cleaner to the duct.
- The duct runs are all in one plane to simplify this introductory problem.
- All branch entries to the main are made at 30° angles. Preentry elbows on branches 2A and 3B must therefore be 60°.
- By convention, the loss due to a branch entry is included in the branch and not in the main.
- All hoods are simple hoods with a single loss factor, as shown in Fig. 9.5.

The calculation sheet for this problem is shown in Example 9.1. The solution of this design problem provides the following information, which is required for the effective installation, testing, and operation of this system:

1. Construction details, including duct layout and the number and type of special duct fittings in the installation.
2. Data for effective choice of the air mover including airflow and fan static pressure (see Chapter 10).
3. The hood static pressure that will exist if the desired hood airflow is achieved (line 20). Once the system is installed this pressure can be used to evaluate hood airflow, as discussed in Chapter 3.

9.3.2 Electroplating Shop (Example 9.2)

In Chapter 6 the common processes releasing air contaminants in electroplating operations were described and a variety of hood designs for effective control were presented. A technique described in the *Ventilation Manual* was used to calculate the minimum exhaust rate for a series of open-surface tanks in a small plating shop. This electroplating shop is shown in Fig. 9.6 and the exhaust rates for each

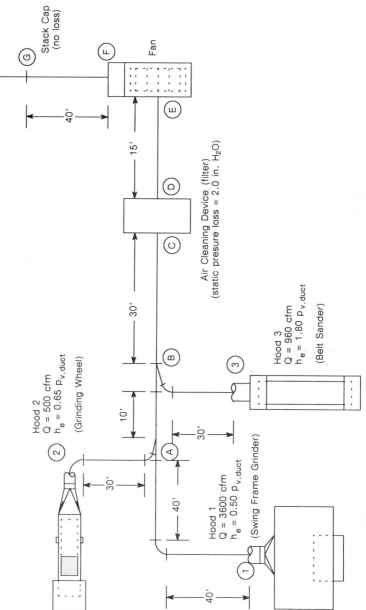

Notes:

1. Maintain duct velocity ≥3500 fpm up to filter; ≥ 2000 fpm beyond

2. Use R/D = 2.0 elbows & 30° entries

Stack Cap (no loss)

Fan

Air Cleaning Device (filter)
(static presure loss = 2.0 in. H₂O)

Hood 2
Q = 500 cfm
h_e = 0.65 $p_{v,duct}$
(Grinding Wheel)

Hood 3
Q = 960 cfm
h_e = 1.80 $p_{v,duct}$
(Belt Sander)

Hood 1
Q = 3600 cfm
h_e = 0.50 $p_{v,duct}$
(Swing Frame Grinder)

Figure 9.5 Layout of ductwork for local exhaust ventilation system in the foundry cleaning room.

VELOCITY PRESSURE METHOD CALCULATION SHEET

(adapted from A.C.G.I.H. COMMITTEE ON INDUSTRIAL VENTILATION)

			1A	2A	AB	3B	BC	CD	DE	EF
1	Duct Segment Identification		1A	2A	AB	3B	BC	CD	DE	EF
2	Volumetric Flowrate	cfm	3600	500	4300	960	5270	5270	5270	5270
3	Minimum Transport Velocity	fpm	3500	3500	3500	3500	3500	→	2000	2000
4	Duct Diameter	inches	13.?	5	14	16			22	22
5	Duct Area	square feet	0.9218	0.384	1.069	0.2873	1.396		2.640	2.640
6	Actual Duct Velocity	fpm	3910	3670	4020	3590	3780		2000	2000
7	Duct Velocity Pressure (p_v)	"H₂O	0.95	0.84	1.01	0.80	0.89		0.25	0.25
8	H S Slot Area	square feet	NA	NA	NA	NA	NA	NA		
9	O Slot Velocity	fpm								
10	O L Slot Velocity Pressure	"H₂O	NA	NA	NA	NA	NA	NA	NA	NA
11	D Slot Loss Factor Section 5 or Fig. 6-10*	0 or 1								
12	T Acceleration Factor	0 or 1								
13	S Plenum Loss per p_v	Items 11 + 12								
14	U Plenum p_s	Items 10 x 13								
15	C Hood Entry Loss Factor Section 5 or Fig. 6-10	1 or 0	0.50	0.65		1.80				
16	T Acceleration Factor	1 or 0	1.00	1.00		1.00				
17	— Entry Loss Factor p_v	Items 15 + 16	1.50	1.65		2.80				
18	I Entry Loss	Items 7 x 17 "H₂O	1.43	1.39		2.24				
19	N Other Loss	"H₂O	0	0		2.24				
20	Hood Static Pressure	Items 14 + 18 + 19 "H₂O	1.43	1.39		2.24				
21	Straight Duct Length	feet	80	30	10	30	30		15	40
22	Friction Factor Figure 6-13 or equation		0.0165	0.054	0.014?	0.035	0.0125		0.0092	0.0092
23	Friction Loss per p_v Items 21 x 22		1.32	1.62	0.14?	1.05	0.375		0.14	0.37
24	No. 90 degree Elbows		1-90,1-60	1-90,1-60	0	1-60	0		0	0
25	Elbow Loss per p_v Item 24 x loss factor		0.27	0.45	0	0.18	0		0	0
26	No. Branches		0	1	0	1	0		0	0
27	Branch Loss per p_v Item 26 x loss factor		0	0.18	0	0.18	0		0	0
28	Special Fittings Loss Factors		0	0	0	0	0		0	0
29	Duct Loss per p_v Items 23 + 25 + 27 + 28		1.59	2.25	0.148	1.41	0.375		0.14	0.37
30	Duct Loss Items 7 x 29	"H₂O	1.51	1.89	0.15	1.13	0.33	2.00	0.04	0.09
31	Duct Static Pressure Loss Items 20 + 30	"H₂O	2.94	3.28	0.15	3.37	C.33	2.00	0.04	0.09
32	Cumulative Static Pressure	"H₂O		3.43	3.43	3.43	3.76	3.76	5.80	5.89
33	Governing Static Pressure	"H₂O	3800	3.28	3.43	3.70	3.76	3.76		
34	Corrected Volumetric Flowrate	cfm				970				
35	Resultant Velocity Pressure	"H₂O	1.06	3.26		0.82				

Example 9.1 Foundry Cleaning Room System Designed by Static Pressure Balance Method

Example 9.1   Design Notes for Calculation Sheet

Branch 1A

Line 6: The velocity is calculated as 3905.4. Round to the nearest 10 fpm, 3910 fpm.

Line 7: Carry to two significant digits.

Line 25: One 90° elbow, loss equals $0.27p_v$.

Branch 2A

Line 25:

90° elbow	$0.27p_v$
60° elbow	$0.18p_v$
Total loss	$0.45p_v$

Line 27: The entry loss is always taken in the branch and not the main; since hood 2 enters at an angle, 2A is the branch and 1A is the main.

Line 31: Calculate the static pressure ratio:

$$\frac{3.28}{2.94} = 1.12$$

Since the ratio is between 1.05 and 1.2 it is necessary to correct the flow in branch 1A.

Line 34: Corrected flow $= 3600(3.28/2.94)^{.5} = 3800$ cfm.

Main AB

Line 1: Add the corrected flow in 1A and flow in 2A to get 4300 cfm in the main.

Line 31: This loss is in series with the loss at junction A.

Line 32: Add the governing static pressure at junction A to the AB loss, $3.28 + 0.15 = 3.43$ in. H_2O.

Branch 3B

Line 25: The loss for 60° elbow is $(60°/90°) \times 0.27p_v = 0.18p_v$.

Line 27: The branch entry loss equals $0.18p_v$.

Line 33: Compare static pressures:

$$\frac{p_{s,1}}{p_{s,2}} = \frac{3.43}{3.37} = 1.02$$

Since the ratio is less than 1.05, it is acceptable.

Line 34: Since the governing static pressure is 3.43 and not 3.37 in. H_2O, an increase in flow will be seen in hood 3. This correction is normally not made when the ratio is less than 1.05 since the correction is small.

Main BC

Line 2: The airflow is 4300 cfm from A to B plus the airflow of 970 cfm from branch 3B, for a total of 5270 cfm.

Line 31: As in main AB, add the loss in main BC in series with junction B.

$$p_s = 0.33 + 3.43 = 3.76$$

Main CD

Line 30: The loss from air cleaner is included here since the loss is expressed in in. H_2O, not as velocity pressure.

Main DE

Line 6: Since the particles were removed in the air cleaner, the minimum duct transport velocity no longer needs to be observed. The velocity can be reduced to approximately 2000 cfm. The duct diameter is chosen to match the fan inlet (see Chapter 10).

Main FG

This calculation could be performed by adding 40 ft to the 15 ft in column DE.

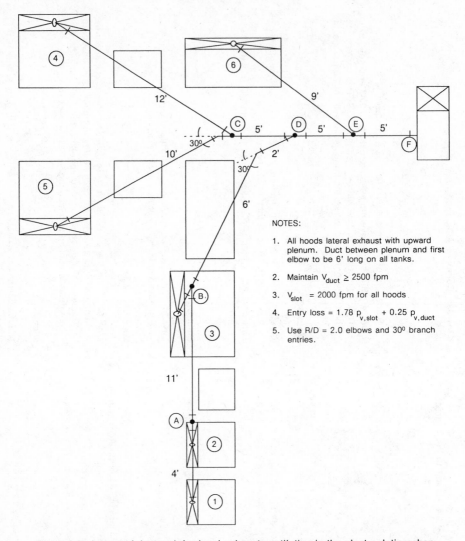

12'
9'
C 5'
30°
10'
2'
30°
6'
D 5'
E 5'
F

NOTES:

1. All hoods lateral exhaust with upward plenum. Duct between plenum and first elbow to be 6' long on all tanks.

2. Maintain $V_{duct} \geq 2500$ fpm

3. $V_{slot} = 2000$ fpm for all hoods

4. Entry loss = $1.78\, p_{v,slot} + 0.25\, p_{v,duct}$

5. Use R/D = 2.0 elbows and 30° branch entries.

11'

4'

Figure 9.6 Layout of ductwork for local exhaust ventilation in the electroplating shop.

tank are presented again in Table 9.3. These data are the basis for the design of a multiple-hood ventilation system using the static pressure balance method and the following design criteria.

- All tanks are provided with lateral slot exhaust using the upward plenum hood design (*Ventilation Manual*, VS 503). This design was chosen because it requires little floor space and the plenum provides an effective baffle. The hood entry loss is

$$h_e = 1.78 p_{v,s} + 0.25 p_{v,d}$$

Tank No.	Solution	Temp.	Class	Tank dimensions L	W	W/L	A	V control (fpm)	Min. Rate (cfm/ft²)	Q (cfm)	A_s (ft²)	Slot Width (ft)	(in)	Min. Plenum Depth
1	Hot Water	200	D-1	3	2	0.67	6	25	130	780	0.39	0.13	1.6	3.2
2	Cr Strip	130	C-3	2⅓	2	0.86	4⅔	50	90	420	0.21	0.09	1.1	2.2
3	Acid Cu	110	P-3	6	3½	0.58	21	25	130	2730	1.36	0.23	2.7	5.4
4	Cr Plate	115	A-1	4	3	0.75	12	150	250	3000	1.50	0.38	4.5	9.0
5	Ni Plate	70	P-2	4	3	0.75	12	100	175	2100	1.05	0.26	3.1	6.2
6	Alk. Cleaner	200	C-1	5	2	0.40	10	100	150	1500	0.25	0.15	1.8	3.6

Table 9.3 Exhaust Rates For Open Surface Tanks in Example 9.2

- All branches include a vertical run of 6 ft from the plenum to the elevation at which the horizontal main duct run will be made. This elevation was required to clear obstructions in the overhead.
- The minimum slot velocity is 2000 fpm. The fishtail design of the hood and the resistance presented by the slot ensure that the exhaust will be relatively constant along the length of the tank (see Chapter 5).
- In this system all 90° elbows have an $R/D = 2.0$ and therefore a loss of $0.27p_v$. The layout of the duct in Fig. 9.6 shows the main running directly over the plenum so that the branches enter the main from the bottom. For this reason all entries except $1A$ will include a 60° elbow and a 30° branch entry.
- All branch entries to main will be 30° and have a loss of $0.18p_v$.
- The submain duct run from hood 1 to junction D appears to present the greatest resistance to airflow; therefore, the sequence of calculations chosen is $1A$, $2A$, AB, $3B$, BD, $4C$, $5C$, CD, DE, $6E$, EF, and stack.
- Table 9.3 is based on Example 6.1 and has been expanded to show the technique used to design the slot and plenum as recommended in the *Ventilation Manual*.

9.4 BLAST GATE BALANCE METHOD

In this design technique the engineer chooses the hood that represents the maximum resistance to flow as the initial point in calculating losses. The system is designed from hood to first junction. At that point the design airflow from hood 1 is added to the airflow from hood 2, the main is sized, and losses are calculated for the section up to the main, where the branch from hood 3 enters the main. When hood 3 joins the main, its design airflow is added to the total of hoods 1 and 2 and the main is sized and losses calculated up to the next branch entry.

Accumulating static pressure losses in this manner assures that if the branch with maximum resistance is properly chosen for the start of the calculation, the static pressure in the main at any junction will be greater than that needed by the entering system branch. The desired airflow distribution can then be achieved by adjusting the blast gates provided on all hoods to adjust the required branch static pressure to equal the static pressure in the main.

This procedure is not as straightforward as it sounds, since the adjustment of a blast gate in one branch will affect the flow in adjacent branches. It is thus necessary to adjust all blast gates in a trial-and-error method until the proper balance is attained. Prior to the design of a system by the blast gate balance method the designer is encouraged to review the limitations of this technique (Caplan 1983).

VELOCITY PRESSURE METHOD CALCULATION SHEET

(adapted from A.C.G.I.H. COMMITTEE ON INDUSTRIAL VENTILATION)

No.		Description		Units	1A	2A	AB	3B(1)	3B(2)	BD	4C	5C	CD
1		Duct Segment Identification			1A	2A	AB	3B(1)	3B(2)	BD	4C	5C	CD
3		Volumetric Flowrate		cfm	740	420	1200	2730	2730	4060	3000	2100	5180
3		Minimum Transport Velocity		fpm	2500	2500	2500	2500	2500	2500	2500	2500	2500
4		Duct Diameter		inches	7½	5½	9	14	13	17	14	12	19
5		Duct Area		square feet	0.306	0.1650	0.441	1.069	0.921	1.376	1.069	0.7854	1.969
6		Actual Duct Velocity		fpm	2540	2550	2720	2550	2960	2390	2800	2670	2630
7		Duct Velocity Pressure (p_v)		"H2O	0.40	0.40	0.46	0.41	0.55	0.42	0.39	0.45	0.43
8	H	S	Slot Area	square feet	0.40	0.20	→	1.36	1.36	→	1.50	1.05	→
9	O	O	Slot Velocity	fpm	2000	2000		2000	2000		2000	2000	
10	O	O	Slot Velocity Pressure	"H2O	0.25	0.25		0.25	0.25		0.25	0.25	
11	D	T	Slot Loss Factor Section 5 or Fig. 6-10*		1.78	1.78		1.78	1.78		1.78	1.78	
12			Acceleration Factor 0 or 1		0	0	NA	0	0	NA	0	0	NA
13		S	Plenum Loss per p_v Items 11 + 12		1.78	1.78		1.78	1.78		1.78	1.78	
14		U	Plenum p_s Items 10 × 13	"H2O	0.45	0.45		0.43	0.43		0.45	0.45	
15		C	Hood Entry Loss Factor Section 5 or Fig. 6-10		0.25	0.25		0.25	0.25		0.25	0.25	
16		T	Acceleration Factor 1 or 0		1.00	1.00		1.00	1.00		1.00	1.00	
17		O	Entry Loss Factor p_v Items 15 + 16		1.25	1.25		1.25	1.25		1.25	1.25	
18		N	Entry Loss Items 7 × 17	"H2O	0.50	0.50		0.51	0.69		0.61	0.56	
19			Other Loss	"H2O				0	0		0	0	
20			Hood Static Pressure Items 14 + 18 + 19	"H2O	0.95	0.95		0.96	1.14		1.06	1.01	
21			Straight Duct Length	feet	10	6	11	8	8	8	16	16	3
22			Friction Factor Figure 6-13 or equation		0.034	0.049	0.0275	0.016	0.016	0.0125	0.015	0.0185	0.0105
23			Friction Loss per p_v Items 21 × 22		0.34	0.29	0.30	0.13	0.13	0.10	0.27	0.30	0.05
24			No. 90 degree Elbows		1-60	1-60	0	1	1	1-30	1-90	1-30 1-90	
25			Elbow Loss per p_v Item 24 × loss factor		0.27	0.27	0	0.27	0.27	0.07	0.27	0.36	0
26			No. Branches		0	0	1	0	0	0	1	0	0
27			Branch Loss per p_v Item 26 × loss factor		0.18	0.18	0.18	0	0	0.18	0.18	0	0
28			Special Fittings Loss Factors		0	0	0.18	0	0	0	0	0	0
29			Duct Loss per p_v Items 23 + 25 + 27 + 28		0.61	0.65	0.48	0.40	0.40	0.35	0.72	0.66	0.05
30			Duct Loss Items 7 × 29		0.24	0.26	0.22	0.16	0.22	0.15	0.35	0.30	0.02
31			Duct Static Pressure Loss Items 20 + 30	"H2O	1.19	1.21	1.43	1.12	1.36	1.64	1.41	1.31	0.02
32			Cumulative Static Pressure	"H2O		1.21	1.49	1.49		1.64			1.43
33			Governing Static Pressure						2860				
34			Corrected Volumetric Flowrate	cfm		1200			2860		1200	2160	5550
35			Resultant Velocity Pressure	"H2O									

Example 9.2 Electroplating Shop System Designed by Static Pressure Balance Method

VELOCITY PRESSURE METHOD CALCULATION SHEET

(adapted from A.C.G.I.H. COMMITTEE ON INDUSTRIAL VENTILATION)

#	Item	Reference	Units	DE	6F	6F	Stack
1	Duct Segment Identification			DE	6F	6F	Stack
2	Volumetric Flowrate		cfm	9610	1500	11,230	11,230
3	Minimum Transport Velocity		fpm	2500	2500	2,500	2,500
4	Duct Diameter		inches	26	10	28	28
5	Duct Area		square feet	3.687	0.5454	4.276	4.276
6	Actual Duct Velocity		fpm	2610	2750	2630	2630
7	Duct Velocity Pressure (p_V)		"H_2O	0.42	0.47	0.43	0.43
8	Slot Area		square feet		0.75		
9	Slot Velocity		fpm		2000		
10	Slot Velocity Pressure		"H_2O		0.25		
11	Slot Loss Factor	Section 5 or Fig. 6-10*			1.78		
12	Acceleration Factor	0 or 1		NA	0	NA	NA
13	Plenum Loss per p_V	Items 11 + 12			0.45		
14	Plenum p_s	Items 10 x 13	"H_2O		0.25		
15	Hood Entry Loss Factor	Section 5 or Fig. 6-10	1 or 0		1.00		
16	Acceleration Factor				1.25		
17	Entry Loss Factor p_V	Items 15 + 16			0.59		
18	Entry Loss	Items 7 x 17	"H_2O				
19	Other Loss		"H_2O				
20	Hood Static Pressure	Items 14 + 18 + 19	"H_2O		1.04		
21	Straight Duct Length		feet	5	15	5	15
22	Friction Factor	Figure 6-13 or equation		0.0074	0.0240	0.006F	0.006F
23	Friction Loss per p_V	Items 21 x 22		0.04	0.36	0.03	0.10
24	No. 90 degree Elbows			0	1	0	0
25	Elbow Loss per p_V	Item 24 x loss factor		0	0.27	0	0
26	No. Branches			0	0	0	0
27	Branch Loss per p_V	Item 26 x loss factor		0	0.18	0	0
28	Special Fittings Loss Factors	Items 23 + 25 + 27 + 28		0	0	0	0
29	Duct Loss per p_V	Items 7 x 29		0.04	0.81	0.03	0.10
30	Duct Loss	Items 7 x 29	"H_2O	0.02	0.38	0.01	0.04
31	Duct Static Pressure Loss	Items 20 + 30	"H_2O	0.02	1.42	0.01	0.04
32	Cumulative Static Pressure		"H_2O	1.66	1.66	1.67	1.71
33	Governing Static Pressure		"H_2O		16.20		
34	Corrected Volumetric Flowrate		cfm				
35	Resultant Velocity Pressure		"H_2O				

Example 9.2   Design Notes for Calculation Sheet

Column 1A

Line 2: The design airflow is entered on this line directly from the Table 9.3 calculations.

Lines 8–14 Lateral slot exhaust hoods have a complex entry loss consisting of two elements

(1) $1.78p_{v,s}$ loss due to air flowing through slot to the plenum
(2) $0.25p_{v,d}$ loss in flowing from plenum to duct

Only the slot entry loss factor is considered in this section of the calculation sheet. The second part of the hood entry loss statement, $0.25p_{v,d}$, is entered on line 15. Since the slot velocity is 2000 fpm for all hoods, the slot entry loss will also be the same.

Line 15: This is the second part of the entry loss, as discussed in line 6. Enter $0.25p_v$.

Line 16: The acceleration of air from outside the hood takes place in two steps. The air is first accelerated to the slot velocity, enters the plenum where the velocity is reduced, and then is accelerated to the duct velocity. The energy for this acceleration can be accounted for in two ways. Acceleration to slot velocity can be entered in line 12 and a second factor accelerating the air from slot to duct velocity

can be entered on line 16. An alternative approach is to add the entire acceleration factor at line 16. This is the technique chosen in this example.

Line 21: The duct length consists of a 6-ft vertical and 4-ft horizontal run.

Line 23: This calculation is $10 \times 0.034 = 0.34p_v$.

Line 29: The individual static pressure losses due to air flowing through the system are summed as fractions of p_v on this line and include losses resulting from friction, elbows, and when appropriate, branch entry and expansions/contractions

Line 30: The losses in line 29 expressed as velocity pressure are converted to in. H_2O in line 30 (line 29 × line 7).

Line 31: Line 30 is added to the hood static pressure loss from line 20 to obtain the total static pressure loss at junction A.

Column 2A

Line 31: In practice, such close agreement of static pressures at a junction is rarely obtained. Since the ratio of the two static pressures is less than 1.05, no correction is needed. The governing static pressure at a junction is always taken as the larger of the two values calculated.

Column AB

Lines 30, 31: Obviously, there is no slot hood in the submain run. Merely calculate the frictional loss due to 1200 cfm flowing through *AB* and add to 1.21 in. H$_2$O to give the required static pressure at junction *B*.

Line 32: The velocity pressure in submain *AB* is higher (0.46 in. H$_2$O) than in either branch 1A or 2A. An additional acceleration factor must be added for the static pressure needed to accomplish this increase in velocity. This correction may be taken merely as the difference between the velocity pressures in the branches (0.40) and the velocity pressure of main *AB* (0.46). This differential (0.06) is included in line 32. This simple correction can be made if the velocity pressures in the two branches are equal. The procedure given in the *Ventilation Manual* Fig. 5–6) must be followed if they differ.

Column 3B(1)

Line 15: This branch was initially sized for a 14-in. duct with a duct velocity of 2550 fpm.

Line 31: The static pressure required by branch 3B at junction *B* is 1.12 in. H$_2$O, but submain *AB* requires 1.49 in. H$_2$O at junction *B*.

$$\frac{p_{s,AB}}{p_{s,3B}} = \frac{1.49}{1.12} = 1.33$$

Since this ratio exceeds 1.2, branch 3B must be resized (see Table 9.2).

Column 3B(2)

Line 1: When a duct is resized it is given the (2) designation.

Line 4: Since the static pressure loss in the branch must be increased, the duct diameter must be reduced. Try the next smaller duct, 13 in. (An equally valid approach is to reduce the slot area, thereby increasing slot velocity and the 1.78$p_{v,s}$ term.)

Line 30: The new static pressure loss at junction *B* is 1.36. Now

$$\frac{p_{s,1}}{p_{s,2}} = \frac{1.49}{1.36} = 1.10$$

According to Table 9.2, junction *B* can now be brought into balance by increasing the airflow through branch 3B:

$$Q_{corrected} = 2730(1.35/1.49)^{0.5} = 2860 \text{ cfm}$$

Column BD

Line 2: The corrected flow in branch 3 is now added to the submain *AB* flow, $Q = 2860 + 1200 = 4060$ cfm.

Line 32: The incremental loss in submain *BD* of 0.15 in. H$_2$O is added to the governing static pressure at *B*: $p_S = 0.15 + 1.49 = 1.64$ in. H$_2$O. The design now proceeds, with the second submain consisting of hoods 4 and 5; and this submain is brought to junction *D*.

Column 4C

Line 25: The static pressure loss of branch 4C is 1.41 and that of branch 5C is 1.31,

$$\frac{p_{s,1}}{p_{s,2}} = \frac{1.41}{1.31} = 1.08$$

Since the ratio is less than 1.20, the static pressure losses at junction C are brought into balance by increasing the airflow in branch $5C$:

$$Q_{corrected} = 2100(1.41/1.31)^{0.5} = 2180 \text{ cfm}$$

Column CD

Line 2: The flow through submain CD is

Branch 5C	2180
Branch 4C	3000
Total	5180 cfm

Line 32: The incremental loss of submain CD is 0.02 in. H_2O. The total static pressure loss at junction D is

Loss at junction C	1.41
Loss CD	0.02
Total	1.43

Junction D must be balanced. The governing static pressure is 1.64 in. H_2O in submain BD.

$$\frac{p_{s,1}}{p_{s,2}} = \frac{1.64}{1.43} = 1.12$$

This junction can therefore be brought into balance by increasing the airflow through submain CD:

$$Q_{corrected} = 5180(1.64/1.43)^{0.5} = 5550 \text{ cfm}$$

Column DE

Line 2: The total air flow in main DE is

Submain CD	5550
Submain BD	4060
Total	9610 cfm

Line 32: The loss of 0.02 in. H_2O in main DE is added to the governing static pressure at junction D of 1.64 in. H_2O to get 1.66 in. H_2O at E.

Column 6E

Line 32: The static pressure loss of branch $6E$ to junction E is 1.42 in. H_2O. The governing static pressure at junction E is 1.66 in. H_2O:

$$\frac{p_{s,1}}{p_{s,2}} = \frac{1.66}{1.42} = 1.15$$

Again, junction E can be brought into balance by increasing the airflow through branch $6E$:

$$Q_{corrected} = 1500(1.66/1.42)^{0.5} = 1620 \text{ cfm}$$

Column EF

Line 2: The total airflow in the main is now

Branch 6E	1,620
From junction D	9,610
Total	11,230 cfm

9.5   OTHER COMPUTATIONAL METHODS

Chapters 8 and 9 have dealt mainly with the manual design of balanced ventilation systems using the velocity pressure method. Although useful in introducing the design process and for the occasional design problem, the manual calculation sheet method is time consuming. During the past decade a number of computer programs for the design of systems have been introduced in the open literature and by engineering consultants and manufacturers. One type of design program has been developed for the programmable hand calculator (Barritt, 1978; Malchaire, 1981; Shotwell, 1984). These have the merit of simplicity and are easily used in the field but may not have advanced features, such as automatic rebalancing. More recently, spreadsheet programs for personal computers have been applied to ventilation design using the velocity pressure method (Rennix, 1987; Koshland and Yost 1987). These powerful programs permit design of complex systems, optimization of design, and correction for nonstandard air.

The widespread application of computer technology to ventilation system design will continue and the person doing routine designs should take advantage of the speed and power of these programs.

Calculation sheet for local exhaust ventilation system (adapted from ACGIH Committee on Industrial Ventilation)

1	Duct Segment Identification	
2	Volumetric Flowrate	cfm
3	Minimum Transport Velocity	fpm
4	Duct Diameter	inches
5	Duct Area	square feet
6	Actual Duct Velocity	fpm
7	Duct Velocity Pressure (p_v)	"H_2O
8	Slot Area	square feet
9	Slot Velocity	fpm
10	Slot Velocity Pressure	"H_2O
11	Slot Loss Factor	Chapter 10
12	Acceleration Factor	0 or 1
13	Plenum Loss per p_v	Items 11 + 12
14	Plenum p_s	Items 10 x 13 — "H_2O
15	Hood Entry Loss Factor	Chapter 10
16	Acceleration Factor	1 or 0
17	Entry Loss Factor p_v	Items 15 + 16
18	Entry Loss	Items 7 x 17 — "H_2O
19	Other Loss	"H_2O
20	Hood Static Pressure	Items 14 + 18 + 19 — "H_2O
21	Straight Duct Length	feet
22	Friction Factor	Figure 5–18 or equation
23	Friction Loss per p_v	Items 21 x 22
24	No. 90 degree Elbows	
25	Elbow Loss per p_v	Item 24 x loss factor
26	No. Branches	
27	Branch Loss per p_v	Item 26 x loss factor
28	Special Fittings Loss Factors	
29	Duct Loss per p_v	Items 23 + 25 + 27 + 28
30	Duct Loss	Items 7 x 29 — "H_2O
31	Duct Static Pressure Loss	Items 20 + 30 — "H_2O
32	Cumulative Static Pressure	"H_2O
33	Governing Static Pressure	"H_2O
34	Corrected Volumetric Flowrate	cfm
35	Resultant Velocity Pressure	"H_2O

(Row grouping labels on the form: items 8–11 "SLOT", items 13–14 "SU", items 15–19 "CTION", collectively under "HOOD".)

LIST OF SYMBOLS

A	area
A_s	slot area
D	duct diameter
h_e	hood entry loss
L	length of tank
$p_{v,d}$	duct velocity pressure
$p_{v,s}$	slot velocity pressure
R/D	ratio of elbow sweep radius to diameter
Q	volume flow
$Q_{corrected}$	corrected airflow
Q_{design}	design air flow
p_s	static pressure
T	temperature
V	average velocity
W	width of tank

REFERENCES

American Conference of Governmental Industrial Hygienists, Committee on Industrial Ventilation (1988), *Industrial Ventilation Manual*, 20th. ed., ACGIH, Lansing, MI.

Barritt, S. L. (1978), "Duct Design by Programmable Calculator," *Heat./Piping/Air Cond.*, December.

Caplan, K. J. (1983), "Balance with Blast Gates—A Precarious Balance," *Heat./Piping/Air Cond.*, February.

Chen, S. Y. S. (1981), "Design Procedure for Duct Calculations," *Heat./Piping/Air Cond.*, January.

Clapp, D. E., D. S. Groh, and J. D. Nenandic (1982), "Ventilation Design by Microcomputer," *Am. Ind. Hyg. Assoc. J.*, **43**:212.

Guffey, S. E. (1983), "An Easier Calculation System for Ventilation Design," *Am. Ind. Hyg. Assoc. J.*, **44**:627.

Guffey, S., and J. Hickey (1983), "Equations for Redesign of Existing Ventilation Systems," *Am. Ind. Hyg. Assoc. J.*, **44**:819.

Koshland, C. P., and M. G. Yost (1987), "Use of a Spreadsheet in the Design of an Industrial Ventilation System," *Appl. Ind. Hyg.*, **2**:204.

Lynch, J. R. (1968), "Computer Design of Industrial Exhaust Systems," *Heat./Piping/Air Cond.*, September.

Malchaire, J. B. (1981), "Design of Industrial Exhaust Systems Using a Programmable Calculator," *Ann. Occup. Hyg.*, **24**:217.

Rennix, C. P. (1987), "Computer Assisted Ventilation Design and Evaluation," *Appl. Ind. Hyg.*, **2**:32.

Shotwell, H. P. (1984), "A Ventilation Design Program for Hand-Held Programmable Computers," *Am. Ind. Hyg. Assoc. J.*, **45**:749.

ADDITIONAL READING

Burton, D. J. (1986), *Industrial Ventilation—A Self Study Companion to the ACGIH Ventilation Manual*, 3rd ed., IVE, Inc., Los Alamos, NM.

10

Fans and Blowers

The fan is an extremely important part of any ventilation system, since it supplies the energy necessary to cause air movement. In a general exhaust ventilation system, the fan frequently *is* the ventilation system, since it is usually just installed in a wall or ceiling without ductwork. The fan used in a local exhaust ventilation (LEV) system, on the other hand, is an integral part of that *system*, and as such must be carefully selected to be compatible with the rest of the system. The network of hoods, ducts, and air cleaners in an LEV system offers a certain inherent resistance to airflow. The amount of air that actually flows through this resistance is a function of the fan attached to the system and the suction it provides to the exhaust system.

The definitions of "fan," "blower," and "compressor" can be the subject of some confusion. Jorgensen (1983) states:

> Any device that produces a current of air by the movement of a broad surface can be called a fan. . . . Broadly speaking, the function of a fan is to propel, displace, or move the air or gas, while the function of a compressor is to increase the pressure, reduce the volume, or compress the air or gas. However, there is always some fluid movement through a compressor, and it is debatable whether that function is less important than the others listed. . . . Some other names for fans and compressors are: ventilator (generally restricted to a very-low-pressure-rise), exhauster (used to signify that gases are being removed from something), and blower (used to signify that gases are being supplied to something).

In this book we call all such devices "fans." Fans can be distinguished from compressors and vacuum pumps, which also move gases but at pressures significantly different from atmospheric (usually, many inches of mercury).

10.1 TYPES OF AIR MOVERS

All fans used in ventilation systems can be categorized into one of two types, depending on the nature of the airflow through the device. In axial flow fans the airflow is parallel to the axis of fan rotation, while centrifugal fans exhaust air radially. Sometimes it is necessary to move air without the use of a fan; one common way to do this is to use an air ejector.

10.1.1 Axial Flow Fans

Axial fans, also called "propeller" fans, find a wide variety of uses in industry. As the name implies, airflow in this type of fan is parallel to the axis of rotation—air does not have to change direction in passing through the fan. There are three basic types of axial fans, each with its special design details and proper area of application to industrial ventilation.

Propeller Fans. The simplest propeller fan, equipped with sheet metal blades, is the common air fan ubiquitous to homes, offices, and hot work environments (Fig. 10.1). The principal advantage of such fans is their ability to move large amounts of air; conversely, their principal disadvantage is their inability to move air against any appreciable resistance. The amount of air that such fans can move is greatly reduced when the static pressure against which they are operating is increased to 0.5 to 1 in. H_2O. If the fan is specially designed with propellers with an airfoil cross section, it can operate in the range 1 to 2 in. H_2O, but no higher.

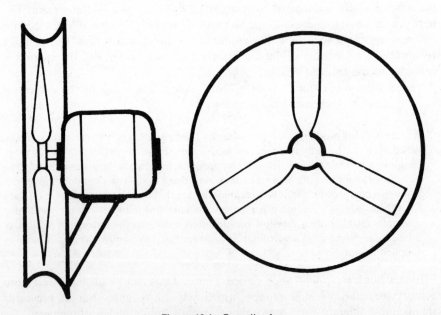

Figure 10.1 Propeller fan.

Despite this limitation, propeller fans are commonly used in industry; their most common use is as wall or ceiling fans in dilution ventilation systems. Since this type of fan cannot develop significant static pressure, care must be taken to supply adequate replacement air, or the fan will not be able to move the amount of air desired due to the negative pressure built up in the room (see Chapter 4). Their poor static pressure capability also precludes their use in local exhaust systems.

Tubeaxial Fans. Tubeaxial fans (Fig. 10.2) are propeller fans which have been modified so that they can be inserted into a duct. The close spacing between the blade tips and the housing improves efficiency and allows the development of greater static pressure than the simple propeller fan. The motor can be located either inside the duct for a direct-drive fan or outside the duct for a belt-driven fan. Such fans are designed to operate against low (0 to 3 in. H_2O) static pressures, and thus have limited application in local exhaust ventilation systems. They are used primarily in simple one-hood systems where space is at a premium and system resistance is low, such as paint spray booths.

Vaneaxial Fans. Vaneaxial fans (Fig. 10.3) are tubeaxial fans which have been modified by the addition of air-directing vanes on the motor housing behind the fan blades. The vanes serve to reduce the whirl in the air as it exits the fan blades, which improves the fan efficiency and allows the development of increased static pressure. Such fans can operate at static pressures up to 10 in. H_2O and are thus suitable for use in local exhaust systems. They must be used with relatively clean air, and thus cannot transport dust-laden gas streams. Such fans are generally more expensive than a comparable centrifugal fan (see section 10.1.2), and are usually employed only when space is at a premium.

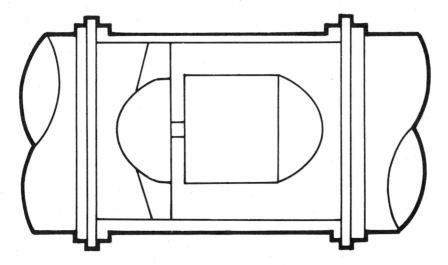

Figure 10.2 Tubeaxial fan.

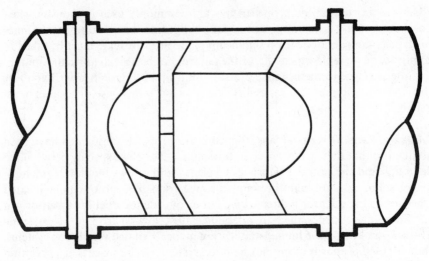

Figure 10.3 Vaneaxial fan.

10.1.2 Centrifugal Fans

Centrifugal fans are distinguished from axial fans in that the air leaves the rotating blades in the radial direction (i.e., perpendicular to the axis of rotation). A simple centrifugal fan is shown in Fig. 10.4. These fans are commonly divided into three performance and application categories, which are determined by the shape of the impeller.

Radial Blade Fans. Radial blade fans are the simplest, and least efficient, type of centrifugal fan (Fig. 10.5a). They are used whenever dusty air or highly corrosive contaminants must be moved, since the blades can be constructed as thick as necessary or replaced to withstand erosion or corrosion. The blades also tend to be self-cleaning, which minimizes dust deposition and subsequent imbalance. Since these fans are less efficient and more expensive than the types described below, they are usually used only for systems transporting dusty or corrosive air through the fan.

Backward-Curved-Blade Fans. Backward-curved-blade fans (Fig. 10.5b) are the preferred choice for local exhaust systems where nondusty air is moved through the fan. This would be the case where the contaminants being exhausted are gases, or where particulate contaminants are being exhausted but an air-cleaning device is located upstream of the fan. This type of fan is more efficient than the radial blade type, thus reducing horsepower requirements. It is not without its disadvantages, however. The backward curve of the blade means that the blade tip must have a higher velocity than the radial blade to impart the same velocity to the air; these fans must thus operate at higher rotating speeds than other centrifugal fans moving the same amount of air. This higher speed

(a)

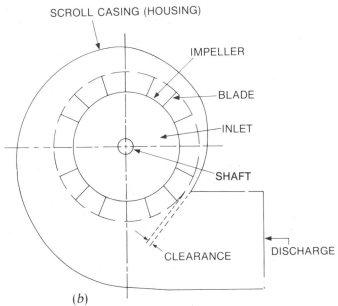

(b)

Figure 10.4 (a) Centrifugal fan; (b) standard centrifugal fan nomenclature.

(a)

(b)

(c)

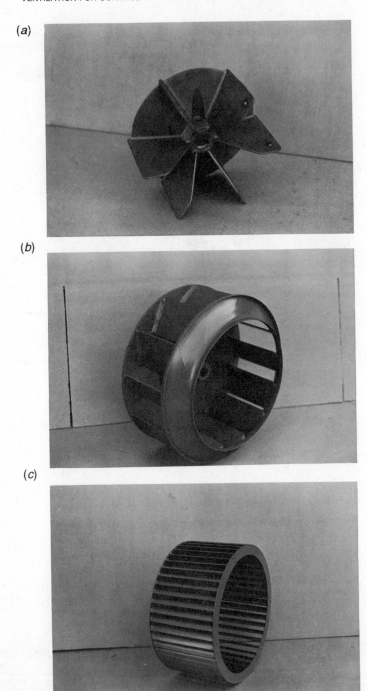

Figure 10.5 Centrifugal fan types: (a) radial blade; (b) backward-curved blade; (c) forward-curved blade.

increases the stresses on the fan and results in a higher capital cost, although this is offset by the higher operating efficiencies attained by this design.

Forward-Curved-Blade Fans. Forward-curved-blade fans are the common "squirrel cage" fans used widely for moving large amounts of air against low-to-moderate static pressures (Fig. 10.5c). This type of fan has many cup-shaped blades which accelerate the air to a high velocity at relatively low rotating speeds. The individual blades and impeller are stamped from sheet steel and assembled by spot welding, so that this design is much cheaper than the heavier impellers used in radial and backward curved blade fans. The high air velocity at low rotating speed results in quieter operation, which is important for many air-moving applications.

This fan type has three main disadvantages: it has a low operating efficiency, it cannot develop a high static pressure, and it cannot be used in corrosive or erosive environments. These limitations preclude its use in most local exhaust systems; its most common use in industry is in air supply and exhaust for building HVAC systems.

Miscellaneous Centrifugal Fans. Centrifugal fans can take several forms other than the classical "stand-alone" variety shown in Figs. 10.4 and 10.5. Examples of such fans are the roof fan (Fig. 10.6a) and "up-blast" roof fan (Fig. 10.6b) used primarily for commercial kitchen range hood exhaust systems, and the "in-line" centrifugal fan (Fig. 10.6c), which is mounted directly in the duct in a manner similar to tubeaxial fans, thus providing the space-saving advantages of an axial fan together with the static pressure capabilities of a centrifugal fan.

10.1.3 Air Ejectors

Air ejectors (also called injectors or jet pumps) are a means of moving air without having it pass through a fan. Typical designs are shown in Fig. 10.7a (a simple ejector) and Fig. 10.7b (a venturi-type ejector). Airflow (secondary air) is induced in the large duct by jet action of the primary fluid leaving the nozzle at high velocity. The ejecting fluid can be water, steam, compressed air, or air from a fan; since the first three involve an excessive use of energy, the usual ejecting fluid is air from a fan.

Ejectors are inherently very inefficient devices, since much of the jet energy is employed in overcoming friction. The maximum efficiency attainable under optimum conditions is only 40%, while typical efficiencies in field conditions range from 15 to 25% (Jorgensen, 1983). Such low efficiencies limit ejector applications to those where for some reason the air cannot come in contact with the fan; examples would include gas streams which are extremely hot, sticky, corrosive, or explosive.

Figure 10.6 Miscellaneous centrifugal fans: (*a*) roof fan; (*b*) upblast roof fan.

(c)

Figure 10.6   (c) In-line centrifugal fan.

10.2 FAN CURVES

A fan's operating characteristics can be described by reference to several important variables. These include the amount of air being moved by the fan (Q, cfm), the static pressure developed by the fan (P_s, in. H_2O), the fan efficiency (η), the fan rotating speed (N, rpm), and the power (H, hp) required to operate the fan. Empirical formulas have been developed relating these variables to each other; such formulas can be displayed both in tabular form, resulting in fan tables, or in graphical form, giving what are known as fan curves.

Manufacturers specify fan performance by means of fan tables, but the performance of various fan types is more easily visualized from the use of fan curves; we will start our discussion using fan curves and then relate the curves to

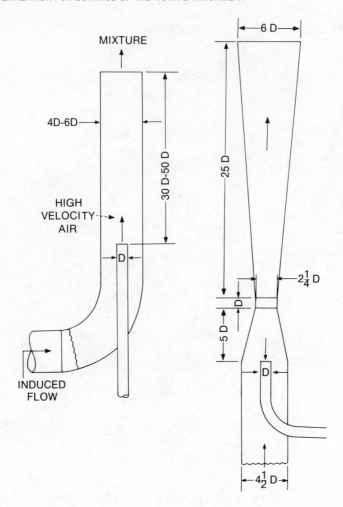

Figure 10.7 Ejectors: (a) simple design; (b) venturi design.

fan tables. Since the end result of using a fan is to deliver a certain amount of airflow to a desired location, it is most common to specify fan performance in terms of airflow. Fan curves then relate all of the other fan variables to airflow. Such curves thus have airflow as the abscissa and the other variables of interest plotted on the ordinate.

10.2.1 Static Pressure Curve

The basic performance of a typical radial blade fan is illustrated in Fig. 10.8a, which plots the amount of air moved by the fan versus the resistance to airflow offered by whatever system is attached to it. As would be expected, maximum airflow occurs when no resistance is presented to the fan; this is the so-called free

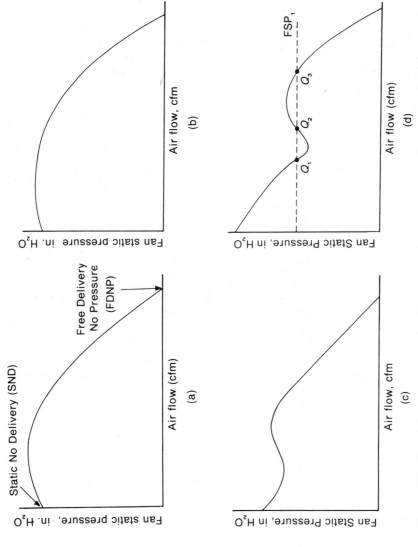

Figure 10.8 Fan static pressure curves: (*a*) radial blade; (*b*) backward-curved blade; (*c*) forward-curved blade; (*d*) vaneaxial fan.

delivery no pressure (FDNP) point. At the other extreme, no airflow occurs when the resistance is infinite (i.e., when the fan inlet or outlet is blocked completely). At this so-called static no delivery (SND) point, the blower develops its maximum static pressure, as shown by the curve. (In actuality, the static pressure increases slightly at very low airflow due to the inefficiency of fan operation in this zone.) The exact shape of the fan pressure curve between these two points depends on the specific fan design, but the curve shown is typical of a radial blade fan.

The static pressure plotted in this curve is called the fan static pressure (P_s). This quantity can be defined with the help of the simple exhaust system pictured in Fig. 10.9. To draw air through this system the fan must develop a certain negative static pressure at the inlet ($p_{s,i}$) and a positive static pressure at the fan outlet ($p_{s,o}$). The total amount of static pressure that must be developed by the fan is thus

$$p_s = |p_{s,i}| + |p_{s,o}| \tag{10.1}$$

Fan static pressure is slightly different from the total static pressure needed to move air through the system. It is defined as

$$P_s = p_s - p_{v,i} \tag{10.2}$$

where $p_{v,i}$ is the velocity pressure of the air at the inlet to the fan. This somewhat arbitrary definition of fan static pressure arises from the standard definition of fan total pressure (P_t); a derivation of Eq. 10.2 is given in Jorgensen (1983).

Static pressure curves for backward-curved centrifugal, forward-curved centrifugal, and vaneaxial fans are shown in Fig. 10.8b, c, and d, respectively. Note the "dip" exhibited by the fan static pressure curves for the vaneaxial and, to a lesser extent, forward-curve blade centrifugal fans. This so-called "stall" zone is a region of unstable fan operation caused by aerodynamic conditions established when these fans are operated at airflows well below the maximum design

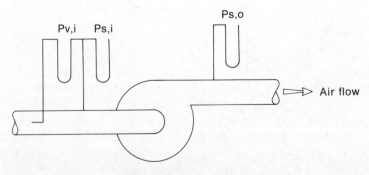

Figure 10.9 Fan pressures.

airflows. Fan operation at or near the stall zone is to be avoided, since the flow will be unstable. This instability is demonstrated in Fig. 10.8 d; at static pressure P_1, three different airflows (Q_1, Q_2, Q_3) can be delivered. Such instability can result in considerable air pulsation and mechanical vibration. These fans should always be operated at airflows high enough to avoid the stall zone, on the downward slope of the fan static pressure curve.

10.2.2 Brake Horsepower Curve

To develop the brake horsepower curve and the mechanical efficiency curve that follows, we must first define the quantity air horsepower (H_a). Air horsepower is a measure of the work done by the fan and is thus defined as the amount of energy added to the moving air per unit time. In general, the work (W) done in moving an airflow Q across a pressure drop Δp is

$$W = (\Delta p)Q \tag{10.3}$$

If the pressure drop is given in pounds per square foot (psf) and the airflow is given in cfm, the work is calculated in the engineering units of foot-pounds per minute (ft-lb/min). To calculate fan air horsepower, the change in pressure that must be used in the equation is the total pressure required to get air through the system, which is usually specified in inches of water. The following conversions can be used with Eq. 10.3 to obtain the desired relationship:

$$H_a(\text{hp}) = \frac{p_t(\text{in. H}_2\text{O})Q(\text{cfm})}{33,000[(\text{ft-lb/min})/\text{hp}]0.192[\text{in. H}_2\text{O}/(\text{lb/ft}^2)]}$$

$$= \frac{p_t Q}{6343} \tag{10.4}$$

This expression describes the amount of energy per unit time being transferred to the moving air. For purposes of designing the fan, however, we need to know the power requirements of the motor driving the fan. If the motor were 100% efficient at transferring energy to the fan blades, and if the fan were 100% efficient at transferring energy of rotation to energy in the moving air, the required motor horsepower, designated by H_m, would be just equal to the air horsepower.

Both of these energy transfers are of course less than 100% efficient, so that the motor horsepower requirement is always considerably higher than the resulting air horsepower. If the efficiency of the motor in converting electrical energy into rotating mechanical energy is designated by η_m and the efficiency of the fan in converting rotating energy into moving air is designated by η_T (standard fan terminology for total mechanical efficiency), we can picture the energy transformation process by Fig. 10.10.

The work done by the rotating fan is called the fan brake horsepower (H_b). Mathematically, H_a and H_b are related by the simple equation

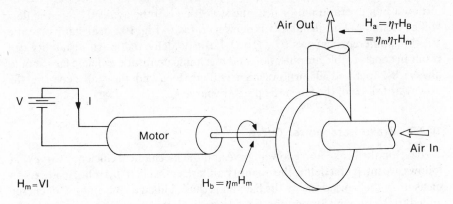

Figure 10.10 Fan-motor horsepower relationships.

$$H_b = \frac{H_a}{\eta_T}$$

$$= \frac{p_t Q}{6343 \eta_T} \qquad (10.5)$$

The electrical horsepower required by the motor is

$$H_m = \frac{H_b}{\eta_m} \qquad (10.6)$$

Combining Eqs. 10.5 and 10.6 gives the relationship between electrical horse-power supplied to the fan and the resulting work done by the moving air:

$$H_a = H_m \eta_m \eta_T \qquad (10.7)$$

Fan motor efficiency η_m is usually about 95%, so that brake horsepower and motor horsepower are nearly equal. Fan horsepower requirements are usually specified in terms of brake horsepower, and the user of fan tables should keep in mind that the motor horsepower used will be slightly higher than the value specified.

This introduction to fan horsepower terminology is a necessary prelude to an understanding of fan horsepower curves. A typical brake horsepower curve for a radial blade fan is shown in Fig. 10.11a. The curve does not intersect the origin, as predicted by Eq. 10.5, because the fan uses some energy to rotate the impeller even when shut tight so that no air flows. The horsepower curve rises steadily with increasing flow, reaching its maximum at free delivery (leading to the possibility that the motor may burn out if such a fan is operated under no load).

Similar curves are shown in Fig. 10.11b for a backward-curved blade fan, in Fig. 10.11c for a forward-curved blade fan, and in Fig. 10.11d for a vaneaxial fan.

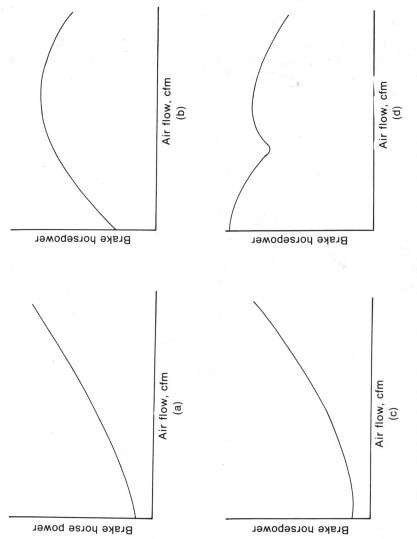

Figure 10.11 Typical fan horsepower curves: (*a*) radial blade fan; (*b*) backward-curved blade fan; (*c*) forward-curved blade fan; (*d*) vaneaxial fan.

Note the downward curve of the brake horsepower curve at high airflows for the backward-curved blade fan. This "load-limiting" characteristic is a desirable feature of this type of fan, since the horsepower requirement does not climb to excessive levels when the fan is operated under low-resistance conditions.

10.2.3   Mechanical Efficiency Curve

Mechanical efficiency refers to the efficiency of the fan in transferring rotating mechanical energy to energy of moving air. Two different efficiencies, called total efficiency and static efficiency, are used in standard fan terminology.

Total efficiency η_T was defined in Eq. 10.5, here rearranged:

$$\eta_T = \frac{H_a}{H_b} \tag{10.5}$$

Substituting Eq. 10.4 into Eq. 10.5 gives

$$\eta_T = \frac{p_t Q}{6343 H_b} \tag{10.8}$$

Note that total efficiency is defined in terms of the total pressure required by the fan. If static pressure is used in place of total pressure, a new efficiency, called the static efficiency (η_s), is defined:

$$\eta_s = \frac{p_s Q}{6343 H_b}$$

$$= \eta_T \frac{p_s}{p_t} \tag{10.9}$$

Five parameters (static pressure, total pressure, horsepower, total efficiency, and static efficiency) can now be plotted as functions of airflow for any particular fan. Typical sets of fan curves for several types of blowers are shown in Fig. 10.12.

10.2.4   Fan Laws

Virtually all centrifugal fans are available in "families" of geometrically similar models, varying only in their size and thus their capacity for moving air. In addition, a particular fan can be operated over a wide range of rotating speeds, which also has the effect of changing the amount of air the fan can move. The changes in a fan's operating characteristics corresponding to changes in fan size and/or rotating speed can be determined by using a series of equations which are known as the "fan laws."

Effect of Rotating Speed. As a blade of a centrifugal fan rotates, it "picks up" a volume of air at the fan inlet and "throws" that same volume out of the fan outlet. Changing the fan rotating speed simply changes the rate at which parcels of air are picked up and thrown out, so that changes in fan rotating speed result in corresponding changes in airflow delivered by the fan. Mathematically, the first fan law can be written as

$$\frac{Q_1}{Q_2} = \frac{N_1}{N_2} \tag{10.10}$$

where N designates fan rotating speed.

As discussed earlier, most of the static pressure losses in a ventilation system are proportional to air velocity squared. Since air velocity is directly proportional to airflow, and airflow is directly proportional to fan speed (Eq. 10.10), it follows that static pressure (and total pressure) varies as the square of the fan speed. This is the second basic fan law:

$$\frac{p_{s,1}}{p_{s,2}} = \frac{p_{t,1}}{p_{t,2}} = \left(\frac{N_1}{N_2}\right)^2 \tag{10.11}$$

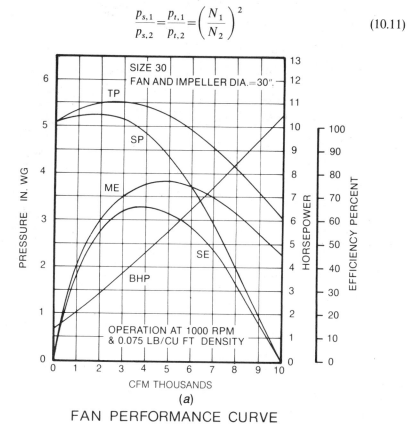

FAN PERFORMANCE CURVE

Figure 10.12 Typical performance curves for several fan designs: (*a*) performance curve for a radial centrifugal fan.

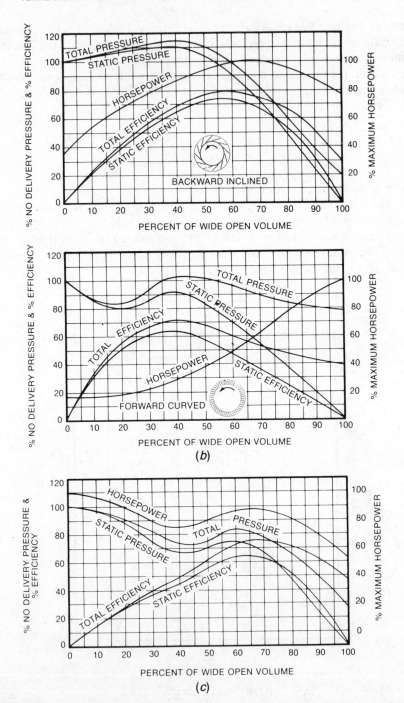

Figure 10.12 (b) Normalized curves for centrifugal fans; (c) normalized curve for an axial flow fan.

Since brake horsepower is directly proportional to total pressure and to airflow (Eq. 10.4), combining Eqs. 10.10 and 10.11 leads to the third basic fan law (i.e., horsepower varies with the third power of fan speed):

$$\frac{H_{b,1}}{H_{b,2}} = \frac{p_{t,1}Q_1}{p_{t,2}Q_2} = \left(\frac{N_1}{N_2}\right)^3 \tag{10.12}$$

These relationships are exact only for the case where the pressure losses through all components of a ventilation system are proportional to the square of the velocity through that component. This is true for most components of typical systems, but the pressure loss through many air-cleaning devices increases more slowly (see Chapter 11). If the pressure loss through an air cleaner is a significant proportion of the total pressure loss in the system, these relationships will have to be modified and a more detailed procedure will have to be followed to determine the new system static pressure and horsepower requirement when fan speed is changed.

Effect of Other Variables. Equations 10.10 through 10.12 describe the behavior of a particular fan when the airflow through that fan is changed. Basic references such as *Fan Engineering* (1983) and the *ASHRAE Handbook* (1985) present many additional fan laws that describe fan behavior when fan size and/or gas density are varied. The reader is referred to these sources for discussion of these laws.

10.2.5 Relationship of Fan Curves to Fan Tables

Graphically, changing the fan speed has the effect of shifting the static pressure curve, as illustrated in Fig. 10.13. For any given fan-motor configuration, there thus exists a "family" of static pressure curves, where each curves corresponds to

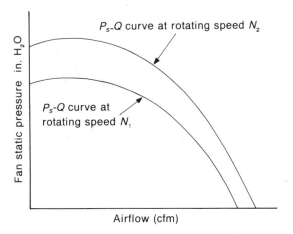

Figure 10.13 Effect of changing fan rotating speed *N* on static pressure–airflow curve. Increasing *N* has the effect of moving the curve upward.

one fan rotating speed; each of these static pressure curves has a set of the other four curves (total pressure, horsepower, total efficiency, and static efficiency) associated with it, so a myriad of curves can be constructed for any given fan. Fan manufacturers have found it easier to present this wealth of data in the form of tables rather than a large set of curves.

A typical fan rating table for a backward-curved blade centrifugal fan is shown in Table 10.1. Each row of the table corresponds to one airflow, and each pair of two columns corresponds to one fan static pressure. The entries in the table show the fan rotating speed necessary to obtain each combination of Q and P_s, and the resulting brake horsepower required to drive the fan. If entries from the table are selected for a constant rotating speed, a fan curve similar to those described above can be constructed. To illustrate this procedure, the entries closest to 600 and 800 rpm have been underlined in Table 10.1 and brought together in the first three columns of Table 10.2. Equation 10.7 can be used to calculate static

Table 10.1   A Section From Typical Fan Rating Table

CFM	Outlet Velocity	1″ S.P.		2″ S.P.		3″ S.P.		4″ S.P.	
		rpm	hp	rpm	hp	rpm	hp	rpm	hp
4000	867	273	0.88						
5000	1084	280	1.09	383	2.24	466	3.48		
6000	1301	291	1.33	388	2.63	469	4.04	539	5.53
7000	1518	302	1.60	396	3.05	473	4.62	541	6.27
8000	1735	315	1.91	406	3.52	480	5.23	546	7.04
9000	1952	327	2.26	417	4.03	489	5.89	553	7.85
10000	2169	340	2.65	429	4.59	500	6.61	561	8.71
11000	2386	353	3.10	441	5.21	511	7.38	571	9.64
12000	2603	368	3.60	454	5.88	522	8.22	582	10.6
13000	2819	383	4.18	466	6.61	534	9.12	593	11.7
14000	3036	398	4.83	479	7.39	547	10.1	605	12.8
15000	3253	415	5.55	492	8.25	560	11.1	617	14.0
16000	3470	431	6.36	506	9.19	572	12.2	630	15.3
17000	3687	448	7.25	521	10.2	585	13.4	642	16.7
18000	3904	465	8.24	536	11.4	598	14.7	655	18.1
19000	4121	483	9.32	551	12.6	612	16.0	668	19.6
20000	4338	501	10.5	567	13.9	626	17.5	680	21.2
21000	4555	519	11.8	583	15.4	640	19.1	693	22.9
22000	4772	537	13.2	600	16.9	655	20.8	707	24.8
23000	4989	556	14.7	616	18.6	671	22.6	721	26.7
24000	5206	575	16.4	633	20.4	687	24.6	736	28.8
25000	5422	593	18.2	650	22.4	703	26.7	751	31.1
26000	5639			668	24.5	719	28.9	766	33.4
27000	5856			686	26.7	735	31.3	782	36.0
28000	6073			703	29.0	752	33.8	797	38.6
29000	6290			721	31.5	769	36.5	813	41.4
30000	6507			740	34.2	786	39.3	830	44.4
31000	6724			758	37.1	803	42.3	846	47.6
32000	6941			777	40.1	821	45.4	863	50.9

Table 10.2 Calculation of Fan Static Efficiencies for the Typical Fan of Table 10.1[a]

P_s (in. H_2O)		Q (cfm)	H_B (hp)	η_s (%)
At 600 rpm				
	1	25,000	18.2	2
	2	22,000	16.9	41
	3	18,000	14.7	58
	4	14,000	12.8	69
	5	7,000	8.0	69
At 800 rpm				
	3	31,000	42.3	35
	4	28,000	38.6	46
	5	25,000	35.6	55
	6	22,000	33.2	63
	7	19,000	30.7	68
	8	15,000	26.2	72
	9	9,000	18.7	68

[a]Data in the first three columns were taken from Table 10.1 at constant fan rotating speeds of 600 and 800 rpm; fan static efficiencies were calculated using these values and Eq. 10.7.

efficiency for each set of conditions in Table 10.2; these values are shown in column 4.

The values in Table 10.2 can now be used to construct fan curves (Fig. 10.14) by plotting fan static pressure (Fig. 10.14a), brake horsepower (Fig. 10.14b), and static efficiency (Fig. 10.14c) for the two rotating speeds. The effect of changing rotating speeds is clearly evident. Note that these curves constructed from an actual fan table resemble the generic fan curves for a backward-curved blade fan shown in Fig. 10.12. The static efficiency values are not normally included in a fan rating table; rather, the point of maximum efficiency for each value of static pressure is indicated by underlining or shading the appropriate value in the table. Thus in our example, according to Table 10.1, the point of maximum efficiency when operating at 4 in. H_2O fan static pressure is attained at a rotating speed of 553 rpm and an airflow of 9000 cfm; at 5 in. H_2O the maximum efficiency is attained at 626 rpm and 11,000 cfm. The maximum efficiency at 600 rpm should thus fall between 9000 and 11,000 cfm, as confirmed by Fig. 10.14c.

10.3 USING FANS IN VENTILATION SYSTEMS

A fan is always used as part of a ventilation system (in some simple dilution ventilation installations it *is* the ventilation system). As such, an understanding of the interaction of the fan and the rest of the system is a crucial part of the system

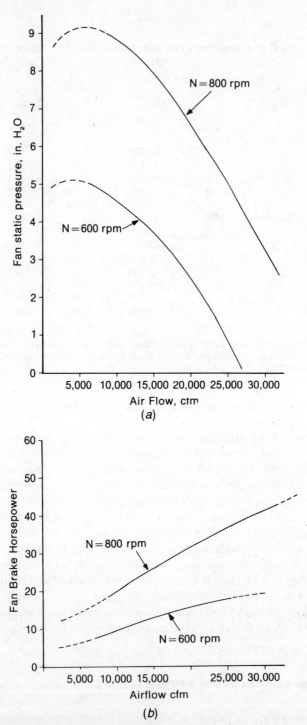

Figure 10.14 Performance curves for the fan of Table 10.1: (*a*) fan static pressure versus airflow; (*b*) brake horsepower versus airflow.

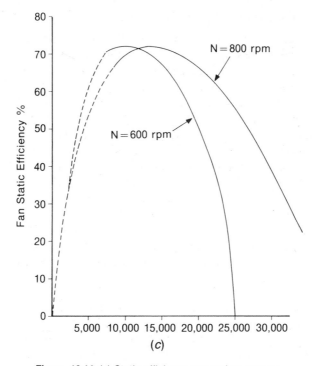

Figure 10.14 (*c*) Static efficiency versus horsepower.

design. The proper fan type, size, and rotating speed must be selected to produce the desired system airflow.

10.3.1 General Exhaust Ventilation Systems

The fan usually used for general exhaust ventilation is the propeller-type axial fan. This fan is designed to move large volumes of air against low resistance, which is what normally is required for general ventilation systems. The selection and installation of general ventilation fans is relatively simple and straightforward. Nonetheless, many such systems operate poorly, for a variety of reasons: the wrong size fan is selected, the fan is installed in the wrong location, insufficient replacement air is provided, and so on.

A properly designed dilution ventilation system will operate at static pressures very close to zero, so that the propeller fan can deliver its maximum airflow. The most common design flaw in such systems, however, is the absence of a proper replacement air system; if an amount of air equal to the exhaust airflow is not provided mechanically to the room being exhausted by a dilution ventilation system, the room will be "starved" for air and the exhaust fan will create a condition of negative static pressure in the room. The design of replacement air systems will be discussed in Chapter 12.

10.3.2   Local Exhaust Ventilation Systems

The proper selection of a fan to be used with a LEV system involves the matching of the fan characteristics to the exhaust system characteristics to obtain the desired airflow. In this regard the exhaust system can be looked on as a fixed resistance to airflow; the amount of air actually flowing through this resistance network depends on the characteristics of the exhaust fan being used. This might be made clearer by use of an electrical analogy (Fig. 10.15). The application of a voltage (V) to a network of resistors results in current flow through each of the network branches. The current through each individual branch (I_1, I_2, etc.) depends on the relative magnitude of each resistance, while the total current (I) depends on the overall resistance of the network. Similarly, the branches of a ventilation system can be thought of as representing a network of resistances to airflow. The pressure difference created by the fan is analogous to the voltage in an electrical circuit; the magnitude of this pressure difference, together with the overall resistance of the ventilation system, determines the total airflow (Q). The airflow in each branch is determined by the relative resistance to airflow presented by that branch.

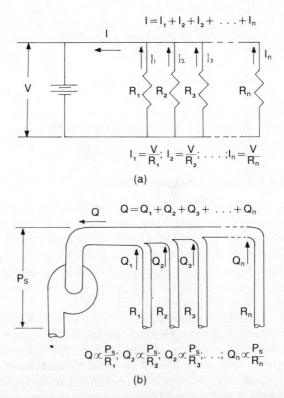

Figure 10.15   Analogy between an electrical network (*a*) and a local exhaust ventilation system (*b*). Voltage is analogous to fan static pressure, electrical resistance to airflow resistance, and current to airflow.

This analogy may be useful to a point, but it is not perfect. In an electrical system it is usually possible to provide a constant voltage, but an exhaust fan develops a variable amount of pressure, as described by its fan curve. The first step in determining the fan's actual operating point is to determine the resistance characteristics of the exhaust system.

Exhaust System Resistance Curve. Chapters 2, 8, and 9 developed the concept that the resistance to airflow of most elements in an exhaust system is proportional to the *square* of the average air velocity through that element. This is true for hoods, ducts, elbows, and miscellaneous other fittings. The principal exception to this rule of thumb is that of air-cleaning devices which might be used in the ventilation system; many of these have pressure drops that vary linearly with velocity and/or vary significantly over time for a constant velocity.

Ignoring air-cleaning devices for the moment, let us consider the airflow resistance of a typical exhaust system, the electroplating shop system designed in Chapters 6 and 9. As detailed in Example 9.2, the electroplating system requires an airflow of 11,230 cfm with a system static pressure of 1.71 in. H_2O. The velocity pressure at the inlet to the fan is 0.43 in. H_2O, so the fan static pressure is

$$P_g = P_g - p_{v,i}$$
$$= 1.71 \text{ in. } H_2O - 0.43 \text{ in. } H_2O \qquad (10.2)$$
$$= 1.28 \text{ in. } H_2O$$

This system has the airflow–resistance relationship shown in Fig. 10.16. This so-called system resistance curve shows the second-power relationship between airflow and static pressure supplied by the fan; that is, to double the air from the design value of 11,230 cfm to 22,460 cfm, the fan static pressure would have to increase fourfold, from 1.28 in. H_2O to 5.12 in. H_2O.

Matching the Fan to the System. The system resistance curve specifies the airflow that will pass through this system for any value of static pressure applied by the fan; what remains in the design procedure is to select the fan that will supply the desired fan static pressure (1.28 in. H_2O) to give the desired airflow (11,230 cfm). The data in Table 10.1 suggest that the fan represented by this fan table might have performance in the proper range for this application; at 11,000 cfm between 1 and 2 in. H_2O the fan is operating near its maximum efficiency, as indicated by the underlined rpm values. The problem is to select the fan rotating speed that will supply 11,230 cfm at 1.28 in. H_2O.

To give an indication of what might happen, the system resistance curve (Fig. 10.16) and the fan static pressure curve at $N = 600$ rpm (Fig. 10.14) can be plotted on the same graph; the results are shown in Fig. 10.17. The fan curve describes the airflow that will be delivered by the fan at different static pressures required to overcome resistance to airflow, while the system curve describes the static pressure needed to overcome the resistance of this *particular* system at different

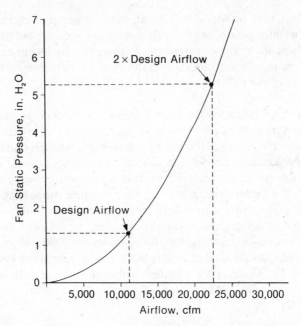

Figure 10.16 Static pressure required to induce different airflows for the electroplating local exhaust ventilation system designed in Chapter 6.

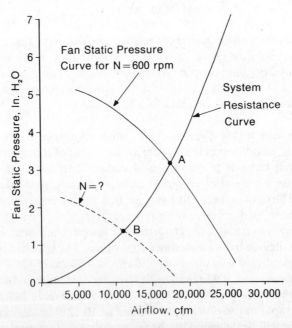

Figure 10.17 Fan static pressure curve at $N = 600$ rpm from Fig. 10.14 and system resistance curve from Fig. 10.16 plotted together. Note point of intersection (A) and desired operating point (B).

airflows. It is evident that the curves intersect at one point, which is an airflow of 17,500 cfm and a fan static pressure of 3.2 in. H_2O. This is the only point common to the fan and to the exhaust system operating characteristics, so that if *this* fan is installed in *this* system of ducts and operated at *this* speed (600 rpm), 17,500 cfm of air will be exhausted and a fan static pressure of 3.2 in. H_2O will be developed.

This is considerably more airflow than is desired for the electroplating system. Recall that there is a *family* of fan static pressure curves, one for each rotating speed. Two of these curves ($N = 600$ rpm and $N = 800$ rpm) are given for the fan in Fig. 10.14a. It should be evident that *one* of these curves, and *one only*, passes through the desired operating point, B (see the dashed line in Fig. 10.17). A key part of the fan selection procedure is to choose the proper fan rotating speed to give the desired intersection point with the system resistance curve. This procedure is described in the next section.

If the proper fan curve is selected, 11,230 cfm of air will indeed flow at a fan static pressure of 1.28 in. H_2O. We have thus met the design requirements for the electroplating operation, and if the hoods and ducts have been designed properly and the proper airflow has been selected, the contaminants given off by the various electroplating tanks should be captured by the exhaust system.

10.4 FAN SELECTION PROCEDURE

Once a desired fan operating point has been selected, it is necessary to select the proper fan which, when inserted in the ventilation system, will furnish the desired combination of fan static pressure and airflow. The process to be followed when using a family of fan curves was described in Section 10.3; one simply selects the fan curve that intersects the system resistance curve at the desired operating point. What is usually available to the designer is not fan curves but a series of fan rating tables similar to that shown in Table 10.1. In this section we describe the procedure to be followed in selecting a fan from such a set of rating tables.

The first step is to select the *type* of fan to be used. In general, dilution ventilation systems will use axial flow fans, LEV systems moving particle-free air through the fan will utilize a backward-curved blade centrifugal fan, and LEV systems that require the fan to move particle-laden air will require a radial-blade centrifugal fan. Once the fan type is selected, it is necessary to consult fan catalogs which contain rating tables for that fan type.

The electroplating ventilation system will once again be used as an example of the procedure to be followed. This system moves particle-free air through a system of ducts, so a backward-curved centrifugal fan should be used. The fan data in Table 10.1 is taken from a catalog of such tables for a family of geometrically similar fans of a range of sizes. Information is given at the top of the table about the dimensions of the particular fan being described; the fans are named according to the diameter of the fan inlet, which in our example in 29 in.

(thus, fan size 429A). This particular series of fans ranges in size from 11-in. inlet diameter to 37-in. inlet diameter.

The first step in choosing a fan from this series is to attempt to match the fan inlet diameter to the diameter of the duct that will enter the fan, as determined by the system design procedure. The diameter of the final duct run for the electroplating design is 28 in. There is no fan in this series with an inlet diameter of 28 in., however; the closest values are 26 in. and 29 in. Since 29 in. is the closest to the desired value, it will be tried first.

Changing the final duct diameter has no effect on system airflow and negligible effect on system static pressure, but the change in velocity will affect the fan static pressure because of the change in inlet velocity pressure. The area of a 29-in. duct is 4.587 ft^2; the inlet velocity is thus

$$V_i = \frac{Q}{A} \tag{2.5}$$

$$= \frac{11{,}230 \text{ ft}^3/\text{min}}{4.587 \text{ ft}^2}$$

$$= 2450 \text{ ft/min}$$

From Eq. 2.14, the corresponding inlet velocity pressure is

$$p_{v,t} = \left(\frac{V_i}{4000}\right)^2 \tag{2.14}$$

$$= 0.38 \text{ in. H}_2\text{O}$$

Equation 10.2 can now be used to calculate the fan static pressure with a 29-in.-diameter inlet:

$$P_s = p_s - p_{v,i} \tag{10.2}$$

$$= 1.71 \text{ in. H}_2\text{O} - 0.38 \text{ in. H}_2\text{O}$$

$$= 1.33 \text{ in. H}_2\text{O}$$

Thus the fan static pressure (1.33 in. H$_2$O) has changed slightly from the value calculated during the system design procedure (1.28 in. H$_2$O).

Once a possible fan size is selected based on inlet diameter, the fan table must be checked to see if the desired combination of airflow and fan static pressure can be attained with this particular fan. Recall that an airflow of 11,230 cfm at a fan static pressure of 1.33 in. H$_2$O is required. In checking the fan table as reproduced in Table 10.1, these values are found to fall between the four entries enclosed in the square in the table; since some calculations will be performed on these numbers, they are reproduced in Table 10.3a. Table 10.3b shows the generalized nomenclature used in the equations that follow.

A quick examination of Table 10.3 reveals that the required airflow is somewhere between the tabular values of 11,000 and 12,000 cfm, and the required fan static pressure is between the tabular values of 1 and 2 in. H_2O; the required fan rotating speed must thus be somewhere between 353 rpm, which would deliver 11,000 cfm at 1 in. H_2O, and 454 rpm, which would deliver 12,000 cfm at 2 in. H_2O. The proper fan operating speed is determined by linear interpolation between the tabular values.

Interpolation must be made between both the airflow values and the static pressure values. The resulting equation for fan operating speed is

$$N = N_{11} + \frac{P_{s,r} - P_{s,1}}{P_{s,2} - P_{s,1}} (N_{12} - N_{11}) + \frac{Q_r - Q_1}{Q_2 - Q_1} (N_{21} - N_{11}) \qquad (10.13)$$

Similarly, the equation for required brake horsepower is

$$H_b = H_{11} + \frac{P_{s,r} - P_{s,1}}{P_{s,2} - P_{s,1}} (H_{12} - H_{11}) + \frac{Q_r - Q_1}{Q_2 - Q_1} (H_{21} - H_{11}) \qquad (10.14)$$

In this example,

$$N = 353 + \frac{1.33 - 1.0}{2.0 - 1.0} (441 - 353) + \frac{11{,}230 - 11{,}000}{12{,}000 - 11{,}000} (368 - 353)$$

$$= 353 + 0.33(88) + 0.23(15)$$

$$= 353 + 29 + 4$$

$$= 386 \text{ rpm}$$

Table 10.3   Fan Interpolation from Rating Tables

(a) Example problem

Air flow (cfm)	Fan Static Pressure				
	1.0 in. H_2O			2.0 in. H_2O	
	Rotating Speed (rpm)	Horsepower		Rotating Speed (rpm)	Horsepower
11,000	353	3.10		441	5.21
12,000	368	3.60		454	5.88

(b) General nomenclature

Air flow	Fan Static Pressure			
	$P_{s,1}$		$P_{s,2}$	
	Rotating Speed	Horsepower	Rotating Speed	Horsepower
Q_1	N_{11}	H_{11}	N_{12}	H_{12}
Q_2	N_{21}	H_{21}	N_{22}	H_{22}

and

$$H_b = 3.10 + 0.33\,(5.21 - 3.10) + 0.24\,(3.60 - 3.10)$$

$$= 3.10 + 0.70 + 0.12$$

$$= 3.92\ \text{hp}$$

Finally, the static efficiency at the operating point can be calculated:

$$\eta_s = \frac{p_s Q}{6343\,H_b}$$

$$= \frac{(1.71)(11{,}240)}{(6343)(3.92)} \tag{10.9}$$

$$= 0.77$$

The fan is thus operating at or near the point of maximum efficiency for this fan type (see Table 10.2 and Fig. 10.14*a*), so this size is an excellent choice for this application. If the efficiency had been lower, the next size fan in the series could have been tried, and a new operating point (rotating speed and horsepower) and efficiency calculated. The fan size that gives the maximum efficiency should be chosen for use.

LIST OF SYMBOLS

A	area
H	power
H_a	air horsepower
H_b	brake horsepower
H_m	motor horsepower
N	fan rotating speed
P_s	fan static pressure
$P_{s,r}$	required fan static pressure
P_t	fan total pressure
p_s	system static pressure
$p_{s,i}$	static pressure at fan inlet
$p_{s,o}$	static pressure at fan outlet
$p_{v,i}$	velocity pressure at fan inlet
Q	airflow
Q_r	required airflow
V_i	velocity at fan inlet
W	work
Δp	pressure drop

η_m electrical-to-mechanical energy conversion efficiency
η_s fan static efficiency
η_T fan total efficiency

REFERENCES

R. Jorgensen, ed. (1983), *Fan Engineering*, 8th ed., Buffalo Forge Company, Buffalo, NY.

American Society of Heating, Refrigerating and Air Conditioning Engineers (1985), *ASHRAE Handbook—1985 Equipment*, ASHRAE, Atlanta, GA.

11

Air-Cleaning Devices

This chapter is not intended to be a comprehensive survey of air-cleaning devices used in conjunction with industrial ventilation systems. Such an effort would take a full volume in itself, and several such texts are available (Licht, 1980; Stern, 1977, 1986; Strauss, 1975; Theodore and Buonicore, 1982). Rather, in this chapter we cover the aspects of air-cleaning device selection and design that are important considerations during the design of local exhaust ventilation systems.

The ventilation system designer must know what *type* of air cleaning device will be used with the system, its *dimensions*, the *configuration* of the inlets and outlets, and the *pressure drop* across the device. In addition, the designer must know if the device changes the psychrometric state point of the airstream in any significant way (e.g., by cooling it, thereby reducing the airflow). Each of these considerations is discussed in this chapter.

The importance of including an air-cleaning device as an integral part of the original ventilation system design cannot be overemphasized. Since a local exhaust ventilation system is being designed to collect a contaminant from the workplace, and this contaminant presumably presents some danger to the workers, it is usually not acceptable just to discharge the contaminant into the general environment. Not only does this trade an outdoor pollution problem for an indoor one, it also creates the potential for contaminant reentry into the workplace.

The air-cleaning device should always be included in the original ventilation system design rather than added on later. Certain characteristics, such as the pressure drop across the device, its size and location, and its power requirements, must be taken into account during the ventilation system design phase.

11.1   CATEGORIES OF AIR-CLEANING DEVICES

The usual method of categorizing air cleaners is by type of contaminant being collected. Particles and gases or vapors require completely different collection mechanisms, and the devices used reflect this difference.

11.1.1   Particle Removers

The principal particle collection methods are gravity settling, centrifugation, filtration, electrostatic precipitation, and scrubbing. In this section we briefly describe the most common types of equipment incorporating these methods, concentrating on the characteristics that are important in selecting an air cleaner for a particular application.

Gravity Settling Devices.   This type of device is the simplest available and has the most limited application. The simple horizontal flow gravity settling chamber (Fig. 11.1) operates by increasing the cross-sectional area of the duct, thereby reducing the gas velocity and increasing the particle residence time inside the settling chamber. Since airborne particles settle at a constant velocity due to the force of gravity, the device is designed to allow particles sufficient time to fall into the hoppers.

In theory, any particle size could be collected in a settling chamber by making the device sufficiently large to give the smallest particles time to fall out of the gas stream. In practical applications, such devices are limited to the collection of very large particles because of the low settling velocities of small particles. This limitation is illustrated in Table 11.1, which lists theoretical gravitational settling velocities for a range of particle sizes.

It should be apparent that only particles with an aerodynamic diameter*

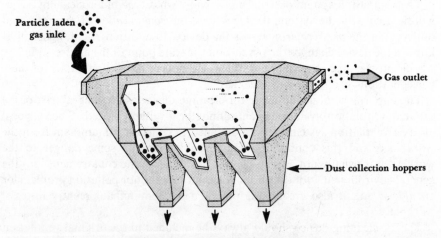

Particle laden gas inlet

Gas outlet

Dust collection hoppers

Figure 11.1   Gravity settling chamber.

---

*Aerodynamic diameter is defined as the diameter of the unit density (density of water, 1 g/cm^3) sphere which settles with the same velocity as the particle under question. Aerodynamic diameter is normally used when describing the properties of air-cleaning devices, since it normalizes for the effects of particle shape and density.

Table 11.1 Terminal Settling Velocities of Particles with Selected Aerodynamic Diameters

Particle Diameter (μm)	Settling Velocity (cm/s)
0.01	0.000007
0.1	0.00008
1	0.004
5	0.08
10	0.3
20	1.2
50	7.5
100	25

larger than about 50 μm will be collected efficiently in such a device. Inhalable and respirable particles are difficult to collect; gravity settlers are thus limited to applications where high concentrations of large particles are generated and present a housecleaning problem which is solved by collecting the particles in a local exhaust ventilation system.

Centrifugal Collectors. These devices, commonly called cyclones, are probably the most common air-cleaning device found in industry. This popularity is due to their simple construction, with no moving parts, which makes them inexpensive to buy and easy to maintain. An offsetting limitation, however, is the large amount of energy required to collect small particles.

Cyclones are similar to gravity collectors in concept, except that they use centrifugal force rather than gravity to separate particles from the airstream. In a conventional cyclone (Fig. 11.2) the air is introduced tangentially into a cylindrical body section and above a conical section leading to a hopper. The gas spins around the cylinder and particles are thrown toward the walls by centrifugal force. Particles that strike the walls fall into the hopper, while those too small to be influenced by the centrifugal force travel with the gas stream into the cone and back up through the outlet.

Since centrifugal forces can be achieved that are many times larger than that due to gravity, these devices can better collect small particles than can gravity settlers. As with gravity settlers, it is theoretically possible to collect particles of any size in cyclones; in practice, excessive energy requirements usually limit their use to particles with an aerodynamic diameter greater than 10 μm.

Cyclones are relatively compact devices and can be integrated into an exhaust system rather easily. Because of the particle-size limitations, they should be used primarily for the collection of nuisance dusts and as preseparators for other more efficient particle removal devices. They can collect both solid and liquid particles, but should not be used for sticky dusts or other materials that would not easily fall down the walls and be collected in the hopper. Cyclones can be made out of almost any material and can thus be used to remove dusts from hot and/or corrosive gas streams.

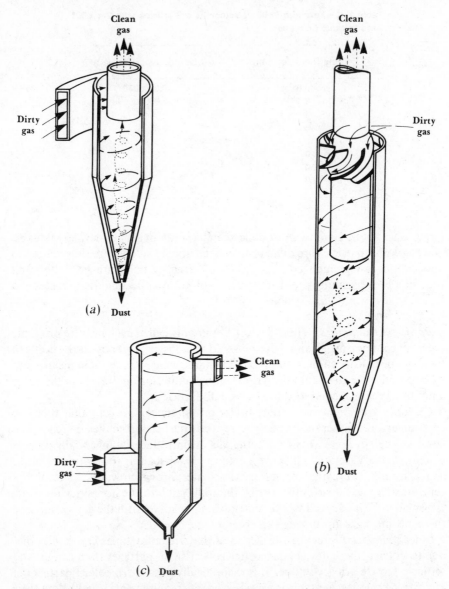

Figure 11.2 Types of cyclones: (*a*) top inlet; (*b*) axial inlet; (*c*) bottom inlet.

Filters. As used in air-cleaning work, "filter" is the general name for three basic types of devices; fibrous and granular ("media") filters, high-efficiency particle air (HEPA) filters, and fabric filters. All three may be used in conjunction with local exhaust ventilation systems, but they have completely different characteristics and applications.

Media Filters These devices are so named because they collect particles by their aerodynamic capture on individual filter elements throughout the depth of the filter medium. The most common media are various types of fibers and granules. Because these filters collect particles throughout the depth of the material, they are generally difficult to clean and reuse. Most fibrous filters, in fact, are designed to be thrown away after they are loaded with dust.

Media filters have two principal uses: as prefilters to remove large particles before a second-stage collector of greater efficiency, and as filters for dusts that cannot easily be collected and cleaned from other collectors. An example of the first use is the common furnace-type filter, which can only collect large particles and is thrown away after becoming loaded with dust. An example of the second type would be a deep-bed fibrous filter used to collect sticky particles.

HEPA Filters   HEPA filters (Fig. 11.3) are used wherever extremely high particle collection efficiency is required. Filters of this type are constructed of a filter paper that is folded to give a very high surface area to the filter. This construction lowers the filtration velocity, leading to the low pressure drop and high collection efficiency characteristic of this filter, but it makes the device impossible to clean. The close spacing between the folds makes the dust-holding capacity of these filters very low. The combination of lack of cleanability with low dust capacity limits the use of these filters to gas streams with very low dust loading. These filters originally were designed to remove radioactive particles from atmospheric air, and should only be used in gas streams where the

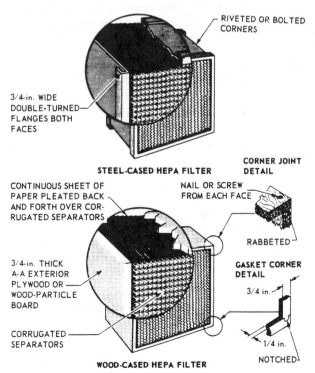

Figure 11.3   High-efficiency particulate air (HEPA) filter construction.

concentration approximates those found in typical outdoor atmospheres (e. g., 50 μg/m^3). Used under these conditions, a HEPA filter should operate at an acceptable pressure drop for at least several years before it must be replaced.

HEPA filters are used extensively wherever ultraclean air is needed, such as in hospital operating rooms and the so-called clean rooms used in electronics and pharmaceutical manufacturing. In these applications the HEPA filters are cleaning the supply air, rather than the exhaust. They are also used where emission levels must be very low, for example as a second-stage collector located downstream from a fabric filter that is collecting lead dust.

HEPA filters are becoming more common as fail-safe devices in exhaust air recirculation systems. Because they have such high inherent collection efficiency, they can serve as second-stage collectors to collect particles if the first stage fails, thus offering protection to the workers (see Chapter 12).

Fabric Filters　Fabric filters (Fig. 11.4) are by far the most common type of filter encountered in the industrial environment and the one most likely to be used in a local exhaust ventilation system. The filter itself consists of pieces of fabric sewn into cylinders or envelopes and mounted in a housing. During operation, exhaust air is drawn through the fabric by the fan; particles either collect in the fabric itself or in a dust cake on the fabric surface and are thus removed from the exiting exhaust stream.

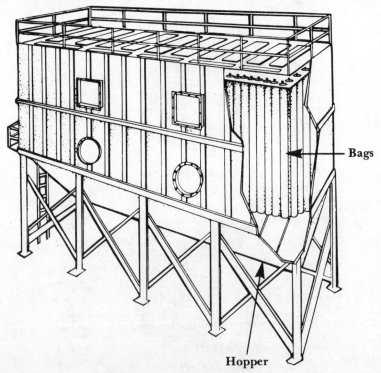

Figure 11.4　Typical pulse-jet baghouse.

As with other filters, fabric filters must eventually be cleaned. In fact, cleaning is so important that fabric filters are classified by the cleaning method used. The three most common cleaning methods are shaking, reverse air flow, and pulse jet air. Each of the three cleaning methods has different applications, which will be described briefly.

Shaking is the oldest and simplest method to clean a fabric filter. The tops of the bags are oscillated in either the horizontal or vertical direction; this motion flexes the fabric and causes the collected dust cake to be dislodged and fall into the hopper. Shaker-cleaned bags are generally constructed of woven fabrics, which collect the dust as a surface cake which is easily dislodged by the fabric flexion.

The shaking mechanism is usually motor driven and can be initiated either automatically (e.g., by sensing excess pressure drop) or manually whenever the operator decides cleaning is necessary. Small units may not employ a motor and are cleaned instead by moving an external lever that oscillates the bags. Exhaust airflow must be shut down before a bag is shaken, to allow the dislodged dust to fall into the hopper. This means that either the process has to be shut down while the baghouse is cleaned, or the baghouse has to be divided into compartments so that one compartment can be isolated and shut down with dampers while the remaining compartments continue filtering the exhaust air. The first alternative is possible for intermittent operations or operations which produce so little dust that the baghouse can be cleaned once at the end of the work shift; most applications, however, will require a compartmentalized baghouse.

In some instances bag shaking is supplemented by a gentle reverse airflow through the fabric; this reverse flow helps to carry the dislodged dust away from the fabric and into the hopper. Reverse flow cleaning requires a system of dampers to shut off the main flow and open the reverse flow; this tends to complicate the baghouse design and operation and increases capital and operating costs. Consequently, reverse air cleaning is usually found only in rather large baghouses filtering a dust that is difficult to remove with shaking alone.

High-temperature filtration requires the use of specialized fabrics, such as fiberglass, which can withstand this severe environment. Woven fiberglass cannot tolerate shaking, so baghouses using these fabrics employ gentle reverse air cleaning without shaking. Such systems usually require bags with special surface treatment to enhance the release of dust.

The last major fabric filter cleaning technique employs blasts of high-pressure air to remove collected dust. A typical system using this method, called pulse-jet cleaning, is shown in Fig. 11.5. A short pulse of high-pressure compressed air is introduced into the top of the bag. This pulse travels down the bag, temporarily reversing flow through the fabric and flexing the bag outward. This sudden flexing dislodges dust from the outer surface of the fabric in the form of agglomerates, which fall toward the hopper. Cleaning pulses can either be applied at regular intervals by incorporating a timer or can be initiated by sensing excess pressure drop across the fabric.

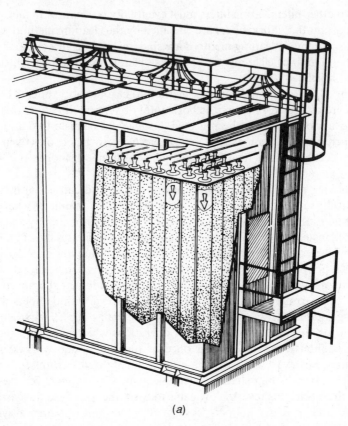

(a)

Figure 11.5 Typical reverse-pulse baghouse during cleaning cycle. (a) general view; (b) close up of pulse travelling down bag

Pulse-jet cleaning requires the use of felted fabrics rather than woven ones, to withstand the violent motion induced by the pulses. Felted fabrics tend to collect dust throughout the depth of the felt rather than at the surface as a cake; this porous dust deposit is more difficult to remove than a surface deposit, so pulse-jet cleaning is less efficient than the other methods and typically must be applied more often.

Pulses can be applied to a bag either *on-line*, where normal airflow is maintained through the system during cleaning, or *off-line*, where a compartment is isolated from the exhaust flow during cleaning. Most pulse-jet systems use on-line cleaning; such systems constitute the simplest type of fabric filter since they require no moving parts, such as isolation dampers or shaking mechanisms. This is a big advantage from the point of view of maintenance, but naturally there are trade-offs involved. On-line cleaning is not as effective at removing dust as off-line cleaning, since the exhaust airflow tends to redeposit dust onto the fabric after the pulse is over. Off-line cleaning allows the removed

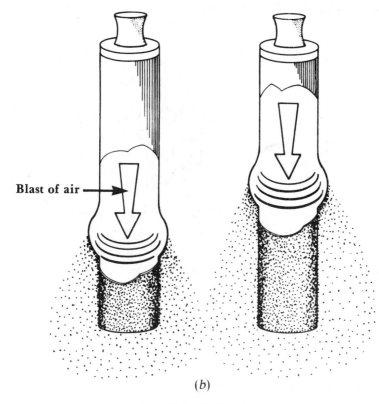

Blast of air —

(b)

Figure 11.5 *Continued.*

dust to fall into the hopper before filtration resumes, but requires the additional complication and expense of isolation dampers.

Electrostatic Precipitators. This class of air cleaner collects particles from an exhaust stream by three distinct steps. Particles are first imparted an electrical charge. Next, the charged particles are attracted toward and collected on grounded plates. Finally, the collected particles are removed from the collecting plates to a hopper. Precipitators are classified as single-stage, where the particle charging and collection occur in the same section of the device, and as two-stage, where the two actions take place in sequential sections.

For economic reasons the application of single-stage precipitators (Fig. 11.6) is limited to large-volume gas streams, such as the exhaust from coal-fired power plants and cement kilns. Such devices will not generally be found on small, multihood local exhaust systems.

Two-stage precipitators (Fig. 11.7), on the other hand, are used for low-volume gas streams with relatively low dust loadings such as atmospheric dust removal in HVAC systems (Chapter 15), smoke and fume removal from general room air, and on single-hood local exhaust systems. The ventilation system designer may encounter suitable applications for such devices from time to time.

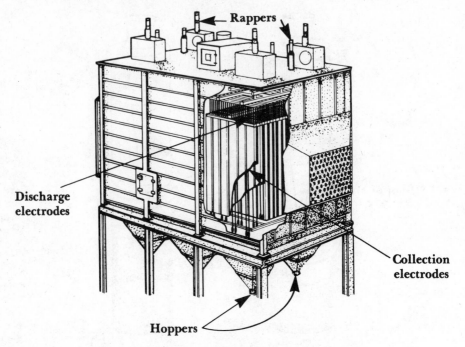

Figure 11.6 Typical single-stage electrostatic precipitators.

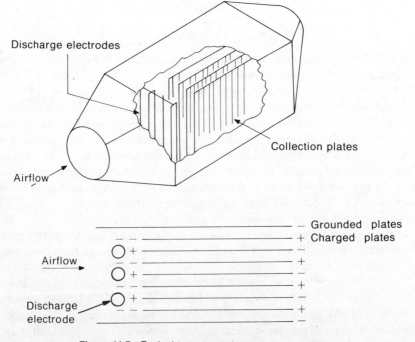

Figure 11.7 Typical two-stage electrostatic precipitator.

Scrubbers. There are several device types that remove particles from a gas stream by contact with a liquid, but the most common is the venturi scrubber. Other scrubbers, such as gravity and centrifugal spray towers, self-induced spray scrubbers, and impingement plate and packed-bed scrubbers, have limited specialized applications for particle collection and are not described here.

The venturi scrubber (Fig. 11.8) is a very simple device, both in construction and operation. Exhaust gases are accelerated to high velocity through the throat of the venturi, where water is introduced. The high-velocity gas shears the water stream into droplets and the high relative velocities cause particles to be collected on the droplets, primarily by impaction. The droplets are subsequently removed from the gas stream in a cyclone or entrainment separator.

Venturi scrubbers have several characteristics that make them useful in particular applications. They can collect particles from high-temperature gas streams, which are then cooled and washed at the same time. This type of scrubber can collect and neutralize corrosive gases and mists, and works well in situations that present a fire or explosion hazard. Venturi scrubbers are very compact because they operate at high gas velocities; they thus do not occupy much space in the plant.

There are inherent disadvantages to venturi scrubbers which limit their application. The primary disadvantage common to all wet devices is the disposal of the waste liquid. The use of liquids creates other problems, such as corrosion, freezing in cold weather, and creation of a high-humidity exhaust gas stream.

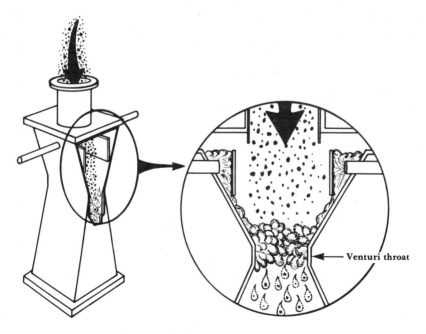

Figure 11.8 Typical venturi scrubber.

Another important disadvantage is the high energy required to collect small particles. Like cyclones, venturi scrubbers trade simple design for difficulty in collecting small particles; both devices require a very high velocity to collect these particles, which entails a high-pressure drop. Scrubbers require additional energy to accelerate the water droplets from rest to the gas velocity. This energy requirement easily outstrips that needed to accelerate the gas. The required pressure drop to operate a scrubber can exceed 60 in. H_2O if small particles are to be collected.

The disadvantages inherent in venturi scrubbers limit their use to applications where other device types have difficulty performing satisfactorily. These include the collection of acid mists and other corrosive liquid aerosols, high-temperature fumes such as those from incineration and foundry and steel furnaces, and sticky, viscous particles such as those found in asphalt production.

11.1.2 Gas and Vapor Removers

The selection of a method to remove gases and vapors from an exhaust stream is somewhat simpler than for particles. There are only three basic methods: absorption in a liquid, adsorption on a solid, and chemical reaction (including incineration) to change a harmful gas or vapor to one less so. With some exceptions for the third, each of these processes involves the diffusion of the gaseous molecules to a surface, where the various removal mechanisms take place. This dependence on a single collection mechanism is in marked contrast to the situation for particles, where a myriad of collection mechanisms operate to remove particles to surfaces. The selection and design of gas and vapor removers should thus present fewer problems than for particle removers.

Absorbers. Absorption involves the diffusion of gaseous molecules to the surface of a liquid, where they are removed by further diffusion into the bulk of the liquid. To be efficient, close contact must be made between the exhaust gas and the liquid. There are two techniques used in the design of practical absorbing devices to ensure such close contact: either the gas is passed through a continuous stream of the liquid or the liquid is dispersed as drops through the gas.

The two most common methods for passing gas through the liquid are plate towers and packed towers (Fig. 11.9). The plates in the former are perforated to allow liquid and gas to flow. The packing in the latter may take the form of simple broken rock or gravel, rings, saddles, or low-density material such as fibers or expanded metal. Complex shapes such as saddles are more expensive but spread the liquid over a larger effective surface area, thereby increasing collection efficiency for a given tower size, but entail additional capital cost over simpler materials.

The diameter of a tower is selected to give the desired gas velocity through the packing material. If the velocity is slowly increased a point will be reached where the gas has sufficient velocity to hold up the liquid and flooding will occur.

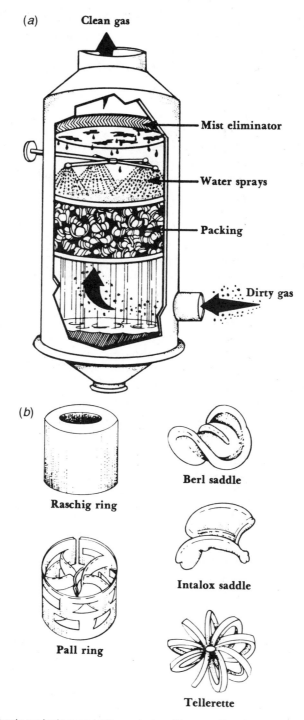

(a)

Clean gas

Mist eliminator

Water sprays

Packing

Dirty gas

(b)

Raschig ring

Berl saddle

Pall ring

Intalox saddle

Tellerette

Figure 11.9 Simple packed tower: (a) tower design; (b) a sampling of typical packing materials.

Scrubbers are usually designed to operate at about 50 to 60% of the flooding velocity; this value and the total gas flow determine the cross-sectional area of the tower. The height of the tower is determined by the inherent efficiency of the device ("height of a transfer unit") and the overall efficiency desired ("number of transfer units").

Adsorbers. Adsorption is similar to absorption in that gas molecules are removed from an exhaust stream by diffusing to a surface—in this case, however, the surface is solid rather than liquid. The mechanisms by which molecules actually adhere to a solid surface are complex, making generalizations about the design of such devices difficult. In general, molecules are held to solid surfaces by physical forces, chemical reaction (chemisorption), or some combination of the two.

Adsorption proceeds in three steps. The first step, the diffusion of the gas to the surface, is identical to the diffusion mechanism discussed for absorption. The second step is the penetration of the molecules to be collected into the pores of the adsorbing material. The third step is the actual adsorption of the molecule into the pore. The third step is very rapid compared to the first two, so it is the rate of diffusion to the surface and into the pores that determines the rate of adsorption in any particular case.

Turk (Stern, 1977) summarizes the factors that control the mass of material that can physically be adsorbed by a given mass of adsorbent.

- The concentration of the material in the space around the adsorbent
- The total surface area of the adsorbent
- The total volume of pores in the adsorbent whose diameters are small enough to facilitate condensation of gas being adsorbed
- The temperature
- The presence of other gases which may compete for sites on the adsorbent
- The characteristics of the molecules to be adsorbed (especially their molecular weight, electrical polarity, size, and shape)
- The electrical polarity of the adsorbent surface

Adsorbents are classified as *polar* or *nonpolar*, depending on the distribution of electrical charges on the surface at the molecular level. The only important nonpolar adsorbent is carbon, which presents an array of neutral atoms that create a homogeneous surface with no potential gradients. Carbon is very efficient at attracting nonpolar molecules in preference to polar ones; it is thus used extensively to collect nonpolar organic molecules (e.g., chlorinated hydrocarbons, alcohols). The form of carbon used for adsorption is activated carbon (also called activated charcoal), which consists of particles of carbon with a very large surface area.

Polar adsorbents are oxides, either of silicon (silica, in the form of silica gel, Fuller's earth, or diatomaceous earths) or metals (most commonly aluminum oxide). These adsorbents are used for the removal of polar molecules such as water vapor, ammonia, formaldehyde, sulfur dioxide, and acetone.

A third category of adsorbents are those which combine adsorption with chemical reaction. This is usually accomplished by the impregnation of activated carbon or alumina with a reagent or catalyst. There is a whole range of such reactions which can be used for the removal of specific gases and vapors. Examples include the impregnation of activated carbon with bromine for the removal of ethylene or other olefins by conversion to dibromide and the impregnation of activated alumina with potassium permanganate for the oxidation of formaldehyde.

Chemical Reaction Devices. One example of chemical reaction, chemisorption, was discussed under adsorption. The other common form of chemical reaction for the removal of gaseous constituents is oxidation, either by flame or by thermal oxidation with or without a catalyst. Both techniques are applicable when the gaseous pollutant can be oxidized. The ideal products of such oxidation of the pollutant are water vapor and either carbon dioxide (for organic material) or sulfur dioxide (for organic sulfides).

If the concentration of the contaminant gas in the exhaust stream is between the lower and upper explosive limits, it generally can be oxidized by direct combustion either in a flare or a direct combustion chamber. Such devices have to be carefully designed and operated, because of the inherent danger of explosion.

Oxidizable contaminants present in concentrations below the lower explosive limit can be removed by thermal oxidation in a device called an afterburner. Afterburners can be either *thermal*, where the contaminant is oxidized by heating in a flame, or *catalytic*, where preheated waste gas is oxidized on a catalyst at temperatures considerably lower than would be possible using thermal oxidation alone. The most popular catalyst is a finely divided precious metal, such as platinum or palladium, supported on another material, such as a crimped nickel-alloy ribbon or activated alumina. The catalytic converter used in automobiles is a common example of such a device.

Catalyst materials are expensive, but are usually employed in small amounts so that the additional cost of a catalytic afterburner over the thermal type is usually not excessive. In any case, the energy savings over the life of the device due to the lower oxidation temperature required would far exceed the additional capital cost. Additional savings can be incurred by recovering the waste energy.

11.2 MATCHING THE AIR-CLEANING DEVICE TO THE CONTAMINANT

11.2.1 Introduction

Possibly the most difficult step in solving an air-cleaning problem is the selection of the device type to be used. This is obviously a complex subject, and can only be summarized in this book. The reader is urged to consult texts on air cleaning such as those listed in the References.

Selection of the proper device type depends on a knowledge of certain basic information about both the contaminant to be removed and the carrier airstream. The most important information about the contaminant is, of course, its physical state. If it is a gas or vapor, further information concerning chemical species, temperature, concentration, and so on, will be required. Particle collection will require information on particle state (solid, liquid, sticky solid), size, shape, density, and surface properties, in addition to concentration in the exhaust stream.

The principal properties of interest concerning the exhaust stream itself are temperature, pressure, and total airflow. In addition, secondary properties such as gas composition, moisture content, combustibility, and viscosity are sometimes of importance.

Information about the contaminant and exhaust air can usually be obtained rather easily if the process is already in operation by sampling the exhaust stream. If it is a new process in the design stage, the required information may be available from another identical or similar plant. Information on many basic industrial operations is available in the literature, and these sources should be consulted (U.S. Environmental Protection Agency, 1973, 1978). For completely new processes, or the application of a new device type to an old process, it may be necessary to construct a pilot-scale air cleaner and test it either on a pilot version of the process or on a side-stream from the actual process to be cleaned.

11.2.2 Device Selection

Many generalizations about the suitability of air-cleaning devices for certain applications should be evident from the discussion in Section 11.1. Much of the basic information about collection device characteristics, summarized in Table 11.2 can be used as an aid in device selection.

Recently, there appears to be increased interest in the fractional collection efficiency of various types of control equipment for collecting particles of various sizes. This is particularly important for the control of respirable particles in the industrial environment. Recent experimental data concerning the fractional efficiency of various categories of particle collection equipment are summarized in Fig. 11.10.

Occasionally, two device types will be equally efficient at collecting the contaminant of interest. When this occurs, selection will depend on the relative annualized operating costs of the two devices. The components of this cost are the capital, operating, and maintenance costs associated with the equipment. Many times a system that is cheaper initially may have higher operating costs (i.e., higher pressure drop) or may be more difficult to maintain. All such factors should be considered carefully when selecting an air cleaner.

11.3 INTEGRATING THE AIR CLEANER AND THE VENTILATION SYSTEM

Previous sections of this chapter have discussed the characteristics of air-cleaning devices and the problems in choosing one for a specific application. In this section we address the problem of integrating the selected device into the local exhaust ventilation system.

From the point of view of the ventilation system designer, the two most important aspects of the air cleaner are its physical characteristics and the pressure drop it will require at the design airflow. Information about both of these should be available from the device manufacturer. The physical character-istics of the device must be known, of course, so that the final layout of the ventilation system can be specified. It is important that the designer obtain detailed information concerning the device size, weight, and location and size of the inlet and outlet ducts.

Many times the designer will develop a preliminary layout of the exhaust system before the air cleaner is selected; this preliminary design will usually have to be modified after a particular device is specified. Such modifications may be minor, or the air cleaner might be so large (for example) that the entire system layout has to be changed. For reasons such as these it is generally a good idea to select the specific air cleaner *before* attempting a detailed design of the LEV system.

The question of pressure drop presents even more problems to the ventilation system designer. The operating pressure drop of the air cleaner must, of course, be known in the design stage to determine the required fan static pressure. For many devices it is difficult if not impossible to obtain good information on the expected operating pressure drop for a particular operation.

This problem may not be particularly important if the pressure drop across the device is a small fraction of the total static pressure required by the exhaust system. In many cases, however, the pressure drop across the air cleaner is the *predominant* pressure drop for the entire system. The designer may spend many hours obtaining the required static pressure of the exhaust system, only to have this precise number overshadowed by a much larger air cleaner static pressure which cannot be specified closer than a factor of 2.

Many factors contribute to uncertainties about air cleaner static pressure requirements. For some devices, such as fabric filters, the pressure drop is strongly dependent on the particular dust being filtered and the cleaning method employed. For others, such as venturi scrubbers, the pressure drop is propor-tional to the collection efficiency; in this case, if the scrubber does not provide the desired collection efficiency, the pressure drop may have to be increased above that selected in the design stage.

The best source of pressure drop data is from an identical air cleaner serving an identical process. The second best source of information is from a pilot-scale device on a pilot plant or side-stream of the exhaust being cleaned. The budget may not allow the collection of such data for small local exhaust systems. In these cases it is best to rely on information supplied by the air cleaner manufacturer.

Table 11.2 Collection Device Characteristics

Collection Device Type	Types of Contaminants	Typical Loading	Collection Efficiency	Pressure Drop (in. H_2O)	Initial Cost	Operating Cost	Serviceability (Durability)
			PARTICLES				
INERTIAL COLLECTORS							
Settling chambers, inertial separators, cyclones, dynamic (fan and separator	Crushing, grinding, machining	0.1–100 g/m³	High for $d_p > 10 \mu m$	2–6	Low	Moderate	Good (erosion)
FILTERS							
Industrial cleanable cloth	All dry powders	0.1–20 g/m³	High for $d_p > 0.1 \mu m$	3–6	Moderate	Moderate	
Paper, Deep bed (HEPA)	Precleaned, atm. air	< 0.001 g/m³	High for all sizes	1–6	Moderate	High	Fair to poor
ELECTROSTATIC PRECIPITATORS							
Single stage	Flyash, H_2SO_4	0.1–2 g/m³	High for $d_p > 0.1 \mu m$	0.5–1	High	Low	Fair

Device	Application	Concentration	Efficiency	Pressure drop			
Two-stage	Welding fume, cigarette smoke	< 1 g/m³	High for $d_p > 0.1\ \mu m$	0.5–1	Moderate	Low	Good
WET SCRUBBERS							
Venturi scrubber	Chemical and metallurgical fumes	0.1–100 g/m³	High for $d_p > 0.25\ \mu m$	20–80	Low	High	Good (corrosion)
Wetted baffles, wetted cyclone, wetted dynamic	Crushing, grinding, machining	0.1–100 g/m³	High for $d_p > 2\ \mu m$	2–6	Moderate	Moderate	Good (corrosion)
GASES AND VAPORS							
Packed tower Absorber	Inorganic gases (HCl, HF, SO_2, Cl_2)	ppm to %	90–99%	4–12	Moderate	Moderate	Poor
Carbon bed (Adsorbers)	Organic gases and vapors (odors)	ppb to %	95–99 + %	2–6	High	Moderate	Fair
Direct flame incinerator	Organic gases and vapors (odors)	ppb to %	90–99	0.2	Low	Extremely high	Good
Catalytic converter	Organic gases and vapors (odors)	ppb to %	90–99	2–6	Moderate	High	Poor

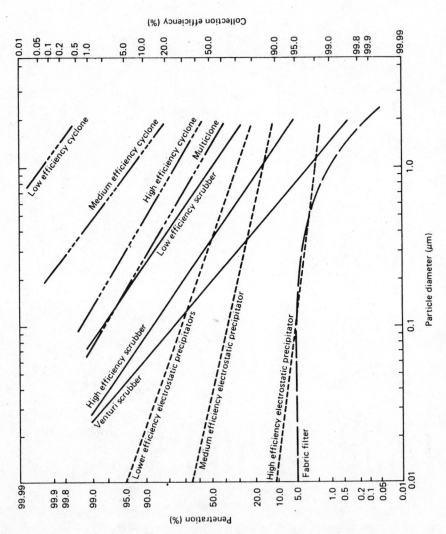

Figure 11.10 Fractional efficiency curves for various air cleaners.

Equations are available that allow the prediction of pressure drop from basic information about the air cleaner and the process being exhausted; such equations are subject to the uncertainties discussed above. They can be useful in the design stage, however, as an aid to device selection.

The remainder of this section contains brief descriptions of the pressure drop characteristics of each basic type of air cleaner. Included are simple equations for predicting pressure drop, and a discussion of any problems peculiar to that device type. A quick review of this material should convince the reader of the difficulty of using general equations to predict operating pressure drop for a specific device and application; the best sources of pressure drop data almost invariably are those mentioned above.

11.3.1 Gravity Settling Devices

As discussed above, these devices are rarely used because of their large size and low collection efficiency for small particles. In any case, the velocity through the chamber itself is very low, so that any pressure drop is due to the inlet and outlet transitions. The *Ventilation Manual* states that a typical value for this pressure drop is 1.5 times the duct velocity pressure.

11.3.2 Centrifugal Collectors

Cyclones have been used widely for many years, and thus a large amount of data have been collected on the pressure drop across such devices. The pressure drop is generally independent of dust type or loading, so it is possible to characterize a cyclone once and use the same data for many applications. Similarly, families of cyclones for different airflows are designed by scaling up dimensions, so pressure drop obtained for one airflow can be used to predict pressure drop at other flows.

Cyclone pressure drops are difficult to predict from theory, so it is best to rely on manufacturer's specifications. Most cyclones are designed to operate at pressure drops in the range 1 to 5 in. H_2O. Pressure drop has been found to vary with the square of the inlet velocity.

11.3.3 Filters

All filters have a pressure drop that varies over time; the nature of this variation depends on the filter design. The pressure drop is lowest when the filter is new. As dust is collected, the pressure drop increases from this baseline value due to the increased resistance caused by the collected dust. This increase in pressure drop cannot continue indefinitely. The system resistance curve becomes steeper, resulting in a new intersection point with the fan curve (Fig. 11.11). The fan and system are now operating at a lower airflow, which may cause dust to start settling out in the ductwork. If this were to continue indefinitely, system airflow would approach zero, so it is necessary for the air cleaner designer to specify a

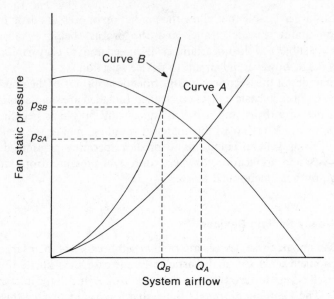

Figure 11.11 Variation of a system resistance curve with change in fabric filter dust loading. Curve *A* represents low filter loading, while curve *B* represents a high-loading condition. Note the different intersection points with the fan curve, resulting in different system airflows and pressure drops.

maximum allowable pressure drop. When pressure drop reaches this value, the filter must either be cleaned or discarded.

As the filter collects dust, the system resistance curve gradually becomes steeper, moving from curve *A* in Fig. 11.11 toward curve *B*. The clean filter and maximum allowable pressure drops define the limiting points of intersection of the system resistance curve and the fan curve. In selecting a pressure drop to assign to the air cleaner, the ventilation system designer should always assume the maximum pressure drop, since this gives the steepest resistance curve. The proper fan can then be selected to give the required airflow (Q_B); when the filter is clean, more air will flow than is necessary (Q_A), thus assuring proper hood operation at all times.

Media Filters. In general, media filters are not cleanable and are changed after a maximum pressure drop, specified by the filter manufacturer, has been attained. The pressure drop across the filter when it is new and at the end of its service life define the limiting values of the system resistance curve.

Media filters vary widely in their design pressure drop, so it is necessary to refer to manufacturers' specifications. Pressure drops across furnace-type filters are generally a few tenths of an inch of water, while large high-velocity filters designed for heavy industrial use can operate at 30 to 40 in. H_2O.

HEPA Filters. Like media filters, high-efficiency filters cannot be cleaned and are discarded after attaining maximum pressure drop. The pressure drop across such devices when new is generally 0.5 to 1 in. H_2O; since such devices are used only in applications with very low dust loadings, the pressure drop should increase very slowly toward the maximum design value (usually 2 in. H_2O). The ventilation designer would do well to choose this maximum value during the design stage, and accept a lower-than-anticipated system pressure drop and higher airflow during early months of operation.

It is important that the pressure drop across all noncleanable filters be monitored continuously during system operation, and that such filters be replaced when maximum allowable pressure drop is attained. If such filters are allowed to operate beyond this point, system resistance will increase to an unacceptable level and airflow will decrease below the amount necessary for efficient hood operation.

Fabric Filters. It is very difficult to predict pressure drop for fabric filters. Several problems are encountered with these devices. The pressure drop is strongly dependent on system variables such as filtration velocity, fabric type, dust type, and cleaning mode. In addition, the pressure drop always *changes* over time, both over the short term during one cleaning cycle and over the long term as the fabric ages and becomes more difficult to clean.

A typical time–pressure drop plot for a fabric filter might look as shown in Fig. 11.12. Given this type of behavior, how is the designer to select one value of

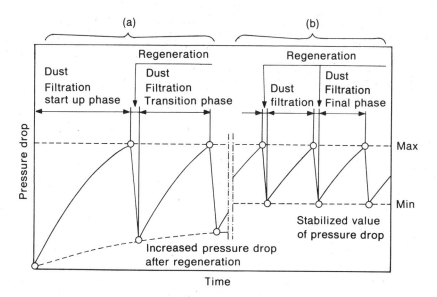

Figure 11.12 Pressure drop versus time for a typical fabric filter. Zone A represents the filter's transient behavior during the startup phase, and zone B represents normal operation after pressure drop has stabilized.

pressure drop to insert in the calculation sheet as the pressure drop across the filter? There are several steps that can be taken to simplify matters. First, the initial transient behavior can be ignored, since it is the long-term steady-state pressure drop that is of interest. One value of pressure drop must be selected which characterizes the behavior of the device in region B of Fig. 11.12, where the pressure drop cycles between two values during the course of one cleaning cycle. The solution here is to select the *maximum* pressure drop for the design, and ensure that a fan is selected that can handle the range of pressure drops expected.

Some simple equations are commonly used to predict maximum pressure drop in a fabric filter. A basic expression for the pressure drop is

$$\Delta p = \Delta p_f + \Delta p_d \tag{11.1}$$

$$= K_1 V + K_2 Vw \tag{11.2}$$

where Δp = pressure drop (in. H_2O)
$\quad \Delta p_f$ = pressure drop across the fabric (in. H_2O)
$\quad \Delta p_d$ = pressure drop across the dust (in. H_2O)
$\quad K_1$ = clean fabric resistance (in. H_2O/fpm)
$\quad V$ = filtration velocity (fpm)
$\quad K_2$ = dust deposit specific resistance (in. H_2O/fpm-lb/ft^2)
$\quad w$ = dust deposit areal density (lb/ft^2)

The difficulty in using this equation is determining values for K_1, K_2, and w for the particular application. The clean fabric resistance K_1 is a function of the fabric used in the filter and is available from the manufacturer. The dust deposit specific resistance K_2 is a complicated function of the dust type, gas viscosity, and dust–fabric interaction and is generally obtainable only from measurements at similar installations or a pilot plant. Similarly, the amount of dust on the fabric per unit area just before cleaning, w, is a complicated function of dust type, fabric type, and cleaning mechanism, and cannot generally be predicted theoretically.

Thus the simple equations given above, while useful in helping to understand what contributes to pressure drop in a fabric filter, are not particularly useful to the ventilation system designer for predicting pressure drop for a particular application. The usual procedure that is followed is to use the manufacturer's data to select a design operating pressure drop, and then adjust the cleaning frequency to attain this pressure drop once the system is operating.

11.3.4   Electrostatic Precipitators

Electrostatic precipitators are among the easiest devices to predict operating pressure drop. From the point of view of the ventilation system designer, electrostatic precipitators are simply large boxes and the pressure drop is low, easy to predict, and will not change significantly during the lifetime of the device. Unfortunately, as was discussed above, these devices are rarely used in local exhaust systems, and thus the designer will not often have the luxury of designing them into a system. Large single-stage precipitators generally operate at a pressure drop of 0.5 to 1.0 in. H_2O, while that across small two-stage precipitators is even less.

11.3.5 Scrubbers

As described above, there is such a wide variety of wet scrubber designs that is it difficult to classify them systematically. One useful approach is to categorize

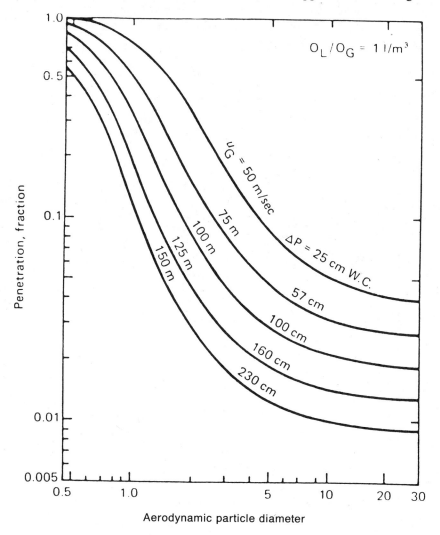

Figure 11.13 Particle penetration versus aerodynamic diameter for a typical venturi scrubber. Penetration curves are presented for a range of velocities in the venturi throat (V_G) at a typical liquid-to-gas flow ratio.

them by pressure drop (Theodore and Buonicore, 1982). Low-energy scrubbers include all devices with pressure drops less than 5 in. H_2O; this would include such devices as simple spray towers. Medium-energy scrubbers have pressure drops between 5 and 15 in. H_2O; most packed bed, impingement plate and wet centrifugal fan scrubbers fall into this category. High-energy scrubbers, then, have pressure drops greater than 15 in. H_2O; the most common such device is the venturi scrubber.

Since these devices operate at a relatively constant pressure drop throughout their lifetimes, and since the desired operating pressure drop is well defined by the device manufacturer, the ventilation system designer's task is relatively straightforward. The major exception to this statement is the venturi scrubber,

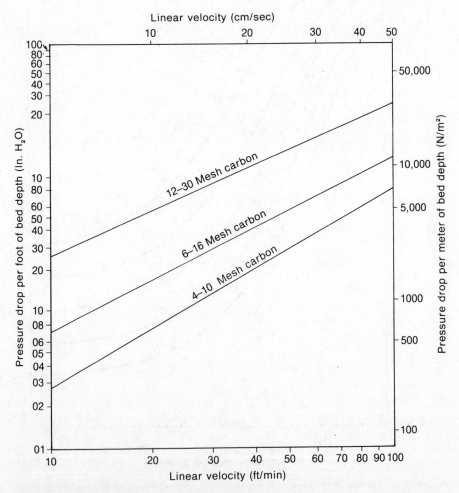

Figure 11.14 Pressure drop versus gas velocity for granular carbon beds with different mesh size granules.

which can operate over a wide range of pressure drops, depending on the dust being collected and the collection efficiency desired.

The pressure drop across a venturi scrubber is primarily a function of the gas velocity in the venturi throat V_G and the ratio of the liquid flow to airflow, Q_L/Q_G. Pressure drop equations are usually of the form

$$p = KV_G^2 \frac{Q_L}{Q_G} \tag{11.3}$$

where K is a constant for a particular scrubber design. Operating pressure drop is selected to give the desired particle collection efficiency. Typical data illustrating the variation of collection efficiency with particle size and pressure drop are presented in Fig. 11.13.

11.3.6   Gas and Vapor Removers

The pressure drop characteristics of gas and vapor collection devices are generally well known, and such devices have the distinct advantage that pressure drop generally does not change over the life of the device. Scrubbers for gas adsorption generally fall into the low- or medium-energy categories, typically having pressure drops of a few inches of water but ranging up to 15 in. H_2O. The pressure drop through an adsorption bed is a function of the bed velocity, bed depth, and size of the granules making up the bed. Typical pressure drop data for granular carbon beds are presented in Fig. 11.14.

LIST OF SYMBOLS

K	venturi scrubber constant (Eq. 11.3)
K_1	clean fabric resistance
K_2	dust deposit specific resistance
Q	airflow
Q_G	gas flow in a venturi
Q_L	liquid flow in a venturi
V	filtration velocity
V_G	velocity in venturi throat
w	dust deposit areal density
Δp	pressure drop
Δp_f	pressure drop across a clean fabric
Δp_d	pressure drop across a dust deposit

REFERENCES

Licht, W. (1980), *Air Pollution Control Engineering: Basic Calculations for Particulate Collection*, Dekker, New York.

Stern, A., ed. (1977), *Air Pollution, 3rd ed., Vol.* 4, *Engineering Control of Air Pollution*, Academic Press, New York.

Stern, A., ed. (1986), *Air Pollution, 3rd ed., Vol.* 7, *Supplement to Measurements, Monitoring, Surveillance, and Engineering Control*, Academic Press, New York.

Strauss, W. (1975), *Industrial Gas Cleaning, 2nd ed.,* Pergamon Press, New York.

Theodore, L., and A. Buonicore, eds. (1982), *Air Pollution Control Equipment: Selection, Design, Operation, and Maintenance*, Prentice-Hall, Englewood Cliffs, NJ. (1982).

U.S. Environmental Protection Agency (1973), *Air Pollution Engineering Manual, 2nd ed.* Publication No. AP-40, J. Danielson, ed., U.S. Government Printing Office, Washington, DC.

U.S. Environmental Protection Agency (1978), *Industrial Guide for Air Pollution Control*, Publication No. EPA-625/6-78-004, U.S. Government Printing Office, Washington, DC.

12

Replacement-Air Systems

Until the mid-1960s little attention was given to the need for replacement air in the design of industrial exhaust ventilation systems. High-quality exhaust systems designed by engineering services would occasionally include a companion design to replace the air exhausted from the workplace, but more commonly a local sheet metal contractor would install the local or general exhaust system with no consideration to a replacement-air system. Further, existing local exhaust systems routinely were modified and expanded without consideration of replacement air. Many of the difficulties encountered in the performance of exhaust ventilation systems in the past could be assigned directly to the lack of adequate replacement air. These chronic problems began to generate increased interest in the 1960s when a number of state and local agencies started to require that replacement-air systems be installed in conjunction with new exhaust systems.

Limited acceptance of the replacement-air concept by industry prior to 1960 was based on the belief that providing replacement tempered air in the northern latitudes would represent an additional cost. It was not realized that even without a replacement-air system the air was drawn into the building by infiltration and heated before it was exhausted. The well-designed replacement-air system supplies heated air in a much more efficient manner (Fig. 12.1).

The importance of replacement air was beginning to receive acceptance in the 1970s, when enthusiasm was dampened by mounting fuel costs. This reluctance to introduce replacement-air systems is still being felt today. Even so, the conscientious designer accepts the necessity of providing replacement air and the availability of a variety of packaged units incorporating filter, fan, and heating sources has made the installation of these systems technically and economically

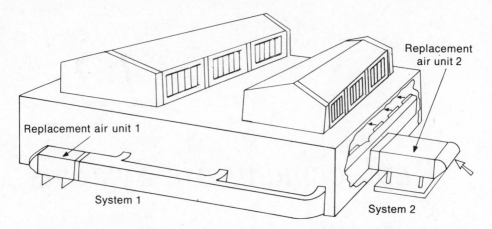

Figure 12.1 Two replacement-air systems consisting of basic unit and distribution duct. In both cases the units are mounted at ground level outside the building. In system 1 the distribution duct is positioned along the building perimeter with three wall penetrations terminating with air diffusers inside the structure. A single wall penetration from the unit in system 2 leads to a distribution manifold on an inside wall with a series of diffusers to provide low-velocity air at working height.

attractive. In addition, the high cost of tempering outside air has encouraged design innovations to permit recovery of heat from large exhaust streams. The recirculation of exhaust streams after suitable air cleaning is also practiced to a limited extent.

12.1 TYPES OF REPLACEMENT-AIR UNITS

A variety of basic HVAC systems (described in Chapter 15) can be applied as replacement-air units, but the typical major installations use packaged systems that are delivered ready for hookup of services and duct connections (Fig. 12.2). These packaged systems are available: (1) for use with hot water or steam coils, (2) as indirect systems fired by gas or oil with venting of products of combustion, and (3) direct gas-fired units with products of combustion mixed directly with the air supplied to the workplace.

The conventional hot water and steam replacement air units are available either as packaged units or they can be assembled from components on-site to meet specific airflow requirements and size constraints. The steam units require significant power plant service and are most popular for larger installations; hot water units are frequently used in smaller systems.

The packaged replacement air unit heated by gas or oil in a heat-exchange section is commonly used for medium and large systems, where the economy and flexibility of an integral unit are important. This design offers relatively high efficiency and, since products of combustion are vented outdoors, are considered safe for all applications, including recirculation of workplace air.

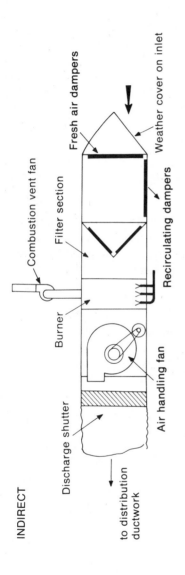

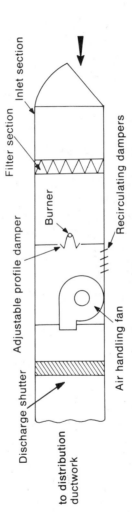

Figure 12.2 Gas-fired replacement-air units: The indirect-fired unit is equipped with a burner chamber and heat exchanger, so the replacement air stream and gas combustion process are separate. These units are equipped with recirculating dampers. The products of combustion of the direct-fired gas units are delivered to the workplace. If air is recirculated from the workplace, it must be introduced downstream of the burner; otherwise, fugitive air contaminants will be thermally degraded in the burner and delivered to the space. The direct-fired replacement air units are equipped with elaborate controls which permit their safe operation.

Where gas is relatively inexpensive, the direct-fired gas units are popular for systems larger than 10,000 cfm. These units are designed with extensive combustion controls to optimize combustion efficiency, thus permitting delivery of the products of combustion to the ventilated space. Recirculation of workplace air is prohibited since thermal degradation products of certain industrial air contaminants may present significant health hazards. As with other types, the direct-fired gas units are usually prepackaged for ease of delivery and installation at the site.

Units are sold by more than a dozen manufacturers. Propeller-type models are commonly available to handle duct resistance up to 0.5 in. H_2O; centrifugal-type models are available to handle external resistance as high as 5 in. H_2O. The duct design is normally calculated in the same manner as HVAC systems (ASHRAE, 1985).

12.2 NEED FOR REPLACEMENT AIR

Today there is general agreement that exhaust and replacement air should be in balance. When exhaust air is removed from the workplace, replacement air should be introduced. This simple guideline requires interpretation in practice. If a large manufacturing facility such as an electroplating and finishing shop operates with a total exhaust of 100,000 cfm, the addition of a small degreaser requiring an additional 2,000 cfm of exhaust should not prompt immediate upgrading of the replacement-air system (RAS). In practice, an RAS is not upgraded until the system is out of balance by more than 10%.

Another way to assess the necessity of an RAS is to compare the room volume to the required exhaust rate. Constance (1983) states that replacement air is necessary when the hourly exhaust volume exceeds three times the building volume. No justification is given for this approach. If this guideline were applied to the installation of a new color printer in a room with a total volume of 10,000 ft^3, the required exhaust of 1000 cfm (60,000 cfh) would warrant a separate RAS or the use of a 1000-cfm supply branch from an existing system.

In an existing facility the need for replacement air is frequently demonstrated by the complaints received from plant personnel. It is useful to describe typical complaints experienced in a number of foundries located in the northern United States as summarized in a hypothetical facility. In this large steel foundry exhaust ventilation totals 230,000 cfm with no replacement-air systems (Fig. 12.3). The range of complaints received by management is consistent with that noted in buildings with significant deficiency in replacement air. Because of the severe air demand in the building, outside air is drawn through all cracks and openings. Workers in small shops or at isolated workbenches on the perimeter walls complain of strong drafts of cold air in the winter. Both pedestrian and material transport doors have significant forces exerted on them under winter conditions due to the negative pressure established by the exhaust fans. It is extremely

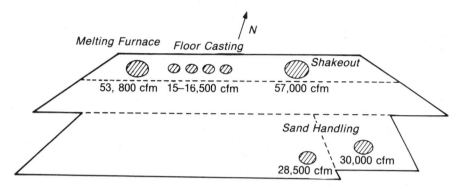

Figure 12.3 Foundry floor layout showing the three building complex with approximate locations of exhaust in four major work areas, as described in Table 12.1. The airflow in cfm is shown directly at the exhaust location. This foundry does not have replacement air.

difficult to open the doors, and this condition leads to frequent back strains. Workers stay clear of closing doors to escape the violent slamming.

Due to the lack of replacement air in the building, the high-pressure blowers on the local exhaust systems create building static pressures approaching -0.1 in. H_2O. This condition causes back-drafting of the stacks on small ovens, resulting both in pollution of the workplace and impaired operation of ovens and furnaces.

The performance of low-pressure-high-volume axial flow roof ventilators used for general exhaust ventilation is also impaired by such a condition. Four roof fans in this foundry pouring area have a rated flow of 18,000 cfm at a static pressure of 0.25 in. H_2O. Under the existing air-starved conditions the actual airflow drops by 10%. Such conditions clearly call for the installation of an adequate RAS.

12.3   QUANTITY OF REPLACEMENT AIR

The first major step in designing a RAS is the selection of the supply airflow. The simplest approach to such a selection is to assume that the supply should equal the exhaust. In practice it is sometimes necessary to generate pressure differentials between major work areas to establish or block airflow from one area to another. This can be done by designing a slight imbalance between supply and exhaust. In the foundry discussed above (Fig. 12.3), the air should flow from the general work area toward the furnace area to avoid fugitive metal fumes from the latter contaminating the rest of the foundry. In this case the replacement air for the electric arc furnace complex should be less than the furnace exhaust, thus creating an air demand in this part of the plant. In general, it is prudent to match the supply and exhaust rates approximately, although supply rates in the range of 90 to 110% of the exhaust capacity are normally used as design boundaries for replacement-air systems.

Determining the required RAS supply rate can be accomplished in a number of ways. In existing installations an estimate of the magnitude of the supply deficiency can frequently be accomplished by a simple procedure. On a day with low wind, one or two large access doors can be opened while the building exhaust and RAS are operating. If the building is fairly tight, any air deficiency will be made up by air flowing through the open doors. If the average inward velocity through the doors is then measured, the product of that velocity and the open area of the door represents the deficiency. This procedure gives results within ± 20% of the values measured by the more accurate methods described below.

A more rigorous approach is based on a review of the original engineering design plans, which specify both the exhaust rates of all installed ventilation and the supply rates of the existing RAS, as shown for the foundry in Table 12.1. The operating shifts and times for all major exhaust are noted so that suitable RAS distribution can be achieved. In our hypothetical foundry there are no replacement-air systems installed.

If engineering data are not available, an alternative approach is to base the projected exhaust on the consensus standards for industrial ventilation applicable at the time the system was designed. In using engineering design standards to estimate exhaust volume the designer's presumption that the original engineer followed the specifications and that a maintenance program was in force to keep the systems operational is probably naive.

The fourth and most accurate way to determine the required replacement air is to evaluate the quantity of air currently being exhausted from the plant by direct airflow measurements on each exhaust system using the techniques described in Chapter 3. Since the building may be air starved, the observed exhaust rate will be less than design if low-pressure exhaust fans are widely used. To minimize this artifact, if weather permits, doors and windows should be opened before making measurements. Measurements of system submain and main exhausts using the Pitot static traverse technique provide excellent results, although hood face velocity and hood static pressure measurements are usually adequate for assessment of airflow. A major problem encountered frequently in establishing such an exhaust inventory is the difficult measurement of the delivery of propeller fans used commonly as roof exhausters, as discussed in Chapter 3. Frequently it is necessary to use manufacturers' performance data to estimate delivery of these fans. In the foundry example, actual airflow measurements were used to provide an inventory of exhaust airflow (Table 12.1).

12.4   DELIVERY OF REPLACEMENT AIR

The general rules regarding delivery of replacement air to the workplace are that (1) the air should be delivered to the active working zone (less than 8 to 10 ft above the floor), (2) the draft on the worker be minimized by maintaining air velocities below 200 fpm, and (3) the replacement air should be delivered in a way that minimizes thermal looping (Fig. 12.4).

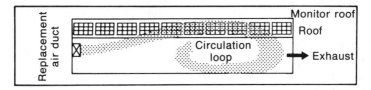

Figure 12.4 Thermal looping occurs when replacement air is heated as it passes over processes. Due to density difference the air flows to the top of the building, frequently "short-circuiting" directly to roof exhausters. (From *Michigan's Occupational Health*, Fall 1960.)

In practice it is often difficult to achieve the desired delivery geometry for the replacement air. In a new facility there is a reasonable opportunity to position the distribution duct to achieve good delivery patterns by close cooperation between the engineer and the architect. The ventilation engineer must obtain a firm commitment on the total exhaust rate, the grouping of local exhaust hoods that operate as one system, and the anticipated time of use. Frequently, it is possible to dedicate RAS to specific exhaust systems and interlock the replacement air units and exhaust fans to operate as integrated supply–exhaust systems.

In a new system the designer can usually locate the RAS so that an airflow path is clearly defined from the RAS delivery grilles to exhaust hoods and not be encumbered by physical obstructions or major draft patterns. As noted above, the replacement air should be delivered at the active occupancy level, not the upper reaches of the building. Frequently, it is possible to choose and position discharge grilles to minimize draft on the worker in the winter but permit cooling air to sweep across the worker in the summer. In many structures the craneway and other obstructions will not permit this placement and the delivery plenum must be positioned at significant heights above the floor and the throw from high-velocity nozzles used to achieve delivery to floor level.

Replacement-air units frequently employ ducts of large cross section and require special collaboration with the architect to provide the necessary space. Since the architect usually does not wish to use valuable interior space for this purpose, RAS ducts are frequently located outside the plant with delivery branches penetrating the structure. The installation of 50,000 to 100,000-cfm replacement-air units is difficult due to their size and mounting requirements. If possible, it is good practice to mount the units on footings at ground level or on an exterior structural wall with suitable brackets. This practice usually permits the designer to use a short distribution manifold. In this case installation costs are relatively low and the accessibility of the unit encourages routine maintenance.

In cases where the availability of space on an exterior wall is restricted, roof location of the replacement air unit may be necessary. Roof installation requires special load-carrying mounting, roof penetrations for electric and fuel services, and of course, penetration for the air delivery duct. Such installations frequently require long and expensive duct runs with involved distribution systems. If the mounting locations for the replacement-air units are at the edge of buildings the units can be positioned by crane; large buildings with replacement-air units located in center bays may require transport by helicopter.

Providing a replacement-air system for an existing large production facility (as typified in the hypothetical foundry example described above) presents many difficulties not seen in new designs. In the foundry example supply volumes for the replacement-air units may be based on both engineering data and actual measurements of exhaust systems. Although the engineering data may provide reasonably accurate information, direct airflow measurement is the preferred method to establish the existing exhaust volumes and the replacement-air requirements. Obviously, the airflow of existing replacement-air units should be measured to determine the replacement-air deficiency.

In most large steel foundries, operations equipped with local exhaust ventilation are grouped by function (e.g., sand handling, molding, melting, pouring, shakeout, and finishing). A few of these operations usually operate continuously, but in our example most operate for specific periods during the three work shifts (Table 12.1). It is possible in this case to couple the exhaust system with individual replacement air units and interlock the units and the exhaust fans to conserve energy and ensure adequate air supply during system operation. In this case, the best approach is to have a "one-on-one" coupling of the RAS and the major exhaust systems.

If two major exhaust systems have approximately the same exhaust airflow, do not operate at the same time, and are located in the same part of the plant, it may be possible to use one RAS to serve both. If the exhaust locations are not suitable for one distribution system, then two alternate duct distribution systems can be supplied by one replacement-air unit using dampers to select the appropriate delivery duct.

Expansion of original buildings occurs over decades, resulting in a myriad of roof structures and bay layouts. In such cases it can be very difficult to find a suitable location for the replacement-air unit. The units should be placed close to the supply location to minimize the length and complexity of the duct run and should not have expensive mounting requirements. In addition, the mounting location for the replacement-air unit should be chosen to minimize the likelihood of reentry of air contaminants from exhaust stacks (see Chapter 16). Ideal locations are rarely found and compromises are required. The solutions chosen for four major replacement-air units (RAUs) in the foundry example are described below. In this example the RAUs are direct gas-fired units delivered by the manufacturer ready for installation.

12.4.1 Replacement-Air System 1 (RAS-1), Melting Furnaces

Two large electric arc furnaces provide the melting capacity for this foundry. To control the emissions of iron oxide fume the furnaces are provided with enclosing hoods (VS-105) with a total exhaust of 53,800 cfm. Since replacement air is not provided the major replacement-air flow path is down the main bay from the floor casting area.

TABLE 12.1 Replacement-Air Requirements in a Steel Foundry Based on Engineering Specifications and Measured Airflow

Process or Operation	Type of Exhaust	Shift	Airflow (cfm)		Code	Location	Airflow (cfm)	Type of Distribution
			From Prints	As Measured				
						Replacement Air		
Melting furnace	Hood	1, 2	60,000	53,800	RAS-1	Ground level, west wall	50,000	Discharge grille on RAU
Floor casting	Roof exhauster	1, 2			RAS-2	Roof, south side	75,000	Three drops
	A		18,000	16,000				
	B		18,000	15,000				
	C		18,000	16,500				
	D		18,000	16,000				
Sand handling	Hoods	1, 2			RAS-3	Roof, south side	60,000	Three drops
	A		35,000	28,500				
	B		40,000	30,000				
Shakeout	Hood	3	—	57,000	RAS-4	Ground level, east wall	60,000	Discharge grille on RAU

In this case the ventilation designer is fortunate since space is available at ground level at the west end of the building to mount a single 50,000-cfm RAU interlocked with the furnace exhaust fans (Fig. 12.5). This large unit is mounted on a structural steel base and the system delivers air directly to the main bay. Since the furnaces are approximately 120 ft away, a plenum and a discharge grille mounted on the wall provided suitable distribution and the airflow path is direct to the exhaust hoods without major obstructions. A minimum residual velocity, less than 200 fpm, resulted at the furnace area workplace.

12.4.2 Replacement-Air System 2 (RAS-2), Floor Casting

The steel casting of large floor molds with hot topping of risers in the main bay is a major source of metal fume contamination in this facility. Since local exhaust ventilation is impossible on these operations the fume rises to the main bay roof and is exhausted by four 18,000-cfm roof fans. Due to the negative pressure developed in the building by the high-pressure centrifugal fans exhausting the furnaces, the roof exhausters do not deliver full capacity, resulting in poor clearance of the fume during floor casting. The fume blankets the floor area and impairs the view of the crane operator and the floorman, making it difficult to conduct operations safely.

A 75,000-cfm RAU is required for this pouring area. The north side of the building adjacent to the pouring area is not available for use. The south side of the building is over 200 ft away from the pouring bay and the building perimeter space at ground level is limited. The building expansion to the south of the main bay is a clerestory design. A 75,000-cfm roof-mounted replacement-air unit is positioned at the southern edge of the building roof with an outside main duct delivering air to three distribution branches entering the clerestory windows (Fig. 12.6). A vertical drop is used to discharge the replacement air at craneway height approximately 40 ft from the main bay. The air demand from the pouring area roof exhausters causes the replacement air to sweep to the bay and clear the

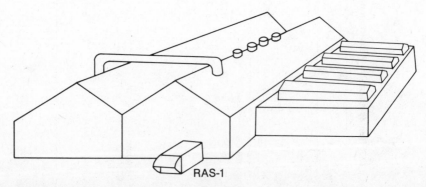

Figure 12.5 Replacement-air system 1 (RAS-1) for the melting furnace area. View from west end of building.

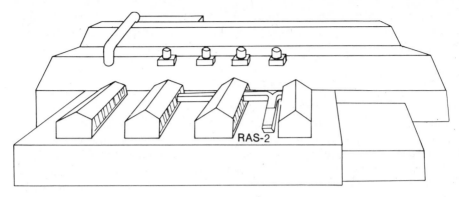

Figure 12.6 Replacement-air system 2 (RAS-2) for the pouring area. View from south end of building complex.

metal fumes. The availability of a clerestory roof design frequently proves useful in permitting practical solutions to the location of RAS distribution ducts.

12.4.3 Replacement-Air System 3 (RAS-3), Sand Handling

A multicomponent sand handling system located in the southeastern corner of the building is equipped with two primary exhaust systems with a total exhaust of 58,500 cfm. The building perimeter space in this area is heavily utilized for foundry support equipment. A roof location for a 60,000 cfm RAU was again chosen as the best solution. (Fig. 12.7). This system is provided with a main duct run along a single clerestory with three drops discharging through diffusers at

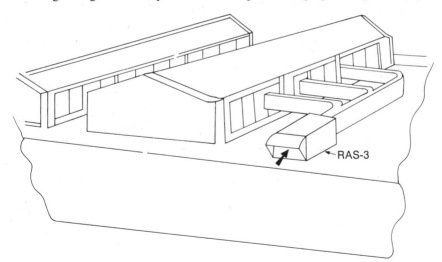

Figure 12.7 Replacement-air system 3 (RAS-3) for the sand-handling area. View from southeast.

the craneway elevation approximately 50 ft from the sand treatment area. This replacement-air unit is electrically interlocked with the two fans in the sand-handling area.

12.4.4   Replacement-Air System 4 (RAS-4), Shakeout

A large shakeout operation with an exhaust of 57,000 cfm conducted only on third shift is located in the main bay. The immediate north wall would have been an ideal location for a 60,000-cfm RAU, in that it would provide cross-bay air sweep, but this space is utilized for a variety of abrasive blasting and heat treatment facilities. The nearest ground-level location is on the east wall of this bay approximately 150 ft from the shakeout (Fig. 12.8). Since there is little other activity under way in the building on the third shift, the possibility of the replacement air short circuiting or thermal looping was minimal and this location was deemed acceptable. The distances from the replacement-air unit to the shakeout exhaust hood preclude the need for an extensive distribution duct; a plenum with a discharge diffuser is mounted directly on the RAU.

Since there is no pouring or sand treatment on the third shift, RAS-2 and RAS-3 will not be operating. A feasible option in this case is not to install a separate RAS for the shakeout but to use RAS-2, as designed, or RAS-3 with an alternative duct to provide replacement air for the shakeout operation.

12.5   REPLACEMENT AIR FOR HEATING

In discussing the application of replacement-air systems earlier in this chapter it was stated that these systems are not designed for primary air heating and thus

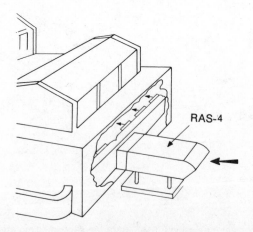

Figure 12.8   Replacement-air system 4 (RAS-4) for the shakeout. View from east end of building.

the delivery air temperature should be approximately room temperature. There are occasions when a packaged direct-fired replacement air unit may provide an inexpensive alternative to a conventional heating system.

Such was the case in an unheated granite cutting shed in New England, where winter workplace temperatures commonly fall below freezing. Granite was transported from a nearby quarry for manufacture of dimensional and architectural stone. Approximately 50 people worked at stone-cutting stations scattered over the main bay (100 ft × 300 ft), many of which were equipped with local exhaust. Dust control at other locations was achieved by wet drilling and polishing using water as a dust suppressant. Since the workstations were not fixed, radiant heating was not appropriate. An expedient solution was to tighten up the building, insulate the walls and roof with sprayed polyurethane foam, and use two 50,000-cfm direct-fired gas RAUs to maintain the air temperatures above 40° F. This installation improved comfort conditions and minimized slips and falls and handling accidents caused by ice formation.

12.6 ENERGY CONSERVATION AND REPLACEMENT AIR

When replacement-air systems were first introduced in the United States, plant managers objected to the cost of heating this "new" air. Management did not easily accept the notion that in those plants without designed replacement air the air that infiltrates the building and is later exhausted is one of the heating loads in a HVAC design. A replacement-air system merely accomplishes this heating in a more efficient manner. This is not to imply that the cost of heating replacement air in northern latitudes is insignificant. The annual cost to heat the replacement air in a steel foundry, such as described above, located in the northern United States was approximately $250,000 in 1986.

The techniques available to minimize this energy cost have been discussed for decades. The energy crisis of the 1970s prompted aggressive action by U.S. industry. The broad approaches to energy conservation in a laboratory setting were discussed in Chapter 7. There are occasions when exhaust air flow is excessive and a reduction in exhaust is possible, permitting a parallel reduction in the replacement-air requirement. When using this approach the engineer must always demonstrate continued effective contaminant control with the reduced exhaust volumes. Another productive approach is to use the exhaust system and its matching RAS only when the process is operating. Interlocking the exhaust fan with the replacement-air units ensures that heating costs are minimized.

In facilities such as hospitals where one-pass ventilation is practiced, major attention has been given to removal of heat from the air before it is exhausted. A common approach is the so-called heat wheel. Warm exhaust air passes through one section of a rotating wheel with a large heat capacity and the heat is transferred to the wheel. As this wheel sector rotates slowly to the outside supply airstream, the heat is supplied to that stream to increase its air temperature. This technique has limited application in industrial exhaust heat recovery, due to its expense and limited capacity. Furthermore, there is potential for contamination of the wheel packing with air contaminants and their subsequent elution by the outside airstream, resulting in workplace contamination. The heat pipe is the basis for another thermal recovery unit (Fig. 12.9). The active element in this design is a bundle of heat pipes; each pipe is split along its length by a partition separating the warm exhaust and cold supply airstreams. Heat energy from the warm exhaust stream is transferred to the other end of the individual heat pipe, where it heats the cold incoming air. The manufacturer states that these units recover 60 to 80% of the difference in sensible heat between the two airstreams.

A major energy conservation approach in industrial exhaust systems has been the removal of the air contaminant from the exhaust stream by suitable air cleaning and the recirculation of the airstream to the workplace. Due to the potential for exposure of employees to toxic air contaminants, this approach has received limited acceptance. NIOSH has sponsored applied research in this area and held several symposia on the application of the concept (NIOSH, 1978). The legal and regulatory aspects of air recirculation in the workplace have been reviewed by Holcomb and Radia (1986). OSHA does not specifically address air recirculation in its regulations but merely states that exposures in the workplace shall not exceed the Permissible Exposure Limit. Certain municipalities, including Chicago, specifically prohibit the practice. Canada permits recirculation but requires that the contaminant concentration in the return airstream not exceed one-twentieth of the TLV; the United Kingdom uses this approach but specifies a factor of one-tenth of the TLV.

The application of recirculation of exhaust air containing toxic contaminants must be based on detailed knowledge of the form and concentration of the chemical species in the workplace and the performance of the air cleaner for these contaminants. The system must include sensors to indicate when a contaminant concentration has exceeded a critical level, and the sensors must activate a system to discharge the exhaust directly outdoors. If possible, the system should operate in a fail-safe mode.

It is the authors' opinion that the present state of air-cleaning technology is such that air recirculation does have application for particle contamination but has limited use for gases and vapors. High-efficiency particle filters used on such applications are "fail-safe" and the drop in air flow as the filter "loads" can easily be monitored. The downstream particle concentration can usually be measured with sensitive and reliable equipment. However, sorbents used for removal of gases and vapors have a finite life before breakthrough occurs. Monitoring the breakthrough of a gas or vapor contaminant is quite difficult and dependent on the specific contaminant. In any case, recirculation should not be used for human carcinogens or for contaminants with low TLV values.

12.7 SUMMARY

It is important that the designer of industrial ventilation systems recognize the importance of replacement-air systems. In this chapter guidelines are presented

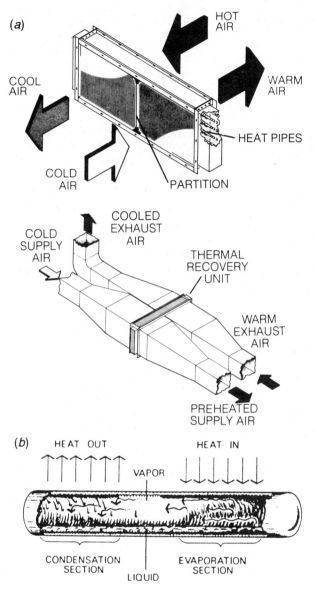

Figure 12.9 The heat pipe coil (*a*) is positioned across the exhaust and supply duct with a partition separating the pipes into warm (exhaust duct) and cold (supply duct) ends (*b*). Each heat pipe consists of an inclined tube filled with a refrigerant. As warm air passes over the exhaust duct end of the pipe, the refrigerant vaporizes. The refrigerant vapor passes to the cold, incoming airstream, where it condenses, releasing the latent heat of condensation, which is transferred to the incoming airstream. The thermal loop continues as the liquid refrigerant flows back to the warm exhaust duct end. The process of refrigerant transfer between ends is facilitated by a grooved inner wall or other capillary wicking technique. The capacity of the heat pipe can be changed by changing the slope or inclination of the pipe coil and hence rate of return of the liquid refrigerant to the evaporation (hot exhaust stream) end. (Courtesy of Q Dot Corporation.)

to aid in determining if replacement air is necessary, procedures are outlined for calculating the supply rate of the replacement-air system, and various options are presented for location of the replacement-air unit and distribution ductwork.

REFERENCES

American Society of Heating, Refrigerating and Air Conditioning Engineers (1985), *ASHRAE Handbook—1985 Fundamentals*, ASHRAE, Atlanta, GA.

Constance, J. D. (1983), *Controlling In-Plant Airborne Contaminants*, Dekker, New York, pp. 127–136.

Hama, G. J. (1974), "Conservation of Air and Energy," *Mich. Occup. Health*, **19**(3): Spring.

Holcomb, M. L., and J. T. Radia (1986), "An Engineering Approach to Feasibility Assessment and Design of Recirculating Exhaust Systems," in *Proceedings of the 1st International Symposium on Ventilation for Contaminant Control*, October 1–3, 1985, Toronto, Canada, H. D. Goodfellow, ed., Elsevier, New York.

U.S. Department of Health, Education, and Welfare, *National Institute for Occupational Safety and Health* (1978), "The Recirculation of Industrial Exhaust Air—Symposium Proceedings," Publication No. 78–141, Washington, DC.

ADDITIONAL READING

Caplan, Knowlton (1985), "Building General Ventilation" in *Industrial Hygiene Aspects of Plant Operation*, Vol. 3, L. V. Cralley and L. J. Cralley, Wiley, New York.

American Conference of Governmental Industrial Hygienists, Committee on Industrial Ventilation (1988), *Industrial Ventilation Manual*, 20th ed. ACGIH, Lansing, MI Sec. 7.

13

Quantification of Hood Performance

It is very important that the performance of exhaust hoods be evaluated once they are installed. Several different tests can be performed, as described below. They are designed to answer one of two important questions: (1) Is the airflow through the hood equal to the design airflow? and (2) Is the hood capturing the contaminants given off by the process it is controlling?

The answer to the first question involves making airflow and static pressure measurements. Such measurements (which in the authors' experience are made on only a small fraction of new systems) basically assess the success of the ventilation system design procedure. That is, such tests measure whether the designer has successfully calculated and balanced all the static pressure losses in the system and specified the proper fan; if this is done properly and the construction follows the design specifications (a very big *if*, in some cases), the measured hood airflow will equal the design flow.

Although it is always necessary to measure airflow in newly installed hoods, the second question is fundamentally of greater importance. The success of a hood design rests ultimately in its ability to capture the contaminants given off by the process. Meeting design airflow while capturing only half of the contaminants is not a satisfactory design; it simply means that the proper airflow was not selected during the design process, or the hood was improperly shaped and positioned, or there was some similar defect in the design process.

Traditionally, the success of a ventilation system in capturing contaminants is assessed by measuring the breathing zone concentration of the air contaminants being captured and comparing the results to a standard (e.g., a TLV or a PEL). Alternatively, if the exposed workers move from one location to another, an area sample may be collected in the vicinity of the hood in question. Such measurements, while important, are at best an indirect measure of hood performance. The airborne concentration at any point outside the hood is a function not only of the hood capture efficiency but of many other factors, such as contaminant generation rate, general room ventilation, operator work practices (for personal samples), and exact sampling location (for area samples). In this chapter we discuss new techniques to measure directly the performance of the hood in capturing contaminants. Measurement of hood capture efficiency should represent the ultimate test of a local exhaust ventilation system.

13.1 HOOD AIRFLOW MEASUREMENTS

The standard procedure used by ventilation engineers and industrial hygienists to evaluate hood performance is simply to measure the total airflow through the hood and compare the measured value to the design value. Once the hood airflow is quantified, its performance can be monitored periodically using the hood static pressure method described in Chapter 3. Techniques for measuring airflow and static pressure are also described in Chapter 3; this discussion will center on the use of such measurements.

The hood airflow measurement method has some great advantages:

1. It is simple and straightforward to perform.
2. It is easy to decide whether or not the performance is adequate—the flow either meets the design value or it does not.
3. The extent of any required "fix" is equally easy to assess—simply increase (or decrease, perhaps) the airflow to the desired level (by adjusting a blast gate or changing the fan speed, for example).

The chief limitation of this method as a means to quantify hood performance lies in the fact that it measures hood velocities, and knowledge of velocity does not in and of itself ensure adequate hood performance. This is especially true for exterior hoods, which rely on the creation of a capture velocity at the point of contaminant generation sufficient to draw the contaminants into the hood. The difficulties inherent in choosing a capture velocity, and the disruptive effects of cross-drafts, are discussed in Chapter 5.

Such difficulties in choosing the proper capture velocity lead to the very real possibility that an exterior hood may have a measured airflow equal to the design airflow and still not capture the contaminants with high efficiency. Realization of this fact has led researchers to investigate the possibility of measuring hood capture efficiency directly.

13.2 HOOD CAPTURE EFFICIENCY

The authors have been among those looking at the concept of capture efficiency as a means to quantify hood performance. As stated by Ellenbecker et al. (1983):

The advances in ventilation control in the last two decades have been limited to empirical improvements in hood design. . . . Although of demonstrated value, the present design guidelines usually result in the construction of a system whose ability to capture contaminants has only been estimated qualitatively; as a next logical step, we propose a system for the *quantitative* evaluation of exhaust system performance.

Such quantitative evaluation involves the determination of exhaust system capture efficiency, defined as the fraction of contaminants given off by a process actually captured by the exhaust hood(s) used to control that process. This concept of capture efficiency as the best measure of hood performance has received considerable attention from researchers in the past few years (Burgess and Morrow, 1976; Conroy *et al.*, 1988; Conroy and Ellenbecker, 1988; Dalrymple, 1986; Ellenbecker et al., 1983; Fletcher and Johnson, 1986; Flynn and Ellenbecker, 1985, 1986, 1987; Jansson, 1980; Regnier et al., 1986). By the above definition, capture efficiency η is simply

$$\eta = \frac{G'}{G} \tag{13.1}$$

where G = rate of contaminant generation (in, e.g., g/min)

$\quad G'$ = rate at which contaminant is captured by the exhaust system, g/min

In practice, G' may be fairly easy to measure by collecting a sample from the duct located downstream from the exhaust hood(s), but in many cases G is very difficult to measure (or estimate) accurately. Four approaches have been used in an attempt to estimate G: (1) perform a material balance to determine the emission rate; (2) measure airborne contaminant concentrations in the workplace and relate them back to G; (3) enclose the contaminant source so that all contaminants are captured and calculate G by measuring the contaminant concentration in the enclosure; and (4) use a tracer to simulate the contaminant emission process.

All four methods have disadvantages. A material balance can rarely be calculated with sufficient accuracy to obtain a realistic estimate of G. The second method is not reliable because of all of the uncertainties involved in translating a generation rate to a resulting airborne exposure. The third method can work well in the laboratory, but is difficult to implement in the field, especially for large contaminant sources. The fourth method is the easiest to use, both in the laboratory and in the field, but care must be taken to ensure that the tracer accurately simulates the actual emission process.

Both aerosol and gaseous tracers have been used; the most common aerosol is an oil mist, which has the advantage of producing a visible tracer but which is fairly difficult to generate and measure. Gaseous tracers generally are easier to generate and measure than aerosols; the most frequently used is sulfur hexafluoride (SF_6), which can be detected at low concentrations with a properly configured gas chromatograph.

Exhaust hood capture efficiency can also be predicted from theoretical or empirical models. Two theoretical approaches, one using analytical solutions to fluid flow equations and the other using numerical solutions, have been developed in recent years. The analytical approach being investigated by the authors and colleagues is described below.

Capture efficiency research is still in its infancy, so that the use of capture efficiency as a system design tool is not widespread. Nonetheless, it represents a powerful new tool for predicting and assessing system performance.

13.2.1 Influence of Cross-Drafts on Hood Performance

As mentioned above, the most important limitation of the hood airflow method of quantifying hood performance is its failure to account for the effects of cross-drafts. Recent work by the authors and colleagues (Conroy and Ellenbecker, 1988; Conroy et al., 1988; Flynn and Ellenbecker, 1985, 1986, 1987; Ellenbecker et al., 1983) has resulted in models that can be used to quantify the effect of cross-drafts on hood airflow patterns and capture efficiency. Models have been developed for round, square, rectangular, and slot hoods. The models are based on analytical solutions to Laplace's equation for frictionless, incompressible, and irrotational fluid flow, and they have been validated using wind tunnel tests. Programs have been written for the personal computer which allow a graphical presentation of the models' airflow predictions.

As an example of the use of these models, consider the simple slot hood shown in Fig. 13.1. This hood has a flanged slot 15 in. long and 1.5 in. wide (an aspect ratio of 10) and might be used to exhaust a small plating tank. The Flynn, Conroy, and Ellenbecker (FCE) model can be used to visualize the effects of cross-drafts on the performance of this hood. The model was programmed in BASIC; an example of the model output is shown in Fig. 13.2. The figure shows an idealized plan view of this hood, including the flange, the hood opening, and

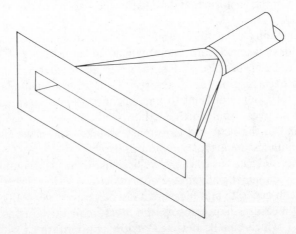

Figure 13.1 Simple flanged slot hood.

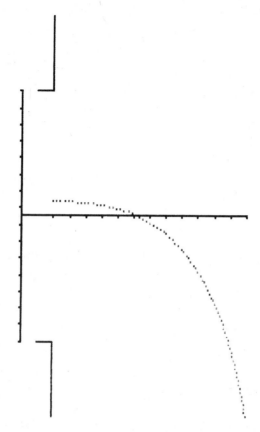

Figure 13.2 Computer model of the hood shown in Fig. 13.1. The slot is seen from above, along with vertical and horizontal distance scales marked 1-in. intervals. L = 15.0 in. w = 1.5 in., V_f = 1000 fpm, V_d = 25 fpm. A single streamline is shown for a face velocity of 1000 fpm and a cross-draft velocity of 25 fpm.

reference scales across the hood face and along the hood centerline. Plotted in dotted lines are the airflow streamlines for a particular combination of hood face velocity V_f and cross-draft velocity V_d. In this case the hood face velocity is 1000 fpm and the cross-draft is relatively low, 25 fpm, and blowing directly across the hood face from the bottom of the figure toward the top. A single streamline, starting at a location upstream and 12 in. in front of the hood, is shown. Such streamlines are useful in visualizing the direction of airflow from any point in front of the hood, and thus are an important feature of the new capture efficiency models. Figure 13.2 and subsequent figures in this chapter show a two-dimensional view of the airflow pattern as it exists on the centerline plane of the hood; any other plane can also be modeled and displayed, so that the entire flow field can be visualized.

Figure 13.3 illustrates the effect of increasing cross-drafts on hood performance; in all cases, the cross-draft is blowing across the hood face (bottom of

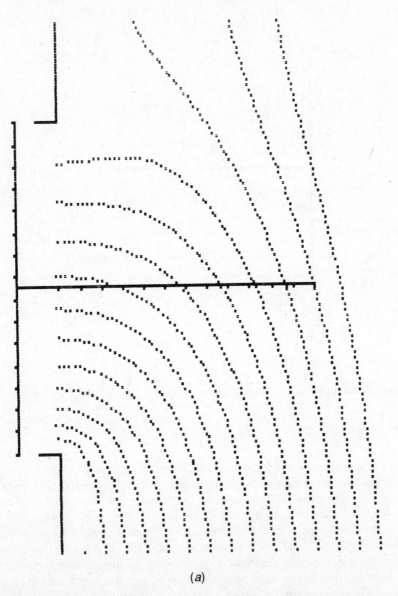

(a)

Figure 13.3 The hood shown in Figs. 13.1 and 13.2, illustrating the effect of increasing cross-draft: (*a*) $V_d = 50$ fpm.

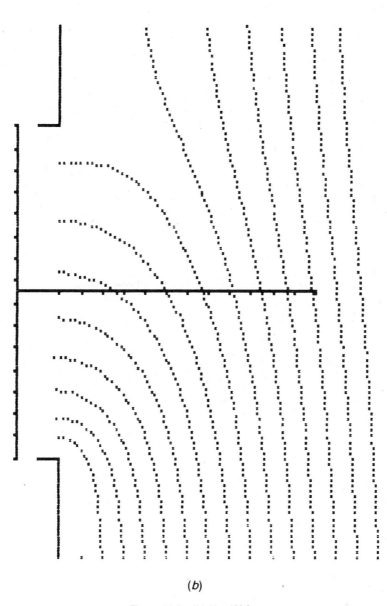

(*b*)

Figure 13.3   (*b*) $V_d = 100$ fpm.

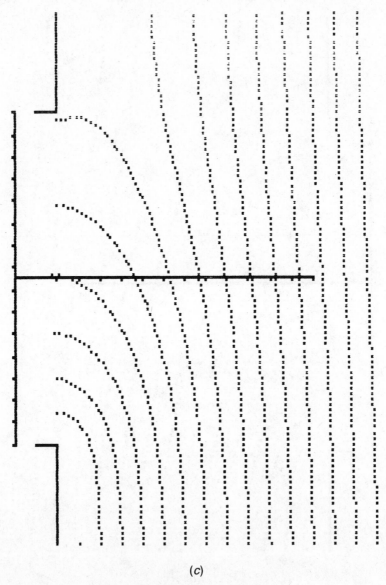

(c)

Figure 13.3   (c) $V_d = 200$ fpm.

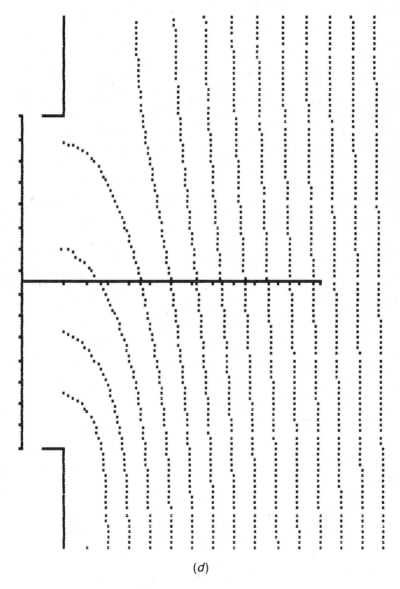

(*d*)

Figure 13.3 (*d*) $V_d = 400$ fpm.

the figure toward the top). As the cross-draft velocity increases, fewer and fewer streamlines enter the hood, so that contaminants would be captured over a smaller and smaller area. When the cross-draft is moderate (e.g., Fig. 13.3*a*, 50 fpm) the reach of the hood extends about 10 in. along the hood centerline; when the cross-draft increases to 400 fpm (Fig. 13.3*d*), the reach is only about 4 in.

13.2.2   Relationship between Airflow Patterns and Capture Efficiency

Knowledge of the airflow patterns created by an exhaust hood in the presence of a cross-draft can be used to estimate hood capture efficiency. Consider the streamlines shown in Fig. 13.3*c*; note that one just curves around and enters the hood at the uppermost edge. This streamline is redrawn in Fig. 13.4, where it is labeled the critical streamline, since all streamlines closer to the hood enter the hood and all streamlines farther away do not. Air flowing through the crosshatched area of Fig. 13.4 thus enters the hood, so that contaminants given off in this vicinity might be expected to enter the hood and be captured, whereas contaminants given off outside this area would escape the hood.

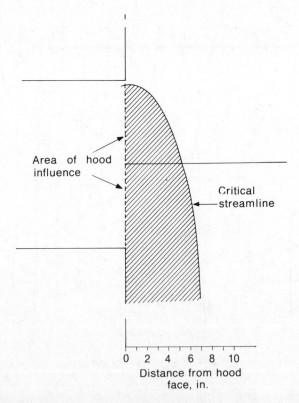

Figure 13.4   The hood of Fig. 13.3 with $V_d = 200$ fpm, showing the critical streamline and the area of hood influence.

The predicted performance of the hood at any position in the flow field can now be described. Using the simplistic model of Fig. 13.4, capture efficiency is a step function; inside the crosshatched area it is equal to 1 and outside the crosshatched area it is equal to 0. If capture efficiency is plotted as a function of distance along the hood centerline x the curve of Fig. 13.5a is obtained. The point at which the critical streamline crosses the hood centerline is termed the critical distance (x_c); at this point the capture efficiency drops from 1 to 0.

In actual practice turbulence is present in the flow field which acts to disperse contaminants off their original streamline. This turbulent diffusion should result in a capture efficiency curve which is sigmoidal (Fig. 13.5b). If diffusion is symmetric about the original streamline, the capture efficiency should equal 50% at the critical distance.

Flynn and Ellenbecker (1986) measured the capture efficiency of several flanged circular hoods, and Conroy and Ellenbecker (1988) did the same for

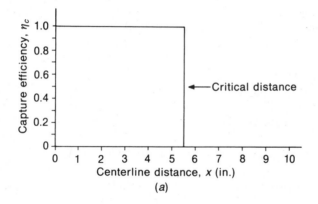

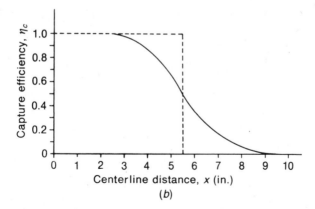

Figure 13.5 Models for hood capture efficiency. (*a*) With no turbulent dispersion of the contaminant, capture efficiency is a step function, dropping from 100% to 0% at the critical distance. (*b*) With turbulent dispersion, the capture efficiency curve becomes sigmoidal and drops more gradually, passing 50% at the critical distance.

flanged slot hoods. They developed models for capture efficiency as a function of hood dimensions (length L and width W or diameter D), hood face velocity (V_f), cross-draft velocity (V_d), position in the flow field, and turbulence intensity (ω). Point sources of gaseous contaminant were used in the experiments so that the behavior of contaminant on individual streamlines could be monitored; the models developed for point sources were then validated for area sources.

Experimental results confirmed that capture efficiency is best described by a logistic function, which results in sigmoidal capture efficiency curves similar to Fig. 13.5b:

$$\eta = \frac{\exp\left(\dfrac{x-\mu}{\omega}\right)}{1+\exp\left(\dfrac{x-\mu}{\omega}\right)} \tag{13.2}$$

where μ is the experimentally measured critical distance, which was found to be very close to the theoretical value x_c. Both μ and ω were found to be functions of the hood dimensions, hood flow, position in the flow field, and cross-draft velocity, as expected.

Typical results using this equation are shown in Fig. 13.6. The four graphs present experimentally derived capture efficiency curves along the centerline in front of a 6-in. flanged circular hood for different values of V_f/V_d. The first curve (Fig. 13.6a) presents results for the lowest ratio tested (0.92). Here the cross-draft quickly predominates and the capture efficiency falls to zero a short distance in front of the hood. The remaining plots illustrate the change in performance as the ratio increases. As the face velocity gets larger relative to the crossdraft velocity the hood has a greater "reach" and capture efficiency remains high at greater distances from the hood face.

Note also that the slope of the curve changes as the reach increases. This is due to the increased time required for the contaminant to travel into the hood, which in turn results in greater turbulent dispersion around the idealized contaminant streamline. The further the contaminant has to travel, the less the capture efficiency curve looks like the step function of Fig. 13.5a.

Flynn and Ellenbecker (1987) found that for the range of hoods and velocities tested, the critical distance along the hood centerline was given by the following empirical relationship:

$$\frac{X}{D} = 0.42\left(\frac{V_f}{V_d}\right)^{0.6} - 0.08 \tag{13.3}$$

where D is the hood diameter. Critical distances can also be calculated at positions in the flow field not located on the centerline, but no simple empirical relationship has been found and the computer model must be used (Conroy and Ellenbecker, 1988).

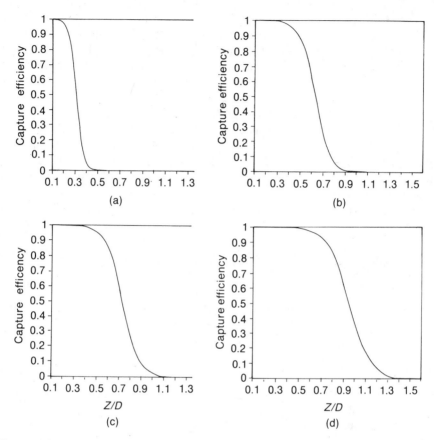

Figure 13.6 Theoretical (solid line) and experimental (cross) capture efficiencies for a 6-in. flanged circular hood at four ratios of face velocity to cross-draft velocity: (a) $V_f/V_d = 0.9$; (b) $V_f/V_d = 2.4$; (c) $V_f/V_d = 3.0$; (d) $V_f/V_d = 4.5$. Note that as the cross-draft ratio increases both the critical distance increases and the shape of the curve becomes flatter.

For a given hood diameter and cross-draft, Eq. 13.3 indicates that the critical distance varies approximately as the square root of the face velocity. This is consistent with the capture velocity findings of DallaValle and others discussed in Chapter 5. The important advance of the capture efficiency method described here over the capture velocity approach now used to describe exterior hood performance is that Eq. 13.2 accounts for the effects of a cross-draft in an explicit, quantifiable manner. That is, if the magnitude and direction of the cross-draft velocity are known, the hood flow required to overcome that velocity can be calculated. This is a considerable improvement over the guidance offered by Table 5.1.

Equation 13.2 can be used to predict capture efficiency for a point source of contaminant located anywhere in front of a flanged exterior hood. For area sources of contaminant release, such as open-surface tanks, the point-source capture efficiency predictions can be integrated over the area to obtain an overall value of capture efficiency for the source. An example of this procedure is shown in Fig. 13.7. The same 15-in. by 1.5-in. flanged hood as used previously is used here to ventilate a 15-in.-square open-surface tank, and the conditions are the same as in Fig. 13.3c ($V_f = 1000$ fpm, $V_d = 200$ fpm). The airflow patterns shown in Fig. 13.3 apply with no source present and must be modified to account for the presence of the tank, which prevents airflow from below the centerline plane. The modification can be accomplished by including a symmetric image suction source located below the surface of the tank. The effect is to increase the reach of the hood and move the critical streamline outward from the hood face. The critical streamline shown in Fig. 13.7 reflects this effect. The integrated area source model for this case predicts a capture efficiency of only about 40%, which is not surprising given the high cross-draft velocity and the resulting critical streamline.

13.2.3   Shortcomings of the Centerline Velocity Approach

The current design procedure for exterior hoods is based on the selection of a proper capture velocity followed by the use of an equation to predict this velocity as a function of hood airflow and distance along the hood centerline. The

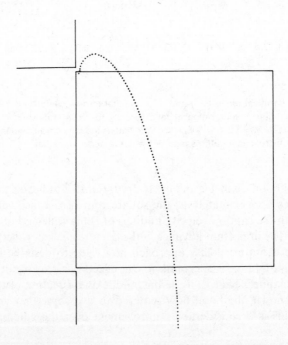

Figure 13.7 Plan view of a 15-in.-square open-surface tank and a 15-in. by 1.5-in. flanged slot hood, showing the critical streamline with $V_f = 1000$ fpm and $V_d = 200$ fpm. The area source capture efficiency in this case is 0.41.

inadequacies of this procedure in the presence of significant cross-drafts can be illustrated using Fig. 13.3c.

Figure 13.3c indicates that the critical streamline crosses the hood centerline about 5.5 in. from the hood face. Since the slot width is 1.5 in., the ratio of critical distance (x) to width (w) in this case is $x/w = 5.5$ in./1.5 in. $= 3.7$. It is instructive to calculate the capture velocity at this distance as predicted by the various models summarized in Table 5.2.

Garrison's empirical model (Eq. 5.60) predicts that

$$V = 0.29 V_f \left(\frac{x}{w} \right)^{-1.1} \tag{5.60}$$

$$= 0.29 \times 1000 \left(\frac{5.5}{1.5} \right)^{-1.1}$$

$$= 69 \text{ fpm}$$

Silverman's equation (Eq. 5.57) predicts that

$$V = 0.36 V_f \left(\frac{x}{w} \right)^{-1} \tag{5.57}$$

$$= 0.36 \times 1000 \left(\frac{5.5}{1.5} \right)^{-1}$$

$$= 98 \text{ fpm}$$

while Flynn, Conroy, and Ellenbecker (FCE) (Eq. 5.61) predict that

$$V = \frac{V_f}{2\pi [0.25 + (x/w)^2]^{1/2} [0.25 + (x/l)^2]^{1/2}} \tag{5.61}$$

$$= \frac{1000}{2\pi [0.25 + (5.5/1.5)^2]^{1/2} [0.25 + (5.5/15)^2]^{1/2}}$$

$$= 69 \text{ fpm}$$

The recently validated Garrison and FCE models predict the same capture velocity, while the much older and widely accepted Silverman model predicts a value that is about 40% too high. Garrison's empirical equation (Eq. 5.60) unfortunately was developed only for x/w ratios less than or equal to 4, so that while it can be used in this example, it cannot be used in cases such as Fig. 13.3b, where the cross-drafts are lower and the ratio of critical distance to hood width is greater than 4. The FCE theoretical model holds for all values of x/w and has been validated experimentally for x/w values up to 7.5.

The actual velocity at the point where the critical streamline crosses the centerline is the vector sum of the capture velocity created by the hood and the cross-draft velocity. The use of the Silverman equation in this case overpredicts velocity toward the hood, and a design based on the Silverman equation would drastically overestimate the performance of the hood.

The current design procedure for open-surface tank hoods also illustrates the shortcomings of the capture-velocity approach. The design procedure, as outlined in the *Ventilation Manual* (ACGIH, 1988), involves first determining the hazard potential and the contaminant evolution rate for the process being performed in the tank; these are used, along with the tank aspect ratio and whether the tank is baffled or not, to determine the necessary airflow, which is specified in cfm per square foot of tank surface area (cfm/ft^2). The specified airflow thus accounts for cross-drafts only qualitatively (the tank is either baffled or it is not) and includes a safety factor to account for contaminant toxicity.

The maximum airflow specified is 250 cfm/ft^2, with the caveat that while 250 cfm/ft^2 "may not produce 150 fpm control velocity at all aspect ratios, the 250 cfm/ft^2 is considered adequate for control." It is instructive to evaluate the adequacy of this airflow for the 15-in.-square source described above, using the area capture efficiency model. With the recommended slot velocity of 2000 fpm and an airflow of 250 cfm/ft^2, the required slot width is 1.88 in. If the cross-draft is 50 fpm, the FCE model predicts a capture efficiency of 91%; this value drops to 61% if the cross-draft is increased to 200 fpm. A more commonly recommended airflow is 150 cfm/ft^2; this airflow gives a predicted efficiency of 84% when the cross-draft is 50 fpm and only 48% when it is 200 fpm. For toxic air contaminants, an efficiency of 84% or 91% may be insufficient to produce safe concentrations at worker breathing zones, while 48% or 61% most certainly would be unacceptable.

13.3 USE OF CAPTURE EFFICIENCY IN HOOD DESIGN

The use of capture efficiency models in the design of exhaust hoods is limited at this time to simple hood shapes, such as rectangles and slots. In addition, the experimental validation of the models has occurred in the laboratory under controlled cross-draft conditions. The models must be tested on real systems in the field, and they must be extended to encompass more complex hood shapes. Much more research needs to be performed on the effects of turbulence on hood performance; all exhaust hoods operate in turbulent fields of varying intensity and scale, and an understanding of such turbulence is key to the quantification of exterior hood performance.

Even at this early stage in its development, the hood capture efficiency concept has proven to be a powerful tool for evaluating hood performance. Although hood capture efficiency has limited practical application at this writing, it is the authors' belief that its incorporation into the design process will result in improved understanding of local exhaust hood performance and the design of better exhaust systems.

LIST OF SYMBOLS

D	hood diameter
G	rate of contaminant generation
G'	rate at which contaminant is captured by an exhaust system
V	velocity
V_d	cross-draft velocity
V_f	hood face velocity
w	slot width
x	distance
x_c	theoretical critical distance
η	capture efficiency
μ	experimental critical distance
ω	turbulence intensity

REFERENCES

American Conference of Governmental Industrial Hygienists, Committee on Industrial Ventilation (1988), *Industrial Ventilation Manual*, 20th ed., ACGIH, Lansing, MI.

Burgess, W., and J. Murrow (1976), "Evaluation of Hoods for Low Volume–High Velocity Exhaust Systems," *Am. Ind. Hyg. Assoc. J.*, **37**(9): 546–549.

Conroy, L., M. Ellenbecker, and M. Flynn (1988), "Prediction and Measurement of Velocity into Flanged Slot Hoods," *Am. Ind. Hyg. Assoc. J.*, **49**(5): 226–234.

Conroy, L., and M. Ellenbecker (1988), "Capture Efficiency of Flanged Slot Hoods under the Influence of a Uniform Crossdraft." Personal communication.

Dalrymple, H. L. (1986), "Development and Use of a Local Exhaust Ventilation System to Control Fume from Hand Held Soldering Irons." in *Proceedings of the 1st International Symposium on Ventilation for Contaminant Control*, October, 1–3, 1985. Toronto, Canada, H. G. Goodfellow, ed., Elsevier, New York.

Ellenbecker, M., R. Gempel, and W. Burgess (1983), "Capture Efficiency of Local Exhaust Ventilation Systems," *Am. Ind. Hyg. Assoc. J.*, **44**(10): 752–755.

Fletcher, B., and A. Johnson (1986), "The Capture Efficiency of Local Exhaust Ventilation Hoods and the Role of Capture Velocity," in *Proceedings of the 1st International Symposium on Ventilation for Contaminant Control*, October 1–3, 1985, Toronto, Canada, H. G. Goodfellow, ed., Elsevier, New York.

Flynn, M., and M. Ellenbecker (1985), "The Potential Flow Solution for Air Flow into a Flanged Circular Hood," *Am. Ind. Hyg. Assoc. J.* **46**(6): 318–322.

Flynn, M., and M. Ellenbecker (1986), "Capture Efficiency of Flanged Circular Exhaust Hoods," *Ann. Occup. Hyg.*, **30**, (4): 497–513.

Flynn, M., and M. Ellenbecker (1987), "Empirical Validation of Theoretical Velocity Fields into Flanged Circular Hoods," *Am. Ind. Hyg. Assoc. J.*, **48**(4): 380–389.

Jansson, A. (1980), "Capture Efficiencies of Local Exhausts for Hand Grinding, Drilling and Welding," *Staub-Reinhalt. Luft* (English ed.), **40**(3): 111–113.

Regnier, R., R. Braconnier, and G. Aubertin (1986), "Study of Capture Devices Integrated into Portable Machine-Tools," in *Proceedings of the 1st International Symposium on Ventilation for Contaminant Control.*, October 1–3, 1985, Toronto, Canada, H. G. Goodfellow, ed., Elsevier, New York.

14

Evaluation and Control of the Thermal Environment

Interest in the evaluation and control of the thermal environment in the workplace stems from both production and health concerns. Certain processes require close humidity control (e.g., textile manufacturing to permit proper handling of the fibers and yarn) and/or temperature control (e.g., precision machining for close dimensional control of parts). In this chapter we do not deal with such production concerns but rather, with the maintenance of a thermal environment suitable for the occupant. The primary concerns are thermal comfort, which may affect performance, and heat stress, which may lead to disease.

Heat stress refers to the environmental thermal conditions imposed on the body; *heat strain* is the result of this thermal stress on the occupant. *Comfort* is defined as the state of mind in which satisfaction is expressed with the thermal environment. The various indices used to characterize the thermal environment and permit the engineer to evaluate its effect on the occupant are reviewed in this chapter, as are the principal control measures applied in the workplace. The air conditioning equipment and appropriate distribution systems to achieve acceptable thermal conditions are described in Chapter 15.

14.1 PROPERTIES OF THE THERMAL ENVIRONMENT

Atmospheric air can be thought of as a binary mixture of dry air and water vapor. Basic properties of air–water vapor mixtures have been determined very accurately and are commonly available in tabular form for standard barometric pressure (14.696 psia). The properties at other than standard pressure can be

371

calculated with good accuracy on hand calculators or by personal computer. A less rigorous method using the perfect gas relationship and Dalton's law results in negligible errors for assessing the health effects of hot environments. The psychrometric chart is the most convenient format for presenting these data for use in solutions to psychrometric problems and the psychrometric chart shown in Fig. 14.1 will be used for that purpose in this chapter.

The thermodynamic state for moist air at a given barometric pressure is defined if two independent properties are known. This state point on the psychrometric chart is usually determined by the dry bulb and psychrometric wet bulb temperatures. These two measurements are commonly made with matched alcohol or mercury-in-glass thermometers. The dry bulb temperature is obtained by an unmodified thermometer: if, however, there is a significant radiation source such as an oven or furnace in the vicinity, the bulb should be shielded with a reflective cover. A cotton wick is sleeved over the bulb of the second thermometer and this thermometer is used to obtain the psychrometric wet bulb temperature. The wick should be washed in distilled water to remove any sizing and then wet with distilled water. If tap water is used, residual salts will build up on the wick, causing a significant error in the reading.

Unsaturated air moving over the wet bulb thermometer evaporates water from the wick, resulting in a cooling of the bulb and a reduction in the temperature reading. The difference between dry bulb and wet bulb temperatures is termed the wet bulb depression; its magnitude depends on the water vapor pressure in the ambient air. An accurate reading of the psychrometric wet bulb temperature requires an air velocity over the bulb in excess of 1200 fpm. This velocity is conveniently obtained by mounting the dry and wet bulb thermometers on a sling (Fig. 14.2) and twirling the sling at a speed that will provide the needed air velocity. The thermometers may also be mounted in an airstream produced by a small blower; these devices are known as aspirating psychrometers (Fig. 14.3). The temperatures are taken after the readings have stabilized and the state point is located on the psychrometric chart at the intersection of the dry and wet bulb temperatures. Once the state point is determined, the other parameters, such as the vapor pressure of water, can be defined with reasonable accuracy.

Once basic familiarity with the psychrometric chart has been attained it is possible to discuss the body's response to the workplace thermal environment and to explore the various methods available to assess the impact of that environment on the comfort and health of the occupant.

14.2 PHYSIOLOGIC RESPONSE TO HEAT

The body displays an impressive ability to maintain a body core temperature between approximately 98.6 and 100.4°F. This temperature equilibrium is achieved by thermal exchange between the environment and the body. The

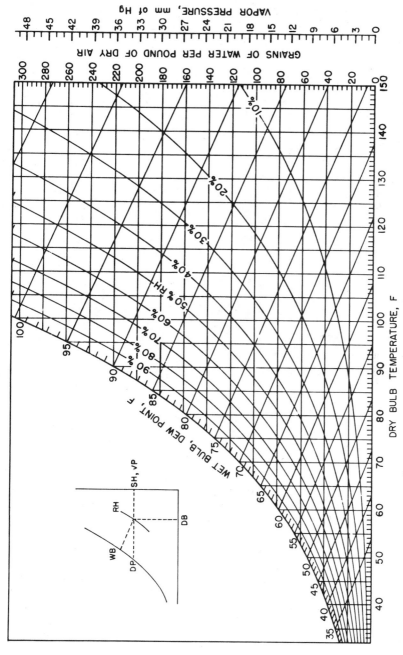

Figure 14.1 Modified psychrometric chart suitable for thermal environment measurements. The dry bulb lines are nearly vertical straight lines which appear to be parallel but in fact are not. The psychrometric wet bulb temperature lines angle upward to the left at approximately 30°. The nest of relative humidity lines curve upward to the right; the boundary lines are the dry bulb lines (relative humidity = 0%) and the saturation line (relative humidity = 100%). Specific humidity of air (gr./lb_a) and vapor pressure (mm Hg) are shown on the vertical axis (From C. H. Powell and A. D. Hosey (1965)).

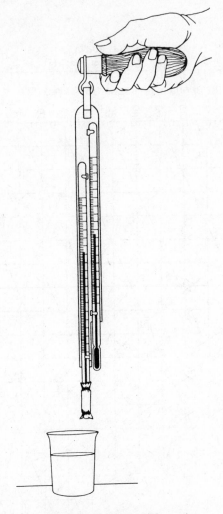

Figure 14.2   Sling psychrometer.

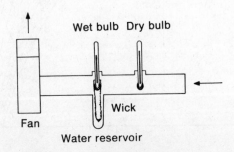

Figure 14.3   Aspirating psychrometer.

exchange is described by the following heat balance equation:

$$(M - W) \pm C \pm R - E = S \tag{14.1}$$

where M = total metabolism
W = external work
C = convective heat transfer
R = heat transfer by radiation
E = evaporative heat loss by sweating
S = change in body heat content

The term $(M - W)$ is the metabolic heat production. When the body is in heat balance, heat production equals heat loss and $S = 0$. To achieve this balance, heat may be lost from the body by convection, radiation, or the evaporation of sweat and Eq. 14.1 then becomes

$$M - W = \pm R \pm C - E \tag{14.2}$$

The deep-body or core temperature is controlled by a complex regulatory system centered in the anterior hypothalamus of the brain. If there is an increase in body temperature, a control mechanism is activated; the peripheral blood vessels dilate and the heart rate increases, resulting in an increased flow of blood from the deep body to the peripheral vessels. The heat associated with this blood flow is transferred by conduction to the epidermis, with a resulting increase in skin temperature and an increased ability to transfer heat to the air by convection if the skin temperature is greater than the air temperature.

If the available convective loss is not adequate to achieve balance, the thermoregulatory control system activates the principal heat control mechanism, evaporative cooling by sweating. The potential for heat transfer from the skin by this method is significant. Approximately 2 thousand sweat glands are distributed over the body, which can provide short-term sweat rates of up to 3 pounds per hour in acclimatized workers. The sweat that wets the warm skin (95°F) will evaporate. The required latent heat of vaporization (1060 Btu per pound of sweat) required to vaporize the sweat is provided from the warm skin. This results in cooling of the skin and the peripheral blood, which exchanges heat with the deep-body tissues. If this mechanism does not control excess heat, storage occurs and heat illness may result (Table 14.1).

Each of the terms of the heat balance equation (Eq. 14.1) must be assessed so that the consequences of the thermal environment on the occupant can be evaluated. The ability to define these heat exchange rates rests principally on work done during World War II by Nelson and his colleagues in the U.S. Army Laboratory at Fort Knox (Nelson et al., 1947) and by the English Medical Research Council and the Royal Naval Tropical Research Unit in England (Bedford, 1946; McArdle et al., 1947).

TABLE 14.1   Heat Disease in Humans

Disorder	Cause	Symptoms	First Aid
Heat cramps	Excessive loss of salt in sweating with inadequate replacement	Painful muscle cramps, body temperature normal or below normal	Rest: administer salt and water
Heat exhaustion	Cardiovascular inadequacy; dehydration	Giddiness, headache; fainting; rapid and weak pulse; vomiting; cold, pale, clammy skin; small rise in body temperature	Rest in shade in recumbent position; administer fluids
Heat stroke	Failure of temperature regulatory center	High body temperature; irritability, prostration, delirium; hot, dry, flushed skin; sweating diminished or absent	Alcohol spray bath or immersion in cold water, medical emergency requiring a physician

14.2.1   Metabolic Heat

During work the body converts chemical energy to mechanical energy. This process is quite inefficient, and large amounts of heat must be rejected.

$$M - W = \text{IBH} \qquad (14.3)$$

where M = total metabolic rate, Btu/hr
  W = external work, Btu/hr
  IBH = internal body heat, Btu/hr

When the worker is at rest the external work is zero and internal body heat is equal to the total metabolism. The minimum heat production possible or the so-called basal metabolic rate is about 240 to 280 Btu/h. Under conditions of work the body is less than 20% efficient. If external work of 200 Btu/h is accomplished, the total metabolic rate is about 1000 Btu/h and the internal body heat generated is 800 Btu/h. It is practice to consider the heat load that must be transferred from the body to ensure that heat storage does not occur as M, the total metabolism. It is important, therefore, that we have a method to estimate the magnitude of this load at work.

The heat load due to body metabolism can be assessed in several ways. The most accurate field technique involves measurement of the worker's oxygen consumption using classical work physiology techniques. In this procedure the worker inhales fresh air and the exhaled breath is collected in a large bag over a short period. The respiratory minute volume and the concentration of oxygen in the exhaled breath are measured with field instruments, and the volume of oxygen consumed is calculated. A liter of oxygen consumed is equivalent to 19.2 Btu of metabolic heat (AIHA, 1971; ISO, 1983).

Data collected on various industrial jobs by work physiologists have been published and are useful in estimating metabolic rates when direct measurements cannot be done (Table 14.2). These data are not comprehensive, and frequently

TABLE 14.2   Energy Expenditures in Different Occupations

Occupation	Btu/h	Occupation	Btu/h
BUILDING		MINING	
Mixing cement	1130	Loading	1280
Plastering walls	980	Hewing	1700
Building wall	960	Packing	1630
Shaping stones	910	Timbering	1540
with hammer		Drilling	1340
Carrying bricks	860		
or cement			
CARPENTRY			
Planing (hardwood)	2180		
Hand sawing (hardwood)	1800		
Chiselling	1370		
Joining floorboards	1060		
Measuring wood	580		
Machine sawing	580		
	Miscellaneous Light Industries		
Loading chemicals	1440	Plastic molding	790
into mixer		Machining	740
		(engineering)	
		Sheet metal working	720
Casting lead balls	1150	Shoe manufacturing	720
in molds (battery		Sewing machine work	670
manufacture)		Shoe repairing	640
Machine fitting	1000	Medium assembly	640
Toolroom workers	940	work	
Pressing	940	Electronics	640
Pressing metal house	910	Printing	530
hold utensils		Hand compositing	530
Turning	880	Electrical armature	530
Joinery	870	winding	
Molding ebonite	860	Drilling	430
Filing (large file)	840	Light assembly work	430
Tool setting	810	Draftsman	430
Light machine work	810	Watch repairing	380
		Typing	340

TABLE 14.2 (Contd.).

Occupation	Btul h	Occupation	Btul h
Steel and Iron Industries			
OPEN HEARTH		HEAVY MILL	
Slag removal	2780	Tending heating	2450
Dolomite	2620	furnaces	
shoveling		Hand rolling	2140
Tipping the	1440		
molds		14-in. MERCHANT MILL	
WIRE ROLL MILL		Merchant mill	2260
Wire bundling	2450	rolling	
Roughing	1970	Forging	1560
		Fettling	1180
		SUNDRY ACTIVITIES	
Shovelling 8-kg load		Tree felling	2570
		with saw	
Distance 1 m		Trimming felled	2450
1–2 m lift	2280	trees	
12 times/min		Digging trenches	2040
		Pushing wheelbarrow	1200
less than 1 m lift	1800	(100-kg load)	
12 times/min			
Distance 2 m			
1–2 m lift	2520		
12 times/min			
less than 1 m lift	2040		
12 times/min			

After Leithead and Lind (1968).

metabolic rates of common tasks of great interest to the occupational health specialist are not available. In such cases, an approach proposed by NIOSH involving detailed industrial engineering study of the work elements of each job may be useful (Mutchler et al., 1975). The individual work elements are identified and the metabolic cost is assigned to each element from a library of data. The average metabolic rate for the work is then calculated from the individual work elements (Table 14.3).

Metabolic rates in industry vary from light electronic assembly (600 Btu/h) to heavy, short-duration tasks in steel making (2000 Btu/h). Frequently, the metabolic load is much greater than both the radiant and convective heat loads; our inability to predict this metabolic load accurately is a major drawback in quantifying the impact of heat on the worker using the heat balance equation.

TABLE 14.3   Estimating Energy Cost of Work by Task Analysis

Task	BTU/h	
A. BODY POSITION AND MOVEMENT		
Sitting	71.5	
Standing	142.9	
Walking	476.4	
Walking uphill	Add 190.6 per meter rise	
B. TYPE OF WORK	AVERAGE	RANGE
Handwork		
Light	95.3	47.6–285.8
Heavy	214.4	
Work one arm		
Light	238.2	166.7–595.5
Heavy	428.8	
Work both arms		
Light	357.3	238.2–833.7
Heavy	595.5	
Work whole body		
Light	833.7	595.5–2143.8
Moderate	1191.0	
Heavy	1667.4	
Very heavy	2143.8	
C. BASAL METABOLISM		
	238.2	
D. SAMPLE CALCULATION[b]	Average	
Assembling work with		
heavy hand tools		
Standing	142.9	
Two-arm work	833.7	
Basal metabolism	238.2	
Total	1214.8	

[a]For standard worker of 70 kg (154 lb) body weight and 1.8 m^2 (19.4 ft^2) body surface.
[b]Example of measuring metabolic heat production of a worker when performing initial screening.
After NIOSH Publication No. 86–113.

14.2.2   Convection

The rate of heat transfer by convection between the worker and the environment is dependent on the area of the body available for convective heat transfer, the velocity of air passing over the body, and the difference between ambient air temperature and the mean weighted skin temperature. Studies conducted by Nelson et al. at Fort Knox during World War II (Nelson, 1947) led to early predictive equations for convective heat transfer which were modified by other investigators, including McKarns and Brief (1966), to provide relationships suitable for field studies. For a lightly clad "standard man" with a body surface area for convective heat transfer of 19.4 ft^2 and a mean skin temperature of 95°F working under conditions of forced convection the equation for heat transfer by convection is

$$C = 0.65V^{0.6}(t_a - 95) \qquad (14.4)$$

where C = heat exchange by convection, Btu/h
  V = velocity of air passing over body, fpm
  t_a = temperature of air or temperature of garment exposed to skin, °F

If the air temperature is above 95°F, convective heat is transferred to the body and the convective heat gain will add to the metabolic heat production. When the air temperature is below 95°F, heat is lost from the body to the environment.

14.2.3 Radiation

The heat transfer by radiation is a complex thermodynamic process; however, an operating relationship suitable for field studies can be devised if the emittance of human skin is assumed to be 1.0, the mean weighted skin temperature is 95°F, and the body area available for radiation exchange is 13.5 ft^2 (Nelson, 1947; Belding and Hatch, 1955; McKarns and Brief, 1966).

$$R = 15.0(\bar{t}_r - 95) \qquad (14.5)$$

where R is the heat exchange by radiation, Btu/h and $\bar{t}_r$ is the mean radiant temperature of the surrounding surfaces, °F.

The mean radiant temperature is the temperature of a "black" enclosure with a uniform wall temperature that would exchange the same amount of heat with the occupant by radiation as the environment under investigation. The mean radiant temperature can be estimated indirectly using the globe thermometer, a 6-in. copper sphere painted black with a thermometer positioned at the center (Fig. 14.4). Placed in the test environment and permitted to equilibrate, the radiation gain on the sphere is balanced by its convective loss. The mean radiant temperature is then calculated.

$$\bar{t}_r = t_g + 0.13V_a^{0.5}(t_g - t_a) \qquad (14.6)$$

where $\bar{t}_r$ = mean radiant temperature, °F
  t_g = globe temperature, °F
  V_a = air velocity at the globe position, fpm
  t_a = dry bulb temperature, °F

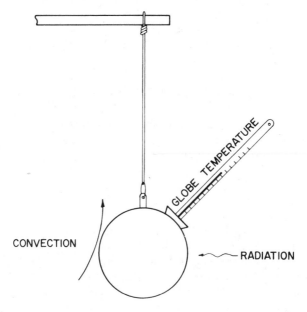

Figure 14.4 Globe thermometer. [From Vernon (1930).]

This globe thermometer device has a significant time lag; the temperature should be noted periodically after 15 minutes of exposure until it stabilizes. Smaller globe thermometers with thin walls providing a shorter response time are now available (see Section 14.4.2).

In high radiative environments such as steel and glass manufacturing operations, the difference between air and globe temperatures may exceed 30°F. Since the mean radiant temperature in such environments will be greater than 95°F, the radiation heat load will be added to the metabolic heat production. In a cold storage meat locker there will be considerable radiative heat loss and the worker may suffer from cold stress.

14.2.4 Evaporation

As noted above, the evaporation of sweat from the skin is a primary method to provide heat loss and maintain thermal balance. If it is assumed that 90% of the total body area of the standard acclimated man is available for sweating and that the saturation vapor pressure of water on the skin at 95°F is 42 mm Hg, Belding and Hatch (1955) have proposed an approximate solution which has, with some modifications, proven useful in field studies.

$$E_{max} = 2.4\, V_a^{0.6}(42 - p_a) \tag{14.7}$$

where E_{max} = maximum evaporative heat loss, Btu/h
$\quad\quad V_a$ = velocity of air passing over body, fpm
$\quad\quad p_a$ = vapor pressure of water in air at the worker's location, mm Hg

The ability of the body to sweat efficiently is governed to a large degree by acclimatization. It is possible for an acclimatized person to sweat a rate of 1 L/h for up to 4 hours, providing a heat loss by evaporation of approximately 2400 Btu/h. This sweat rate cannot be maintained for an extended period. The maximum sustained rate for an acclimatized worker has been proposed as 2.5 liters for a 4-hour period (McArdle et al., 1947).

14.2.5 Storage

To prevent heat disease resulting from an increase in body temperature, the total of the metabolic, convective, and radiation loads must be balanced by evaporative sweating, as noted in Eq. 14.1. If this balance is not obtained, the body will store heat. Relatively small levels of stored heat will result in significant changes in body temperature; storage of 125 Btu by a 150-lb man results in an increase in body temperature of approximately 1°F. To ensure that a temperature excursion will not exceed 2.0°F, a target recommended by the International Labor Organization, the heat storage, cannot exceed 250 Btu. This target can be monitored by indices of comfort, indices of heat stress, and by physiologic measurements of heat strain of the exposed worker. The varied approaches widely used in the United States to monitor the thermal environment are described in this chapter.

14.3 INDICES OF COMFORT

The sensory perception by occupants of the thermal comfort afforded by the environment has been given major attention for over fifty years. The major indices of comfort currently in use in the United States are reviewed in this section and sample calculations are presented in Example 14.1.

14.3.1 Original Effective Temperature

In the early 1920s Houghten and Yaglou (1923) under sponsorship of ASHRAE studied the thermal impressions of observers to various thermal test environments. From this study they devised a sensory scale of warmth called the *effective temperature* (ET) which combined air temperature, humidity, and air velocity into a single numerical index (Yaglou, 1927). For more than fifty years the effective temperature was the most widely used index of comfort in the world. Over the years it has been modified to improve its predictive capability, the most important modification being Bedford's use of the globe temperature to include radiation effects in the index (Bedford, 1946). Since it continues to be encountered in discussions of thermal comfort, the development of the original effective temperature and its limitations will be discussed as an introduction to the more accurate empirical indices now available.

The experimental protocol on which Houghten and Yaglou based their index is described in Fig. 14.5a. Male subjects passed from a control chamber to an adjoining test chamber and compared the relative feeling of warmth in the two atmospheres. Two series of tests were run—one with subjects at rest and stripped to the waist and another with subjects at rest but lightly clothed (1 clo insulation). The numerical value of the *effective temperature index* was defined as the air

Control Room	Test Room		
$t_a = 70\,°F$	Test	$t_a(°F)$	$t_{wb}\,(°F)$
$t_{wb} = 70\,°F$	1	70	70
RH = 100%	2	72	67.5
ET = 70°F	3	74	64.0
	4	76	60.5

(a)

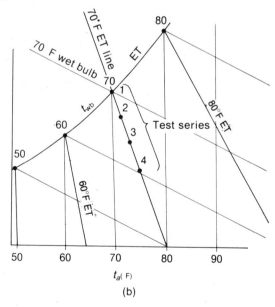

(b)

Figure 14.5 (a) Schematic of adjoining test chambers used by Houghten and Yaglou in the experimental development of the effective temperature index. In the study illustrated in this figure the control room was maintained at a saturated condition of 70°F. Conditions in the test chamber were varied. In the test series the subjects identified the four noted dry bulb–wet bulb conditions as inducing the same sense of warmth as the control room. (b) In this psychrometric chart the loci of these four tests represent the loci of the ET 70°F line.

temperature of the still and saturated (relative humidity = 100%) air in the control chamber which induced the same sense of warmth or cold as the still but nonsaturated test environment.

The results of a test series at a given effective temperature (in this example 70°F) plotted on a psychrometric chart give a series of air temperature and relative humidity state points the observers felt were equivalent to the control room saturated at 70°F (Fig. 14.5b). The loci of these "equivalent" conditions is a straight line with negative slope displaced from the wet bulb lines. These data were transferred to charts to depict comfort zones; the charts showed that there are many combinations of air temperature and humidity that produce the same feeling of warmth, namely all points on a given effective temperature line. The application of the chart was limited to United States males occupants living or working in spaces heated by convection, not radiation, and it could only be applied to situations where the exposure periods allowed the occupants to become fully acclimatized. Comfort charts based on the original effective temperature index were published in the ASHRAE Fundamentals Volumes through 1967.

Bedford devised a procedure to correct the effective temperature for radiation (Bedford, 1946). In this procedure the globe temperature, not the air temperature, is used with a pseudo wet bulb to determine the effective temperature graphically (Fig. 14.6). This scale, known as the *corrected effective temperature* (CET), is valid for increases in radiant load up to 32°F.

Graphical solutions were later devised for the calculation of effective temperature for various values of air movement for both normally clothed subjects and those stripped to the waist (Fig. 14.7). Although this extended the use of the effective temperature from still air, the index still did not permit a correction for metabolic rate although Yaglou (1947) did note that as the metabolic rate

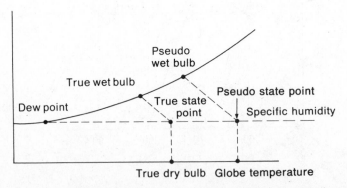

Figure 14.6 Calculation of pseudo-wet-bulb temperature for determination of corrected effective temperature. Determine the state point from the dry bulb and wet bulb temperatures. Establish a pseudo state point from the specific humidity and at an air temperature equal to the globe temperature. Determine the pseudo wet bulb for this state point. The pseudo wet bulb and the globe temperature are used to determine the corrected effective temperature (Fig. 14.7). See Example 14.1 for this calculation.

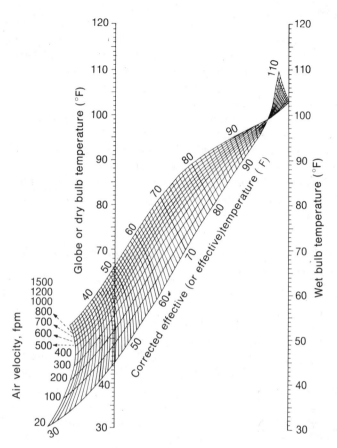

Figure 14.7 Chart showing corrected effective or effective temperature for normally clothed persons.

increased, the effective temperature lines approached the wet bulb temperature lines.

As indices of comfort, both the effective temperature and corrected effective temperature have shortcomings. Yaglou stated that the effective temperature index overestimates the effects of humidity at low air temperatures and conversely underestimates humidity effects at high air temperatures. This problem can be traced to the test protocol used in the original investigation. Since the subject's ability to differentiate small differences between environmental conditions in the control and test chamber degrades after a few minutes, the characterization of the test atmosphere as "cooler," "warmer," or "the same as" the control chamber had to be done rapidly. When the subject entered the control chamber the exposed clothing and skin adsorbed water vapor and the heat of adsorption generated a sense of warmth; when the subject then immediately went from the control to the test chamber, this adsorbed moisture evaporated, causing a brief cooling effect which the subject could sense.

Houghten and Yaglou did not recognize sorption of moisture as the basis for this phenomenon; they believed that it was due to adaptation. The effective temperature scale is, therefore, based on an incorrect belief that all combinations of air temperature and relative humidity felt to be equally warm within the first minute or so would result in the same degree of adaptation after extended exposure.

In later years Yaglou cautioned that care should be used in applying the effective temperature due to the above effect. He stated that the air temperature is a better index of comfort than the effective temperature for ordinary conditions with minimum air movement. However, he felt that the effective temperature was useful in evaluating warm, moist conditions where the adsorption effect does not occur since the skin is wet with perspiration. In addition to its use in assessment of comfort, the effective temperature has been a widely used predictor of heat stress, as discussed later in this chapter.

14.3.2   American Society of Heating, Refrigerating, and Air Conditioning Engineers Indices

The original effective temperature scale was reevaluated by ASHRAE in the 1950s. This work led to an improved index of thermal comfort identified as the new effective temperature (ET*). The ET* loci are plotted on the psychrometric chart in the same manner described for the original effective temperature (Fig. 14.5). The ET* scale is labelled as the dry bulb temperature at the intersection of loci of constant physiologic strain with the 50% relative humidity line, not the saturation line as is the case with the original effective temperature. The ET* index was used to define the comfort zone recommended in the ASHRAE Standard 55–74 Thermal Environmental' Conditions For Human Occupancy (ASHRAE, 1974). This standard applied to occupants who were lightly clothed and seated or engaged in sedentary activity with an average air velocity below 40 fpm. Although these qualifications may seem restrictive, it is true, as ASHRAE indicated, that perhaps 80% of the indoor work population are covered in this application realm.

ASHRAE 55–74 was later replaced with ASHRAE 55–1981, which used another index, operative temperature, with humidity ratio to define the comfort envelope (Fig. 14.8) (ASHRAE, 1981). Operative temperature is a calorimetric scale devised to measure the combined thermal effects of air temperature, mean radiant temperature, and air movement on unclothed subjects. The operative temperature index represents the temperature of a uniform black enclosure in which the occupants would exchange the same amount of heat by convection and radiation as in the given nonuniform environment. Under limited conditions (air movement less than 80 fpm and a mean radiant temperature below 120°F) the operative temperature, t_o, can be approximated by the average of the air temperature and the mean radiant temperature. Since this condition exists in most indoor environments covered by ASHRAE Thermal Comfort Standards, the operative temperature was utilized as a basis for defining acceptable conditions in the ASHRAE 55–1981 standard.

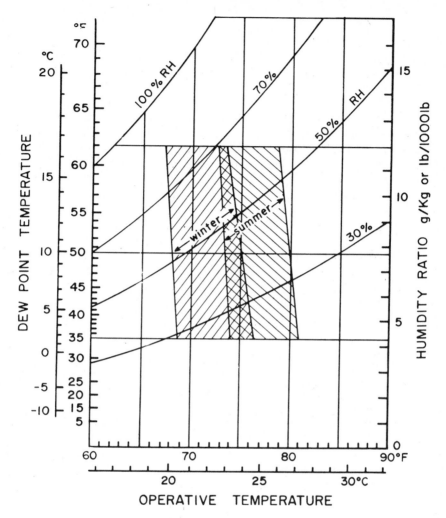

Figure 14.8 Comfort zone from ASHRAE 55–1981 based on operative temperature (Copyright 1981, American Society of Heating, Refrigerating, and Air Conditioning Engineers).

The principal proposal in ASHRAE 55-1981 was to define summer and winter thermal conditions acceptable to 80% of the population based on current clothing practice in the United States. The issue of clothing had not been addressed adequately in previous indices of comfort or in published thermal comfort standards, even though it obviously has a significant effect on a person's sense of warmth and comfort.

The insulation property of clothing is identified by the *clo*. As defined by Gagge, the clo is the amount of insulation necessary to maintain comfort and a measured skin temperature of 92°F in a room at 70°F with air movement not over 10 fpm, humidity not over 50%, and with a metabolism of 360 BTU/h. In physical terms it is the amount of insulation that will allow the passage of 1 cal/m^2/h with a temperature gradient of 0.18°C between the two surfaces. The range of clo values encountered in the United States is shown in Table 14.4. The ASHRAE 55-1981 standard lists clo values to achieve various levels of acceptance during sedentary activity (Fig. 14.9).

In addition to the comfort zones, ASHRAE Standard 55-1981 includes additional information on the control of the thermal environment necessary to achieve acceptance by occupants. Average air movement in the summer may be extended above 50 fpm dependent on air temperature (Fig. 14.10). Adjustments to the comfort zone are cited based on the activity level of the occupants. (Fig. 14.11).

TABLE 14.4 CLO Values for Various Clothing Systems

Mens' Clothing	clo Value		Womens' clothing	clo Value
		Underwear		
Sleeveless	0.06		Bra and Panties	0.05
T-shirt	0.09		Half slip	0.13
Underpants	0.05		Full slip	0.19
		Torso		
Shirt			Blouse	
Light				
Short sleeve	0.14		Light	0.20[a]
Long sleeve	0.22		Heavy	0.29[a]
Heavy				
Short sleeve	0.25		Dress	
Long sleeve	0.29		Light	0.22[b]
(plus 5% for			Heavy	0.90[a,b]
tie or turtleneck)				
Vest			Skirt	
Light	0.15		Light	0.10[b]
Heavy	0.29		Heavy	0.22[b]
Trousers			Slacks	
Light	0.26		Light	0.26
Heavy	0.32		Heavy	0.44
Sweater			Sweater	
Light	0.20[a]		Light	0.17[a]
Heavy	0.37[a]		Heavy	0.37[a]
Jacket			Jacket	
Light	0.22		Light	0.17
Heavy	0.49		Heavy	0.37
		Footwear		
Socks			Stockings	
Ankle length	0.04		Any length	0.01
Knee-high	0.10		Panty hose	0.01
Shoes			Shoes	
Sandals	0.02		Sandals	0.02
Oxfords	0.04		Pumps	0.04
Boots	0.08		Boots	0.08

[a]Less 10% if short sleeve or sleeveless.
[b]Plus 5% if below knee length, less 5% if above.
After NIOSH 1986.

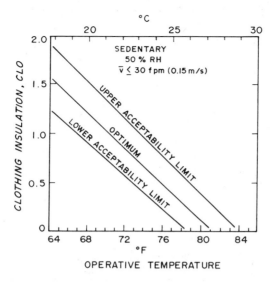

Figure 14.9 Range of clothing insulation to achieve comfort at various operative temperatures (Copyright 1981, American Society of Heating, Refrigerating, and Air Conditioning Engineers).

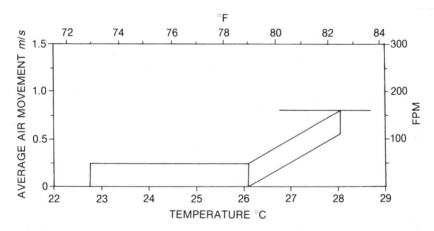

Figure 14.10 Average air velocity permitted in the summer and extended summer zones under ASHRAE Comfort Standard 55–1981 (Copyright 1981, American Society of Heating, Refrigerating, and Air Conditioning Engineers).

14.3.3 Fanger Comfort Equation

As noted in Section 14.1, the heat balance equation must balance $(S = 0)$ to ensure that heat storage does not occur with resulting discomfort and heat disease. The environmental conditions permitting heat balance are broad; within those extended boundaries there is a narrow range of conditions that provide thermal comfort. In a series of studies Fanger (1982) has provided a method to quantify the degree of comfort presented by a given thermal environment (ASHRAE,

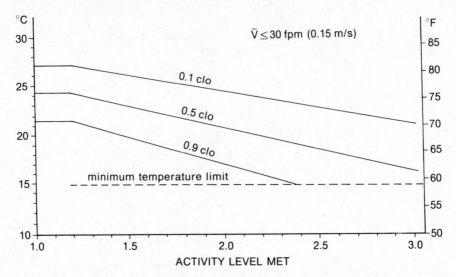

Figure 14.11 Optimum operative temperatures for occupants under varying metabolic rates and clothing in spaces with air velocity less than 30 fpm. (Copyright 1981, American Society of Heating, Refrigerating, and Air Conditioning Engineers).

1985). The comfort charts devised by Fanger and validated in a series of studies have received major acceptance in the United States and Europe.

Fanger states that at a given level of activity the only physiologic parameters that influence heat balance are the mean skin temperature and sweat loss and that there is a narrow range of skin temperature and sweat loss consistent with the comfort zone. The relationship between mean skin temperature and net metabolic rate is defined by Fanger (1982) in one equation and the relationship between evaporative heat loss and net metabolic rate in a second.

Combining these relationships with the basic heat balance equation resulted in a new function that permitted Fanger to propose combinations of environmental parameters (air movement, mean radiant temperature, dry bulb temperature, and vapor pressure of water in air) that provide a comfortable thermal environment for a person with a known metabolic rate and dressed at a given clo level. These data are translated into a set of comfort charts. To apply the charts, metabolic rate and clo values are estimated; the combination of four acceptable thermal environmental factors for thermal comfort are defined by the charts (ASHRAE 1985).

None of the indices of thermal comfort described previously permit an estimate of the percentage of the occupants who are satisfied with the given thermal environment. The Fanger Comfort Charts permit the investigator to make such a calculation. The degree of discomfort in a given environment can be estimated using a predicted mean value (PMV) – predicted percentage of dissatisfied (PPD) index. The PMV is based on the following psychophysical scale:

-3 cold
-2 cool

−1 slightly cool
 0 neutral comfort
 1 slightly warm
 2 warm
 3 hot

The PMV can be calculated from an equation proposed by Fanger (1982) or obtained from tables of the type shown in Table 14.5. Once the PMV is obtained it is used to establish the predicted percentage of occupants dissatisfied (PPD) with the given thermal environment from Fig. 14.12 (Example 14.1). A thermal comfort meter is now available for direct determination of PMV and PPD values in the workplace (Fig. 14.13).

The ISO (DIS 7730) standard on thermal comfort proposes that the PMV be maintained within +0.5 of the neutral point (ISO, 1984); this would limit the percentage of dissatisfied persons to 10%.

14.4 INDICES OF HEAT STRESS

In addition to indices of thermal comfort of the work environment, it is important that indices be available to predict the ability of the worker to handle severe thermal stress without progression to heat disease. Since the turn of the century approximately fifty indices of heat stress have been proposed. The indices that are in wide use or have the potential for extensive application are covered in this section and calculations are shown in Example 14.2.

14.4.1 Corrected Effective Temperature

Yaglou (1947) considered extending the application of the *corrected effective temperature* (CET) index, originally devised for thermal comfort applications, to heat stress, and during the period 1950–1970 it was widely used for this purpose. As noted in Section 14.3, this index has significant limitations, including its inability to correct for clothing and metabolic rate. The heat stress application of the CET during the 1950s and 1960s utilized the guidelines in Table 14.6. Environmental conditions exceeding these guidelines could not be tolerated by the worker.

The most extensive critique of the value of the CET index for heat stress was conducted by investigators in England during World War II (McArdle et al. 1947). As noted earlier, this study revealed that the CET exaggerated the stress produced by environments characterized by high dry bulb temperatures in the air movement range 100 to 300 fpm, while underestimating the stress of hot and humid environments with low air movement. The most significant finding was that there was not a linear relationship between CET and the observed physiological indices of strain, including elevation of rectal temperature, pulse rate, sweat loss, and work tolerance time. Different climatic conditions with the same CET did not produce the same physiologic strain.

Table 14.5 Fanger Predicted Mean Values (PMV) for selected combinations of clothing I_{cl}, metabolic rate M, air movement V_{ar}, relative humidity RH, and air temperature. The PMV values are included in the range $+3$ hot to -3 cold. The application of these tables in conjunction with Figure 14.12 to obtain the predicted percentage of dissatisfied index (PPD) is described in Example 14.1.

PMV	M = 1,0 met v_{ar} = < 0,1 m/s RH = 50%						
	I_{cl} , clo						
t_a °C	0,1	0,3	0,5	0,8	1,0	1,5	2,0
10						−2,2	−1,4
12						−1,8	−1,0
14					−2,5	−1,4	−0,7
16				−2,5	−1,9	−1,0	−0,3
18				−1,9	−1,4	−0,5	0,0
20			−2,3	−1,3	−0,9	−0,1	0,4
22		−2,3	−1,5	−0,7	−0,3	0,4	0,8
24	−2,3	−1,4	−0,8	−0,1	0,2	0,8	1,1
26	−1,2	−0,5	0,0	0,6	0,8	1,2	1,5
28	−0,1	0,4	0,8	1,2	1,4	1,7	1,9
30	1,0	1,3	1,6	1,8	1,9	2,1	2,3
32	2,0	2,2	2,3	2,4	2,5	2,6	2,6

PMV	M = 1,2 met v_{ar} = < 0,1 m/s RH = 50%						
	I_{cl} , clo						
t_a °C	0,1	0,3	0,5	0,8	1,0	1,5	2,0
10					−2,7	−1,6	−0,9
12				−2,8	−2,2	−1,2	−0,6
14				−2,3	−1,8	−0,9	−0,3
16			−2,8	−1,8	−1,3	−0,5	−0,0
18		−2,9	−2,1	−1,2	−0,8	−0,1	0,3
20		−2,2	−1,5	−0,7	−0,4	0,2	0,6
22	−2,3	−1,4	−0,8	−0,2	0,1	0,6	0,9
24	−1,4	−0,7	−0,2	0,3	0,6	1,0	1,3
26	−0,5	0,1	0,4	0,8	1,0	1,4	1,6
28	0,4	0,8	1,1	1,3	1,5	1,7	1,9
30	1,3	1,5	1,7	1,8	1,9	2,1	2,2
32	2,0	2,1	2,2	2,3	2,3	2,4	2,4

PMV	M = 1,6 met v_{ar} = 0,2 m/s RH = 50%						
	I_{cl} , clo						
t_a °C	0,1	0,3	0,5	0,8	1,0	1,5	2,0
10				−2,0	−1,5	−0,7	−0,2
12			−2,6	−1,6	−1,2	−0,4	0,0
14		−2,9	−2,1	−1,3	−0,9	−0,2	0,3
16		−2,4	−1,7	−0,9	−0,5	0,1	0,5
18	−2,8	−1,8	−1,2	−0,5	−0,2	0,4	0,7
20	−2,1	−1,3	−0,7	−0,1	0,2	0,6	0,9
22	−1,4	−0,7	−0,2	0,3	0,5	0,9	1,2
24	−0,7	−0,2	0,2	0,7	0,8	1,2	1,4
26	−0,0	0,4	0,7	1,1	1,2	1,5	1,6
28	0,7	1,0	1,2	1,5	1,6	1,8	1,9
30	1,4	1,6	1,7	1,9	1,9	2,0	2,1
32	2,1	2,2	2,2	2,3	2,3	2,3	2,4

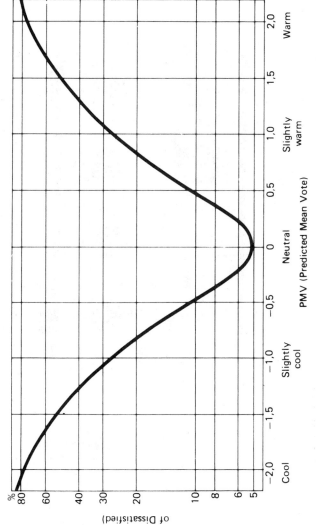

Figure 14.12 Fanger predicted percentage of dissatisfied (PPD) index (Courtesy of Brüel & Kjaer).

Figure 14.13   Thermal comfort meter. The values for clothing, activity, and humidity are set and a heated transducer integrates the effects of air temperature, mean radiant temperature, and air velocity. The instrument calculates the PMV and PPD values. (Courtesy of Brüel and Kjaer.)

TABLE 14.6 Guidelines for Interpreting CET Data

Type of Activity	Maximum Full Shift CET Permissible (°F)
	Unacclimatized
Sedentary	86
Moderate work	82.5
Hard work	80
	Acclimatized
Sedentary	89.6
Moderate work	86.1
Hard work	83.6

14.4.2 Wet Bulb–Globe Temperature

In the early 1950s a number of U.S. Marine recruits exposed to severe heat conditions during training died of heat stroke. The U.S. Navy Bureau of Medicine and Surgery requested that Professor Yaglou, then at the Harvard School of Public Health, devise a simple field procedure to monitor climatic conditions and provide guidelines under which training activities could safely be conducted. A single number index was devised based on passive monitoring of dry bulb air temperature, natural wet bulb temperature (no aspirating air), and the globe temperature (Fig. 14.14). Both indoor and outdoor indices of wet bulb–globe temperature (WBGT) were proposed.

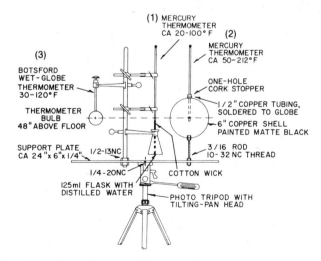

Figure 14.14 WBGT survey instruments for manual measurement of indoor workplaces, including (1) natural wet bulb temperature (2) globe temperature, and for comparison purposes, (3) the wet globe discussed in Section 14.4.3.

Indoors:
$$WBGT = 0.7t_{nwb} + 0.3t_g \qquad (14.8)$$

Outdoors:
$$WBGT = 0.7t_{nwb} + 0.2t_g + 0.1t_a \qquad (14.9)$$

where t_{nwb} = natural wet bulb temperature, either °C or °F
  t_g = globe temperature, either °C or °F
  t_a = air temperature, either °C or °F

The WBGT index proved valuable to the military and continues to be used by the U.S. Armed Forces (Yaglou and Minard, 1957). During the 1950s and 1960s the WBGT was adopted by industry, and in 1973 the WBGT index was adopted by the ACGIH-TLV Committee (Fig. 14.15). The NIOSH Criteria Document (1972) and the revised NIOSH Criteria Document (1986) on hot environments recommend that the WBGT be adopted for industry (Fig. 14.16).

Manual measurements of WBGT such as described in Fig. 14.14 can now be complemented by recording instruments for fixed-location monitoring (Fig. 14.17). Since it does correlate well with indices of strain, the WBGT is now the most broadly applied heat stress index in the world.

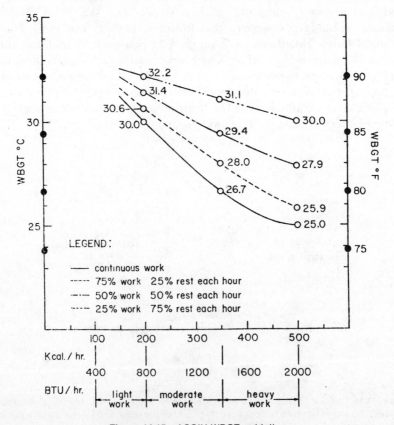

Figure 14.15 ACGIH WBGT guidelines.

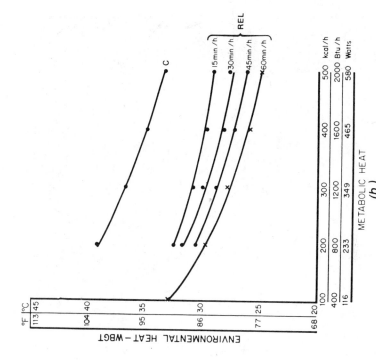

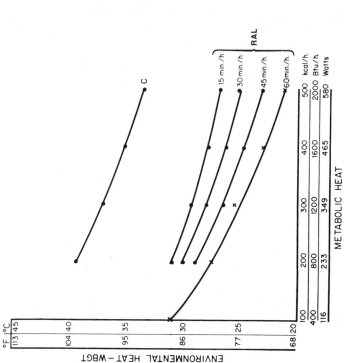

Figure 14.16 NIOSH heat stress alert limits for (a) heat-unacclimatized workers and (b) heat acclimatized workers. C, ceiling limit; RAL, recommended alert limit (From NIOSH Publication 86–113).

(a)

(b)

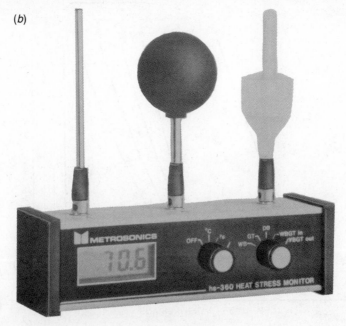

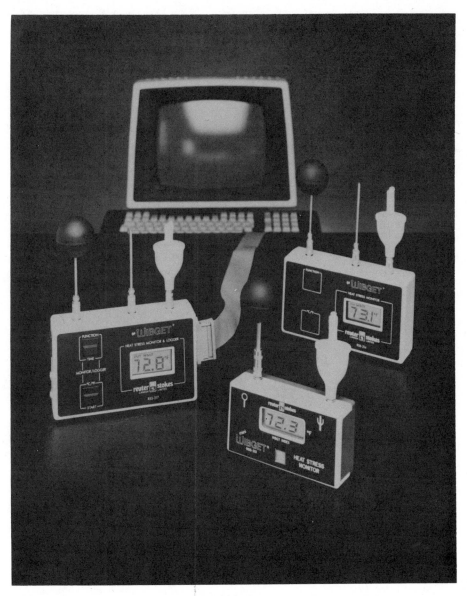

(c)

Figure 14.17 WBGT instruments: (a) Brüel & Kjaer WBGT Heat Stress Monitor Type 1219 (with WBGT Transducer MM 0030) (Courtesy of Brüel & Kjaer); (b) Metrosonics h_s 360 Heat Stress Monitor (Courtesy of Metrosonics, Inc.); (c) Reuter Stokes Heat Stress Monitor (Courtesy of Reuter Stokes).

14.4.3 Wet Globe Temperature

The ideal index for field evaluation of the thermal environment is a single number index based on a simple instrument, which accurately predicts heat stress in all environments. The WBGT index comes close; the wet globe devised by Botsford (1971) comes closer. This device (Fig. 14.18), also called the Botsball, is based on a 2 1/2-in.-diameter thin copper sphere covered with a black cotton mesh cloth. A dial thermometer is inserted in a plastic tube fastened to the sphere, with the sensing head positioned at the center of the sphere. A second concentric tube provides an annular space which acts as a water reservoir to wet the cotton mesh. This instrument is placed in the workplace and the wet globe temperature (WGT) is read when evaporative cooling and heating by convection and radiation equilibrate.

A number of studies have shown excellent correlation between the WGT and the WBGT. Caution is needed since the relationship may not hold for all combinations of thermal environments. The revised NIOSH Criteria Document (1986) recommends WGT as a preliminary screening instrument using the correlations noted below with the calculated equivalent WBGT values compared directly to the proposed WBGT threshold limits shown in Fig. 14.16.

Indoor: equivalent WBGT = WGT + 3°C   (14.10)

Outdoor: equivalent WBGT = WGT + 1°C   (14.11)

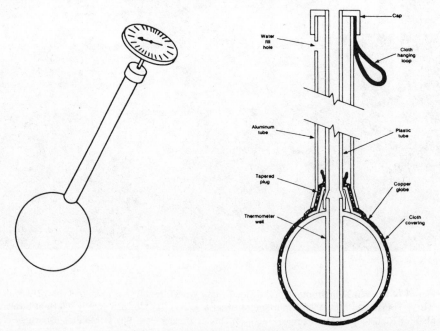

Figure 14.18 Wet globe showing (a) the general appearance of the wet globe and (b) detailed construction of stem and globe without dial thermometer in place. (Courtesy of Brüel and Kjaer.)

14.4.4 Predicted Four-Hour Sweat Rate

During World War II investigators in England found that sweat rate was the best index of strain (McArdle et al., 1947). Since their evaluation of heat stress indices such as Corrected Effective Temperature revealed significant weaknesses, they proceeded to devise an index based on sweat rate. This index, the *predicted 4-hour sweat rate* (P4SR), may be the most accurate of all available heat stress indices. The P4SR is certainly the most flexible since it permits correction for variations in clothing, air movement, work rate, and radiation. The *basic 4-hour sweat rate* (B4SR) is obtained from the initial nomograph (Fig. 14.19) based on observed dry bulb, wet bulb, globe temperatures, and air velocity. The B4SR is then corrected for clothing, work rate, and radiation to give a *predicted 4-hour sweat rate* (P4SR), as shown in Fig. 14.19 and Example 14.2. The investigators state that an acclimatized person may safely work at levels up to a P4SR of 2.5 liters.

This index is rarely used in the United States, apparently due to its complexity, but the authors have found it helpful in conjunction with other heat stress indices.

14.4.5 Heat Stress Index

A rational *heat stress index* (HSI) was proposed by Belding and Hatch (1955). The index is derived from the basic heat balance equation (Eq. 14.1). In this index, the stress to the worker is represented by the required heat exchange rate achieved by evaporation of sweat necessary to maintain body heat balance, that is,

$$E_{req} = M \pm C \pm R \qquad (14.12)$$

where the convective and radiative heat loads are given by Eqs. 14.4 and 14.5.

This required evaporation rate is compared to the maximum evaporative capacity of the body under the stated environmental conditions (from Eq. 14.7) or 2400 Btu/hr, whichever is the lesser value. The 2400-Btu/hr limit is proposed based on an acceptable sweat rate of 1 liter (2.2 lb) per hour for an 8-hour day. The heat stress index is defined as

$$\text{HSI} = \frac{E_{req}}{E_{max}} \times 100 \qquad (14.13)$$

and can be interpreted as shown in Table 14.7.

The heat exchange rates for convection, radiation, and evaporative loss initially were based on early versions of the equations that overpredicted the stress presented by the environment. The modified coefficients and exponents introduced in Eq. 14.4, 14.5, and 14.7 (McKarns and Brief, 1966) have resulted in improved accuracy. Due to its complexity, the index is not widely used by U.S. industry to predict the outcome of work in hot environments, although it is a

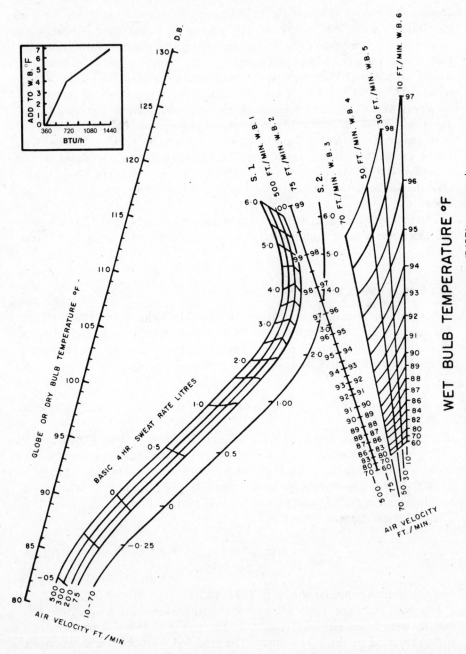

Figure 14.19 McArdle predicted four-hour sweat rate (P4SR) nomograph.

Instruction for Figure 14.19

Calculation of P4SR

The P4SR is calculated in three stages as follows:

Stage 1: Calculation of modified wet-bulb temperature

a If t_g differs from t_a, the wet bulb temperature is corrected: the correction is $0.4(t_g-t_a)$.

b If the energy expenditure exceeds 360 BTU/h a correction is obtained from the inset chart in the upper left-hand corner of the nomograph.

c. If the workers are clothed in regular garments, add 1.8°F.

Stage 2
A line is drawn from the appropriate point on the globe thermometer scale on the left of the nomograph to the modified wet bulb temperature (from stage 1) on the wet bulb scale for the wind speed. Read the B4SR from the cluster of curves in the center of the nomograph for the correct air speed.

Stage 3
The P4SR is obtained by adding to the B4SR an amount dependent on activity and clothing. For men working in light clothing: $P4SR = B4SR + 0.25 + 0.004 \ (M - 370)$.

---

rational approach and it is effective in identifying the major sources of heat load so that effective control measures can be implemented.

14.5 INDICES OF HEAT STRAIN

The best measurement of heat strain, or the impact of heat on the body, is mean body temperature. Unfortunately, this cannot easily be measured. However, it

TABLE 14.7 Evaluation of Heat Stress Index

HSI	Physiologic and Hygienic Implications
-20	Mild cold strain.
0	No thermal strain.
10–30	Mild to moderate heat strain. Little effect on heavy physical work, but possible decrement on skilled work.
40–60	Severe heat strain, involving threat to health unless workers are physically fit. Accimatization required. Some decrement in performance. Medical selection necessary.
70–90	Very severe heat strain. Personnel should be selected by medical examination and trial on job. Ensure adequate water intake. Control of conditions desirable.
100	Maximum strain tolerated by fit acclimatized young men.
Over 100	Exposure time limited by rise in deep-body temperature.

From Olesen (1985).

can be approximated from deep-body or core temperatures. Additional indices of strain are sweat rate and pulse rate. Each of these three potential indices is discussed with emphasis on the ability of the nonphysiologist to use these indices of strain under field conditions.

14.5.1   Core and Skin Temperature

The core or deep-body temperature of tissue and organs can be approximated by peripheral temperatures such as skin, tympanic, esophageal, oral, and axilla skin temperatures. Skin temperature varies considerably over the body, and a multipoint sampling harness must be used for an effective measurement. Tympanic and esophageal temperature measurements involve medical intervention and are difficult to implement in field studies. Oral and axilla temperatures are easy to obtain, and of these, oral temperature is the most commonly used. If the thermometer is carefully held under the tongue for five minutes a reasonably accurate measurement can be made. Oral temperature is usually 0.8°F lower than the core temperature.

Invasive techniques include the rectal temperature, which can be measured accurately with miniature thermocouple or thermistor probes. If rectal temperature is measured in conjunction with mean skin temperature, the core temperature may be determined with good accuracy; obviously, this represents a difficult procedure for field studies.

A simple procedure for in-plant use is the measurement of urine temperature, which is a reasonable approximation of core temperature (Gempel, 1979). In a heat stress study in tire plants, workers coming off work were asked to void into a thermally insulated bottle conditioned with water at 98 to 99°F. The water was dumped, the workers urinated in the bottle, and measurements of urine temperature as a prediction of core temperature were made immediately with a quick-response thermistor probe. It is inadvisable to permit body temperatures to exceed 100.4°F during prolonged heavy work in hot environments; work should be terminated if deep-body temperatures exceed 102.2°F.

14.5.2   Heart Rate

As noted earlier in the chapter, heart rate is a good index of heat strain. Brouha proposed an oral temperature-recovery heart rate technique to explore the impact of heat on the worker. In this procedure work is stopped, the worker is seated, and an oral thermometer is placed under his tongue. The heart rate is measured during the last one-half of the first three minutes of seated recovery. Using simple criteria, one can determine the ability of workers to continue their duties in a hot environment without ill effect.

A modification of this technique, using only heart rate, has been proposed by Fuller and Smith (1981) (Table 14.8) to evaluate hot environments. This is a simple approach that can easily be completed by plant personnel without special equipment.

Table 14.8 Heart Rate Recovery Criteria

Pattern and Comments	Criteria[a]	
	p_3	p_1-p_3
Satisfactory: Satisfactory pulse rate pattern; no excessive stress.	<90	
High: High recovery pulse rate patterns indicate work at a high metabolic level with little or no accumulated body heat. Individual jobs showing this condition require further study. Marginal stress.	>90	>10
No recovery: No recovery patterns. May indicate high stress.	>90	<10

[a]Heart rate measurement period: p_1 $\frac{1}{2}$ to 1 min; p_2, $1\frac{1}{2}$ to 2 min; p_3, $2\frac{1}{2}$ to 3 min.

14.5.3 Sweat Rate

This procedure requires that a mass balance be conducted on the worker by monitoring all liquid and food intake and urine and fecal excretion. If these data are known, the difference in body weight over the work shift corrected for material balance is used to calculate sweat rate. Short-term sweat rates on acclimatized workers should not exceed 2.5 liters per 4 hours. Due to the difficulty in assessing the mass balance, this technique is rarely applied in industry.

14.6 CONTROL OF THE THERMAL ENVIRONMENT

To this point this chapter has stressed the evaluation of both thermal comfort and thermal stress. Of equal importance is control of the thermal environment. The control of comfort conditions in the workplace is based on the prudent application of heating, ventilating, and air conditioning technology, described in Chapter 15. The control of heat stress can be more difficult and can require a range of vigorous controls, including engineering, work practice, and personal protective clothing. The effective application of these controls requires that the engineer be familiar with the heat balance equation and the origin of the individual heat transfer terms. In this section we outline briefly the controls that must be implemented jointly by plant and facilities engineers and occupational health specialists.

14.6.1 Engineering

Air Temperature. If the analysis described in Section 14.3 shows that convective load is a major component of the thermal burden to the worker, the two critical terms that must be controlled are air velocity and air temperature. In the most common situation the air temperature is greater than the skin temperature of 95°F and the heat transfer term for convection is positive, that is, it represents a heat gain by the body.

The initial concern is to reduce the air temperature to minimize this term. First consideration should be given to the outdoor temperatures during days of high stress. In many industrial settings, the indoor and outdoor temperatures track one another. In other locations, although the outdoor temperatures are lower than the indoor temperatures, the high radiant solar load on the building manifests itself as building structure heating and then conversion to a convective heat load on occupants. This condition can be minimized by intercepting the radiant load to the building.

The solar radiation load on roofs and windows is often an important factor in the overheating of the workplace, which may easily be overlooked in a heat stress control program. This solar load may account for 75% or more of the total heat penetration to a building with an uninsulated roof or in buildings with many windows on the southern exposure. In the northern United States, the intensity of solar radiation on a flat roof varies from 160 $Btu/h/ft^2$ at 8:00 A.M. and 4:00 P.M. to a maximum of 300 $Btu/h/ft^2$ at noon. The radiant heat from the sun passing through the east or west windows at 8:00 A.M. or 4:00 P.M., where the sun's rays are approximately perpendicular to the windows, amounts to about 210 $Btu/h/ft^2$ of glass surface.

Reflective roof insulation is a good preventive measure; one-half of the total energy contained in solar radiation is propagated by wavelengths lying within the "visible" range (0.4 to 0.7 μm) and the extent to which these wavelengths can be reflected depends on the brightness of the surface. A new aluminium roof will reflect about 90% of the visible radiation, fresh white paint about 75%, and black asphalt shingles only 10%. Good roof insulation will reduce solar load by 50%. Window awnings will reduce solar load through glass by two-thirds and new-technology reflective glass will provide impressive reductions.

Water can either be used as a pool on flat roofs or sprayed onto the roof. The solar load is reduced by reflection from a continuous water surface, by the evaporation of the water from the roof, and finally by the heat storage ability of the water. If water is sprayed on the roof and the rate of spraying is controlled so that most of the water evaporates as it touches the roof, the roof surface will approach the outdoor air wet bulb temperature, resulting in a roof temperature somewhat below outdoor air temperature. Water spray is an expensive approach since water must be pumped through spray nozzles with extensive piping and equipment in addition to water cost.

In many hot, dry regions where the outside wet bulb temperature is below 70°F, heat relief can be obtained at low cost by evaporative cooling. As noted in Chapter 15, when dry air is passed through a spray air washer, or a thick bed of wet excelsior, evaporation will cool the air to within a few degrees of the initial wet bulb temperature, although the humidity will approach the saturation point. Thus air at 100°F dry bulb and 68°F wet bulb can be cooled to 70°F by the use of a well-designed air washer. If this cool air is supplied to the working area, each pound of air will remove about 5.0 Btu when its temperature is allowed to rise to 90°F, or 3.5 Btu if the rise is limited to 85°F. Assuming that no significant amount of moisture is released from plant processes the relative humidity in the working space will be about 50% in the former case and 60% in the latter. The air distribution ducts for evaporative cooling systems must be insulated and duct velocities should be high to reduce heating as the air traverses the plant. This method of cooling is particularly suited to textile mills requiring artificial humidification during much of the summer.

In most of the United States evaporative cooling is not feasible, and frequently it is not appropriate for the manufacturing operations. In such cases air conditioning by mechanical refrigeration must be considered. A general discussion of mechanical air conditioning processes is presented in Chapter 15.

Mechanical refrigeration systems provide a flexible method of handling both sensible and insensible heat loads to provide an optimum thermal environment. However, the initial and operating cost of such systems for hot heavy industries such as glass and steel precludes their general plant use and limits their application to spot cooling and recovery or rest stations.

In many instances direct or "spot cooling" is the most practical method of air distribution. This is true in buildings having extremely high ceilings or with relatively few workers in a large area where it would be impractical to attempt to cool the entire space. Such spot cooling stations may be located directly at the workstation (e.g., both stationary and marine power plants) or at a work recovery station (e.g., glass and steel facilities). A good deal of information on the acceptable air temperature range and air velocities is available in the literature. In general, the cooling air should not be more than 10°F below ambient air temperature and air velocity should not exceed 400 fpm since convective or evaporative transfer rates are not increased by velocities above this range.

Mean Radiant Temperature. In hot, dry industries such as metal, glass, or ceramics the heat load is predominantly radiant and can be controlled by insulating surfaces of furnaces and ovens, thereby reducing the emitting surface temperature, by reducing the emissivity of the surface by covering it with appropriate materials such as aluminum on steam lines, and by providing highly reflective barriers between the emitting source and the worker. Each of these approaches, if properly implemented, will reduce the mean radiant temperature. Certain precautions must be taken when using reflective shields. If furnaces and ovens are screened from the general working areas with partitions of highly reflective material such as aluminum, openings must be provided at the bottom and top for circulation of air. The chimney effect produced by this temperature differential is often sufficient to remove this convective heat directly to the out-of-doors. If air circulation is not maintained, the radiant energy may be reflected back to the emitting source or to ancillary equipment such as motors and cause overheating and resulting damage. In addition, the reflecting panel may assume the surface temperature of the furnace or oven and become a significant radiation source. In any case the impact of any of these techniques can be assessed by globe temperature measurements at the workstation and the reduction in radiation load calculated (Eq. 14.5).

Moisture Content of Air. As discussed earlier, the evaporation of sweat is the most important method the body uses to achieve thermal equilibrium. When the water vapor content of air is high, the efficiency of sweating is seriously impaired. In such situations it is important to consider engineering control methods to prevent excess humidities.

In hot, moist industries such as food preparation and laundries, where large amounts of water vapor are released to the working space, the preferred control technique is to capture water vapor at the source by local exhaust ventilation. In special circumstances moisture can be removed from workplaces by desiccators that use silica gel or activated alumina. Dual desiccant beds are used in such systems; one bed is operational and removes moisture from the air while a second bed is reactivated. The reactivation process uses 300°F air or dry steam to expel the adsorbed water. The bed is then allowed to cool to normal temperature, after which it is again put into service. The cost of these systems preclude their general use in the prevention of heat disease. Chill coils used in conjunction with mechanical refrigeration systems are more commonly used to reduce moisture content as described in Chapter 15.

Air Velocity. The movement of air over the worker's body is probably the least understood control method in industry, yet its proper application can be easily understood if one reviews the heat balance equation and the equations defining the heat transfer rates. Convective heat transfer is a function of velocity and air temperature (Eq. 14.4). If the air temperature is above 95°F, the transfer is positive, that is, there is a gain in body heat. Conversely, if the air temperature is below 95°F, the term will be negative, representing a heat loss from the body. Therefore, if the air temperature is above skin temperature, the velocity should be lowered to minimize convective load; if it is below skin temperature, the heat load is reduced by increasing air velocity. In the case dictating a reduction in air velocity the impact of this reduction on the available sweating capacity should be evaluated using Eq. 14.7. In low humidity environments the advantage of high air velocity on evaporative loss may outweigh a smaller penalty from convective gain. At any rate, no advantage is obtained at velocities above 400 fpm, so that higher velocities should be avoided.

Mechanization of Process. Sometimes it is possible to mechanize the process to reduce heat stress, and excellent examples are seen in bottle manufacturing plants, the steel industry, foundries manufacturing small castings, and in certain malting and brewing processes. Mechanization is particularly effective in reducing human exertion and therefore the resulting metabolic load.

14.6.2 Work Practices

The preferred method to prevent heat stress is the application of the engineering control methods described above. If this is not possible, work practice controls may be the only solution. The principal control of this type is a work/recovery cycle. Where the worker is exposed to heat beyond the upper limit for continuous work, short periods of work in the heat should be alternated with rest periods, preferably in a cooler environment. Symptoms of heat distress, including heart palpitation, dyspnea, dizziness, nausea, headache, and mental confusion, normally do not appear in healthy persons until the heart rate rises well above 140 beats per minute, and the rectal temperature above 101°F. The length of the safe working period is determined by the time it takes for the pulse and the rectal temperatures to rise to critical levels, after which the worker must rest until the pulse and rectal temperature drop to near normal before work is resumed.

The length of the recovery period will depend on the rate and duration of work, the severity of heat exposure, the environment of the recovery location, and the person's condition. In any case, the worker should rest before fatigue has developed, as it requires a much longer time to recuperate after becoming thoroughly tired. The rest period should be long enough to reduce the heart rate to 100 beats per minute or less and to prevent the core temperature from rising to 102°F.

A number of investigators have proposed work recovery regimens for hot work, and the revised NIOSH Criteria Document (NIOSH, 1986) defines acceptable time periods based on metabolic load and work conditions.

It is also important to train the worker in the hazards of work in hot environments. Self-monitoring by the worker must frequently be used to guard against heat strain, and the onset of symptoms must be clearly defined.

Acclimatization of workers to hot environments is outside the scope of this book. The importance of acclimatization in optimizing the ability to handle the heat load, described well elsewhere, must be a part of any heat stress control program.

14.6.3 Protective Clothing

On first glance it may seem that the loss of heat from the body would be maximized by wearing a minimum amount of clothing. This is true for convective losses, but it is not the case for predominantly radiant environments. It is important that clothing be worn in the presence of radiant heat, since it intercepts the radiation load and transfers it by convection before it reaches the skin. A variety of aluminized garments are available for production operations such as steel making, and for short-term repair operations when severe radiant loads are encountered.

If the worker is stationed at a fixed work location, an air-supplied hood, helmet, or suit with suitable air distribution harness can be used. These systems should be operated at high airflow rates to achieve convective cooling while removing moisture from the envelope surrounding the worker, thus permitting heat balance. The required airflow varies with the degree of enclosure and the air temperature of the supply air.

When additional cooling is necessary, air line equipment can be used with a Vortex tube that requires 12 to 20 cfm at 75 to 100 psig. Two thermal streams are obtained. A cold stream of 6 to 8 cfm of 40 to 50°F air is delivered to the garment. A hot stream is discharged to the ambient atmosphere. A conventional mechanical refrigeration system can also be used to cool the air line supply.

The approaches described above require the use of an air hose and therefore restrict worker mobility. Self-contained systems for direct body cooling use either a short or a long garment fitted with a number of water-filled pouches which are frozen. The garment is donned over an undershift or mesh underwear and provides direct cooling of the trunk and arms. This garment has been given extensive application in short-term emergency work applications in the United States and deep mines in South Africa.

In another system a fluid cooled by passing through a bed of ice or dry ice is pumped through a tubing matrix in a tight-fitting helmet. These systems are used fairly extensively in industry and in car racing, where cooling of the driver is critical.

The control of the thermal environment to provide comfort and eliminate heat disease must be based on the environmental and physiologic measurements outlined in this chapter. Once the problem is defined, a variety of controls are available. A principal control is the proper application and maintenance of the range of heating, ventilation, and air conditioning facilities described in Chapter 15.

Example 14.1   Calculation of Thermal Comfort Indices

Lightly clothed office personnel (0.5 clo) are conducting sedentary desk activities (1 met) in an environment where the dry bulb temperature is 72°F, the relative humidity is 40%, the wet bulb temperature is 58°F, the globe temperature is 74°F and air velocity is 50 fpm.

Calculate the original effective temperature, ASHRAE operative temperature and determine the Fanger *predicted mean vote* (PMV) and *predicted percentage dissatisfied* (PPD).

Original Effective Temperature, ET

From Fig. 14.7, assume that t_a is approximately equal to $\bar{t}_r$. Enter the chart with $t_a = 72$ and RH $= 40\%$. ET $= 66.5$.

ASHRAE Operative Temperature

Calculate t_o. If V is less than 90 fpm and $\bar{t}_r$ is less than 120°F, then

$$t_o = \frac{t_a + \bar{t}_r}{2}$$

$$= \frac{72 + 75.8}{2} = 73.9°F$$

where

$$\bar{t}_r = t_g + 0.13V^{0.5}(t_g - t_a)$$
$$= 74 + 0.3(50)^{0.5}(74 - 72)$$
$$= 75.8°F$$

From Fig. 14.8, enter with $t_o = 73.9°F$. This condition lies just inside the summer comfort zone.

Fanger Predicted Percentage Dissatisfied

From Table 14.5, enter with

$$\begin{cases} M = 1.0 \text{ met} \\ V = 50 \text{ fpm } (0.25 \text{ m/s}) \\ t_a = 72°F \ (22 \text{ C}) \\ I_{cl} = 0.5 \\ RH = 40\% \end{cases}$$

Table is available only for RH $= 50\%$; complete tables are available from Fanger, 1982

PMV from the top element of Table 14.5:

$$PMV = -1.5$$

This predicted mean vote indicates that the environment is slightly cool to cool for 0.5 clo. If clothing is increased to 1.0 clo, PMV $= 0.3$ PPD.

From Fig. 14.12

For 0.5 clo: PPD $= 55\%$
For 1.0 clo: PPD $= 7\%$

Example 14.2 Calculation of Heat Stress Indices

The following measurements were made at a workstation in a foundry where an operator wearing light clothing works at an average rate of 800 Btu/hr.

$$t_a = 98°F$$
$$t_{wb} = 69°F$$
$$t_{nwb} = 70°F$$
$$t_g = 106°F$$
$$V = 100 \text{ fpm}$$

Calculate CET, WBGT, HSI, and P4SR for this worker.

Corrected Effective Temperature, CET

Step 1: Calculate the pseudo-wet-bulb temperature as shown in Fig. 14.6. The psychrometric chart (Fig. 14.1) is used in this step. The pseudo wet bulb (t_{pwb}) is 72°F from this calculation.

Step 2: Enter the CET chart (Fig. 14.7) with $t_{pwb} = 72°F$, $t_g = 106°F$, and $V = 100$ fpm. Read a CET value of 84°F.

Wet Bulb Globe Temperature, WBGT

Since this workplace is indoors, Eq. 14.8 applies:

$$WBGT = 0.7t_{nwb} + 0.3t_g$$
$$= 0.7(70) + 0.3(106) = 80.8°F$$

Predicted Four-Hour Sweat Rate, P4SR (Fig. 14.19)

Step 1: Calculation of modified wet bulb temperature

(a) $0.4(106 - 98) = 3.2$.
(b) From inset chart, for $M = 800$, correction is 4.0.
(c) Light clothing over shorts: correction is 1.8.

Total wet bulb correction $= 3.2 + 4.0 + 1.8 = 9.0$
Modified $t_{wb} = 69 + 9 = 78°F$

Step 2: Calculation of B4SR (Fig. 14.19)
Enter chart with $t_g = 106°F$ and modified $t_{wb} = 78°F$. Use WB 2 line for 75 fpm (closest to $V = 100$): read B4SR = 1.25.

Step 3: Calculation of P4SR
Worker is wearing light clothing over shorts, so

$$P4SR = B4SR + 0.25 + 0.004(M - 370)$$
$$= 1.25 + 0.25 + 0.004(430) = 3.22$$

Heat Stress Index, HSI (Eqs. 14.12 and 14.13)

$$HSI = \frac{E_{req}}{E_{max}} = \frac{M + R + C}{E_{max}}$$

$$M = 800 \text{ Btu/hr}$$

$$R = 15(\bar{t}_r - 95) = 15(116 - 95) = 315 \text{ Btu/hr}$$
$$\bar{t}_r = t_g + 0.13V^{0.5}(t_g - t_a)$$
$$= 106 + 0.13(100)^{0.5}(106 - 98) = 116°F$$

$C = 0.65V^{0.6}(t_a - 95) = 0.65(100)^{0.6}(98 - 95) = 31$ Btu/hr

$E_{max} = 2.4V^{0.6}(42 - p_a) = 2.4(100)^{0.6}(42 - 10)$
$= 1213$ Btu/hr

Value of p_a of 10 mm Hg from Fig. 14.1

$$HSI = \frac{M+R+C}{E_{max}} = \frac{800+315+31}{1213} \times 100 = 95$$

Interpretation of Results

CET = 84°F   (From Table 14.6)

If a worker is unacclimatized, he or she must be restricted to sedentary activities if full shift work is planned. If acclimatized, the worker can conduct moderate work for a full shift.

WBGT = 80.8°F   (From Fig. 14.16)

If acclimatized, the worker can conduct work at M=800 Btu/h for a full shift.

(From Fig. 14.16)
If unacclimatized, full shift work is also acceptable.

P4SR = 3.22

Interpretation varies. Conservative guideline is that an acclimatized worker may be safety employed at P4SR up to 2.5. Therefore, this condition is unacceptable for continuous work.

HSI = 98   (Table 14.7)

An HSI of 100 is the maximum strain tolerated by fit acclimatized young men. This condition represents severe heat stress and requires vigorous controls.

LIST OF SYMBOLS

C	heat exchange by convection
CET	corrected effective temperature (for radiation)
E_{max}	maximum evaporative cooling which can be achieved by sweating under the prevailing climatic conditions
E_{req}	total evaporative heat loss by sweating required to maintain thermal equilibrium
ET	effective temperature
HSI	Heat Stress Index
M	metabolic rate
P4SR	predicted 4-hour sweat rate
p_a	water vapor pressure of ambient air
RH	relative humidity
R	heat exchange by radiation
S	change in body heat content
t_a	ambient air dry bulb temperature
t_g	black globe temperature
t_{nwb}	natural wet bulb temperature
t_o	operative temperature
$\bar{t}_r$	mean radiant temperature
t_{pwb}	pseudo wet bulb temperature
t_{wb}	psychrometric wet bulb temperature
V_a	air velocity
W	external work
WBGT	wet bulb globe temperature
WGT	wet globe temperature

REFERENCES

American Conference of Governmental Industrial Hygienists (1973), "Threshold Limit Values for Chemical Substances and Physical Agents in the Work Environment," ACGIH, Cincinnati, OH.

American Industrial Hygiene Association Technical Committee on Ergonomics (1971), "Ergonomics Guide to Assessment of Metabolic and Cardiac Costs of Physical Work," *Am. Ind. Hyg. Assoc. J.*, **31**:560–564.

American Society of Heating, Refrigerating and Air Conditioning Engineers (1974), "Thermal Environmental Conditions for Human Occupancy," ASHRAE Standard 55-1974, ASHRAE, Atlanta, GA.

American Society of Heating, Refrigerating and Air Conditioning Engineers (1981), "Thermal Environmental Conditions for Human Occupancy," ASHRAE Standard 55-1981, ASHRAE, Atlanta, GA.

Bedford, T. (1946), "Environmental Warmth and Its Measurement," Medical Research Council, War Memo No. 17, H.M. Stationery Office, London.

Belding, H. S., and T. F. Hatch (1955), "Index for Evaluating Heat Stress in Terms of Resulting Physiological Strain," *Heat./Piping/Air Cond.* **25**:129–135.

Botsford, J. H. (1971), "A Wet Globe Thermometer for Environmental Heat Measurement", *Am. Ind. Hyg. Assoc. J.* **32**:1–10.

Brouha, L. (1960), *Physiology in Industry*, Pergamon Press, New York.

Candas, V., J. P. Libert, and J. J. Vogt (1979), "Influence of Air Velocity and Heat Acclimation on Human Skin Wettedness and Sweating Efficiency," *J. Appl. Physiol.,* **47**:1194–1200.

Fuller, F., and P. E. Smith (1981), "Evaluation of Heat Stress in a Hot Workshop by Physiological Measurements," *Am. Ind. Hyg. Assoc. J.,* **42**:32–37.

Gempel, R. (1979), *Proceedings of the American Industrial Hygiene Conference.*

Houghten, F. C., and C. P. Yaglou (1923), "Determination of the Comfort Zone", *J. Am. Soc. Heat Vent. Eng.,* **29**:515.

International Organization for Standardization (1983), "Determination of Metabolic Rate," ISO/TC 159/SC 5 N&E, ISO, Geneva.

International Organization for Standardization (1984), "Moderate Thermal Environments—Determination of the PMV and PPD Indices and Specifications of the Conditions for Thermal Comfort," ISO DIS 7730, ISO, Geneva.

Leithead, C. S., and A. R. Lind (1964), *Heat Stress and Heat Disorders*, F.A. Davis, Philadelphia.

McArdle, B., et al. (1947), "The Prediction of the Physiologic Effects of Warm and Hot Environments: the P4SR Index," Medical Research Council, R. N. P. Report 47/391, H. M. Stationery Office, London.

McKarns, J. S., and R. S. Brief (1966), "Nomographs Give Refined Estimate of Heat Stress Index," *Heat./Piping/Air Cond.,* **38**:113–116.

Mutchler, J. E., D. D. Malzahn, J. L. Vecchio, and R. D. Soule (1975), "An Improved Method for Monitoring Heat Stress Levels in the Workplace," U. S. Department of Health, Education, and Welfare, Public Health Service, Center for Disease Control, National Institute for Occupational Safety and Health, Publication No. 75–161.

Nelson, N., L. Eichna, S. N. Horvath, W. B. Shelley, and T. F. Hatch (1947), "Thermal Exchanges of Man at High Temperatures," *Am. J. Physiol.,* **151**:626.

Powell C. H., and A. D. Hosey, eds. (1965). The Industrial Environment—Its Evaluation and Control. 2nd ed., Public Health Services Publication No. 614.

U. S. Department of Health, Education, and Welfare, Health Services and Mental Health Administration, National Institute for Occupational Safety and Health (1972), "Criteria for a Recommended Standard—Occupational Exposure to Hot Environments," Publication No. HSM-10269, Washington, DC.

Vernon, H. M. (1930), "The Measurement of Radiant Heat in Relation to Human Comfort," *J. Physiol. Proc.,* **70**:15.

Yaglou, C. P. (1927), "Temperature, Humidity, and Air Movement in Industries, the Effective Temperature Index," *J. Ind. Hyg.,* **9**:297.

Yaglou, C. P. (1947), "A Method for Improving the Effective Temperature Indices," *ASHVE Trans.,* **53**:307.

Yaglou, C. P., and D. Minard (1957), "Control of Heat Casualties at Military Training Centers," *Arch. Ind. Health,* **16**:302.

ADDITIONAL READINGS

American Society of Heating, Refrigerating and Air Conditioning Engineers (1985), *ASHRAE Handbook—1985 Fundamentals*, ASHRAE, Atlanta, GA.

Fanger, P. O. (1982), *Thermal Comfort*, R. E. Krieger, Melbourne, FL.

Olesen, B. W. (1982), "Thermal Comfort," *B&K Tech. Rev.* No. 2.

Olesen, B. W. (1985), "Heat Stress," *B&K Tech. Rev.*, No. 2.

U. S. Department of Health and Human Services, National Institute for Occupational Safety and Health (1986), "Criteria for a Recommended Standard—Occupational Exposure to Hot Environments, Revised Criteria 1986," Publication No. 86–113, Washington, DC.

15

Air Conditioning for Comfort and Health

Over twenty million workers in the United States work in air-conditioned buildings. Once installed principally for worker comfort and health, air conditioning is now frequently required in modern high-technology factories for production reasons. The heating, ventilating, and air conditioning (HVAC) systems in industry use central air processing units which can regulate the thermodynamic state of the air (psychrometric state point) supplied to the space to control the external and internal heat and moisture loads. Although occupational health specialists are often required to investigate complaints arising from HVAC systems installed in offices and factories, they frequently do not have sufficient understanding of such equipment. This chapter is devoted to a brief review of conventional HVAC systems, their impact on worker comfort and health, and techniques to evaluate the adequacy of the air supply for these systems.

15.1 HVAC PROCESSING EQUIPMENT

In the simplest configuration, a constant-volume central HVAC system takes return air from the occupied building area and mixes it with outside air to replace the air exhausted from the space (Fig. 15.1). This blend of outside and return air is passed through a filter and heat transfer coils. Winter heating is provided by direct-resistance heaters or coils provided with hot water or steam. Coils are operated with cold water or brine or more frequently with a refrigeration fluid using a direct expansion coil to achieve summer cooling. The individual elements of such an HVAC system are described in this section.

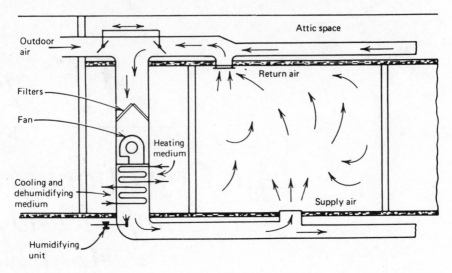

Figure 15.1 Major components of an air conditioning system serving an occupied space. (F. C. McQuiston and J. D. Parker, *Heating, Ventilating, and Air Conditioning* (1982) Reprinted by permission of John Wiley and Sons, New York.)

15.1.1 Mixing Boxes and Dampers

Mixing two airstreams with different thermodynamic conditions to produce a third stream is one of the most common processes encountered in HVAC systems. In Fig. 15.2a, streams A and B are mixed to produce stream C. The psychrometric state point C must be located on the line joining A and B, and its position is defined by the mass ratio of the two streams (Fig. 15.2b). If the streams are of equal mass, the final point will be at the midpoint of line AB. If the mass ratios are 1 part A and 2 parts B, the final state point will be one-third of the difference from B to A.

The airflows of streams A and B are regulated by dampers. Mechanical dampers used to regulate flow through ducts in HVAC systems are usually vaned dampers operated by pneumatic cylinders or solenoids (Fig. 15.3). Dampers are commonly used to regulate the quantity of outside or replacement air taken into the system, the quantity of air bypassing a HVAC system component, and the

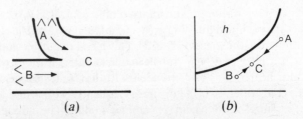

Figure 15.2 Mixing two airstreams in an HVAC system.

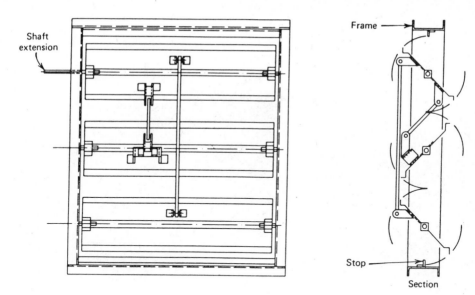

Figure 15.3 Common damper design used in HVAC systems. (J. L. Threlkeld, *Thermal Environmental Engineering* 2nd ed., (1970) Reprinted by permission of Prentice Hall, Inc. Englewood Cliffs, N.J.)

airflow supplied to an occupied space. In addition to vaned dampers, variable throat, butterfly, and blastgate devices are used in HVAC systems.

15.1.2 Filters

The most common HVAC filters are based on fiberglass or nonwoven synthetic materials mounted in flat panels, in a bag or envelope configuration for increased area and reduced resistance to airflow, or as a moving screen (Fig. 15.4). The efficiency of these filters ranges from 30 to 80% by weight when collecting atmospheric dust.

Dust filters are normally placed in front of heating or cooling coils. In addition to removing dust from the airstream to prevent plugging of coil fins, the filters are effective in distributing the airflow across the coil bank, thereby improving the performance of the coils.

15.1.3 Heating Coils

Heating coils provide sensible heat transfer, that is, heat transfer which results in an increase in air temperature, to airstreams. These coils handle heat loss during fall and winter conditions and as reheat coils to help achieve a desired psychrometric state point in summer operation. In its simplest form the heating coil is a bundle of tubes carrying hot water or steam. Air passes over the tubes, resulting in heat transfer. To improve their efficiency the surface area of the

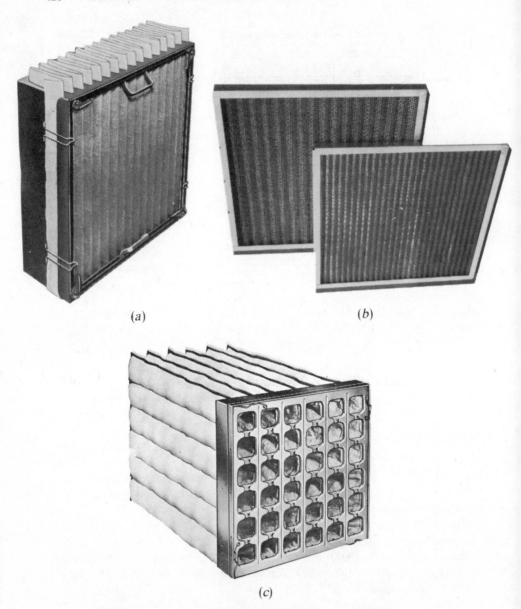

(a)   (b)

(c)

Figure 15.4 Variety of particle filters for HVAC systems (a) disposable pleated media; (b) cleanable filter; (c) replaceable bag type media. (Courtesy of Trane Company)

copper or steel tubes is usually greatly extended by covering the surface with spiral fins, flat fins, or plates (Fig. 15.5a). A coil bank is a parallel array of such individual finned tubes (Fig. 15.5b).

The psychrometric process performed by the coil is shown in Fig. 15.5c. Since no moisture is added, the state point A moves horizontally to the right along the

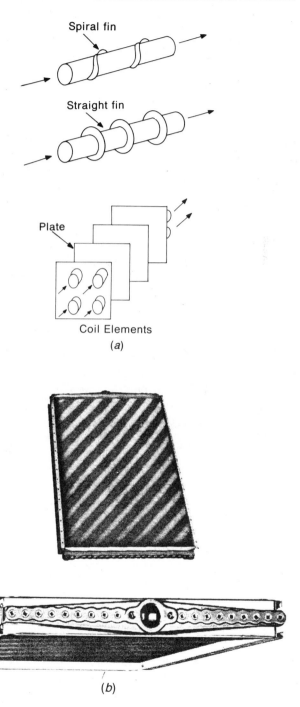

Figure 15.5 (*a*) Extending the heat exchange surface of tubes in heating coils. (*b*) Large coil bank in an HVAC system.

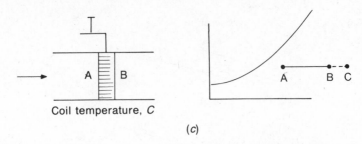

Coil temperature, C

(*c*)

Figure 15.5 (*c*) The psychrometric process for a simple heating coil showing the "bypass" ratio, the ratio of AB to AC.

constant-moisture-content line. The position of the final state point B depends on the energy input to the steam or hot water coil. If the heat exchange were ideal, the air temperature exiting the coil would be equal to the surface temperature of the coil. In fact, it is not and the efficiency of the coil depends on many factors, including fin design, cleanliness of the fins, temperature gradient between coil temperature and air, and the velocity of the air passing over the fins. The efficiency of the coil is usually rated by its hypothetical bypass ratio; air enters the coil at temperature A, exits at B, and the actual coil surface temperature is C. The "bypass" ratio, or the ratio of AB to AC, is a convenient way to describe coil performance (Fig. 15.5*c*).

Unless a properly maintained filter is placed upstream of the coil, airborne dust will collect in the space between the fins plugging the coil. The dirt acts as a insulator and seriously impairs heat transfer. Additionally, since the air must now pass through a reduced area the velocity increases, resulting in poor heat transfer and increased resistance to flow. Occasionally, the debris between fins may release, and if the duct velocity is adequate, it is discharged from the supply duct to the occupied space, much to the discomfort of the occupants.

Since the typical heating coil experiences a pressure drop of 0.1 to 0.2 in. H_2O at a velocity of 500 fpm, it has a reasonably flat velocity profile and thus is a good location for measuring the total system airflow.

15.1.4 Cooling Coils

A cooling coil is similar in appearance to a heating coil but they can easily be distinguished by the type of insulation and controls on the piping manifold serving the coils (Fig. 15.6*a*). The cooling coil may provide heat transfer for both sensible and insensible cooling, that is, direct reduction in air temperature or removal of water vapor from the air. In a simple exchange coil, well water may be adequate for modest cooling loads, while low-temperature brine or refrigerant is needed for major reductions in air temperature or moisture content. If the surface temperature of the coil is above the dew point of air at state point A, sensible cooling occurs with the state point moving to the left along a constant

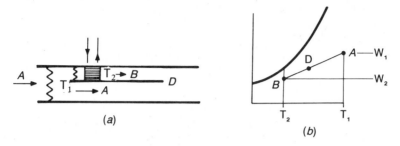

Figure 15.6 (a) Schematic diagram of a simple cooling coil which provides both sensible cooling and dehumidification. A bypass path may also be used in this equipment. (b) The psychrometric plot of the cooled stream shows a reduction in specific humidity ratio ,W, lb water vapor per lb of dry air, and a temperature drop from T_1 to T_2. The bypass stream at state point A is then mixed with stream at state point B to provide exit air at D.

humidity ratio line AB. If the coil surface temperature is below the dew point at A, both sensible cooling and latent or insensible cooling will occur (Fig. 15.6.b) and the state point of the exit air at A can be driven to B. If only a fraction of the stream passes over the coil, a new state point D will result.

15.1.5 Cooling System and Refrigeration Plant

Cooling coils are usually operated directly or indirectly by a refrigerant fluid from a central refrigeration system (Fig. 15.7). The principal components of a refrigeration plant are a liquid receiver, evaporator, compressor, and condenser (Fig. 15.8). In many large systems the cooling is done indirectly; that is, the refrigeration fluid cools an intermediate heat transfer fluid such as water or brine, which is pumped through the cooling coil to provide cooling. In a direct-expansion system (Fig. 15.9) the refrigeration fluid flows through an expansion valve and the latent heat of vaporaization is provided by the flowing air. As air is cooled, moisture may condense on the cold coil surfaces. When such insensible cooling occurs, the water condensing on the coil drains to a condensate tray. This tray must be cleaned periodically to prevent organic growth, with its resulting odor and biological air contamination.

After expansion the refrigeration gas is compressed, cooled, and liquefied in the condenser using a cold fluid or air to remove heat from the refrigeration fluid at the condenser. In a large central system the cooling source is usually water supplied by a cooling tower (Fig. 15.10). The warm water from the refrigeration plant condenser is discharged through nozzles at the top of the tower and cascades downward through baffles or packing material. Ambient air moves countercurrent to water flow, resulting in evaporative cooling of the water. This water is recirculated to the refrigerating plant condenser to provide cooling of the refrigeration gas. In summary, the primary heat transfer takes place at the cooling coil in the duct. This energy is transferred to the cooling water at the refrigeration plant evaporator and condenser. Finally, the energy is released

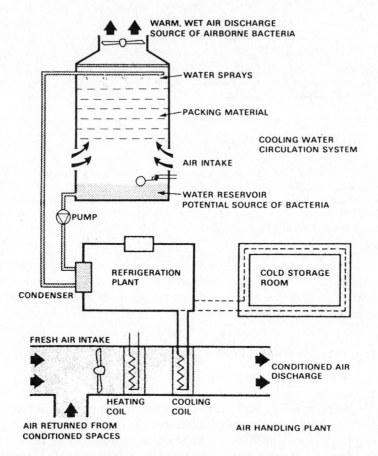

Figure 15.7 The major components of a cooling system include the cooling coil positioned in the duct, a refrigeration plant providing the cooling medium, and a cooling tower which finally strips the heat energy from the condenser water. (Ager, B. P. and J. A. Tickner (1983) *Ann. Occup. Hyg.* **27**, 341–348. Reprinted with permission Pergamon Journals, Ltd.)

from the cooling water to the air at the tower and subsequently to the environment.

The cooling tower is provided with replacement water to compensate for loss due to evaporation and mist carryover. The daily makeup rate usually amounts to about 5% of the total recirculating water supply. To minimize bacterial growth and the formation of salts in the tower, the water is continually treated with biocides and chemicals. Drift eliminators are installed to minimize the release of water droplets from the tower.

Several hygienic issues are of concern relative to cooling towers. Legionnaires' disease has been traced to bacteria originating in the warm, moist environment of cooling towers. Chemical processing plants may pump wastewater streams containing low levels of volatile solvents through the towers, which can result in very low concentrations of solvent vapors in the cooling tower plume. Baffles are

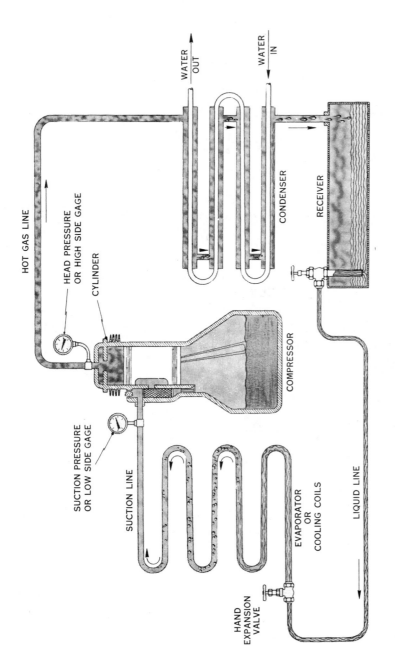

Figure 15.8 Schematic of a mechanical refrigeration plant utilizing an evaporator in the duct to absorb heat from the air, a compressor, a condenser, and a liquid receiver. (Courtesy of Trane Company.)

WATER OUT

WATER IN

HOT GAS LINE

HEAD PRESSURE OR HIGH SIDE GAGE

CYLINDER

CONDENSER

RECEIVER

COMPRESSOR

SUCTION PRESSURE OR LOW SIDE GAGE

SUCTION LINE

LIQUID LINE

EVAPORATOR OR COOLING COILS

HAND EXPANSION VALVE

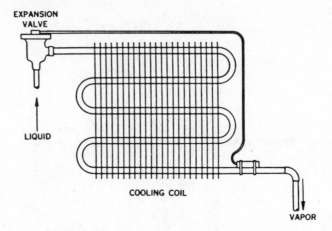

Figure 15.9 A direct expansion coil can be identified by the unique tubing and liquid traps on the manifold side of the coil. (Courtesy of Trane Company.)

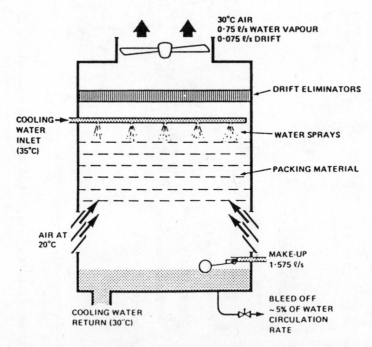

Figure 15.10 Schematic diagram of a small cooling tower. (Ager, B. P. and J. A. Tickner (1983) *Ann. Occup. Hyg.* **27**, 341–348. Reprinted with permission Pergamon Journals, Ltd.)

frequently made of Transite®, a composite board of cement and asbestos. Repairs involving sawing of the baffles should be done with the necessary controls to minimize worker exposure to asbestos. Fiberglas®-based cement board is now available and should be specified for such applications.

In the system described earlier the heat from the hot refrigeration gas is transferred to the cooling water, which is then processed through the cooling tower. A second common system uses an air-cooled condenser, where the intermediate water cooling loop is eliminated. The hot gas is directly cooled by the condenser (Fig. 15.11), where the Freon is liquefied and accumulated for return to the evaporator in the duct.

15.1.6   Spray Humidifiers

Spray humidifiers, commonly called air washers, are designed to humidify, with or without heating or cooling. In their simplest form (Fig. 15.12a) a pump discharges water to a bank of spray nozzles in the airstream, where evaporation occurs. Excess water returns to an open tank at the base of the HVAC enclosure, where the pump takes suction. Replacement water is supplied to replace that lost through evaporation and mist entrainment. Since the water has achieved the wet bulb temperature of point A (Fig. 15.12b) the process is one of adiabatic saturation; that is, no heat is added or removed. The heat to vaporize the water is taken from the air and causes sensible cooling. Ideally, the sensible cooling of the airstream is equal to the latent heat required to vaporize the water so that there is no change in the total enthalpy of the air–water vapor mixture and the process line is coincident with a constant-enthalpy line (Fig. 15.12b) Impingement plates typically are positioned downstream of the spray humidifier to minimize droplet carryover. The spray humidifier is frequently used in conjunction with a preheater and an afterheater (Fig. 15.13).

Since the air washer presents a warm, moist environment it encourages bacterial and mold growth, which can lead to odors. If contained in a droplet, biological particles may be released to the airstream and travel to the occupied space. Documented cases of respiratory disease have been traced to such sources (Ager and Tickner, 1983, CDC MMWR, 1984).

15.2   AIR CONDITIONING SYSTEMS

The individual pieces of processing equipment described above can be assembled in a number of different configurations to provide conditioned air and establish the comfort conditions described in Chapter 14. These systems have evolved over the past several decades and the occupational health specialist should be familiar with the designs encountered most frequently. In this section we review the basic systems and describe their operational characteristics.

The initial step in designing an HVAC system is to define the load the system must handle. The winter heating load is primarily external, that is, the loss of heat from the building. The summer cooling load is both internal (from the occupants, equipment, and lighting) and external (solar). Calculation of these loads is beyond the scope of this book, but a number of computer software programs are now available to assist the plant engineer in this task (Sun, 1986).

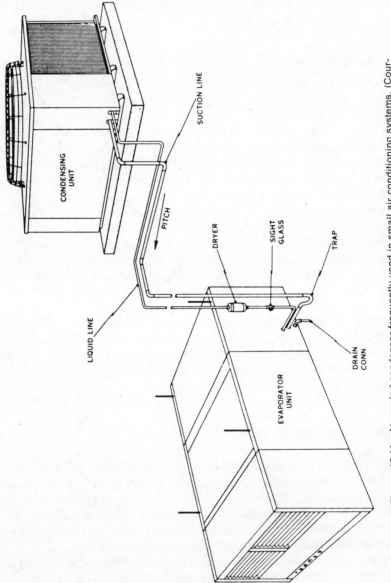

Figure 15.11 Air-cooled condenser frequently used in small air conditioning systems. (Courtesy of Trane Company.)

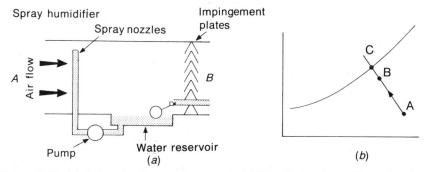

Figure 15.12 (a) Schematic diagram of a spray humidifier. (b) Psychrometric process for a spray humidifier

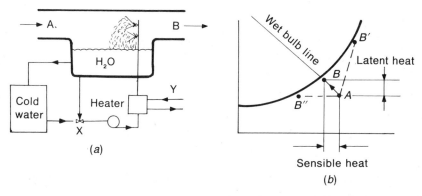

Figure 15.13 (a) Schematic diagram of an air washer with a heater. (b) Psychrometric chart for air washer. If operated without heat under adiabatic conditions, the process line will be AB. If heated, there is an increase in both sensible and insensible heat, as noted in AB'. If cold water is used in the spray nozzles, sensible cooling will be achieved while the humidity ratio remains constant at AB''.

15.2.1 Winter (Heating) Season

The conventional system for winter application includes heating and humidification (Fig. 15.14a). As air circulates through a room from the supply to the return air grilles, both heat and water vapor are lost from the space. The shift in psychrometric state point from 1 to 2 (Fig. 15.14b) demonstrates both a total enthalpy loss (Btu/lb dry air) and a shift in humidity ratio (lb H_2O/lb dry air).

A simple winter HVAC system for an office space is shown in Fig. 15.14a. The ratio of return air to outside air during severe winter conditions is usually chosen to meet the minimum outside air requirements to be described in Section 15.4. In this example the mixing ratio of streams 4 and 2 results in air at state point 5. The outside air duct is frequently equipped with a preheat coil to minimize the chance of freeze-up of the main coil bank. The state point of the air discharged from the preheat coil point 6 determines the amount of moisture that can be added to the

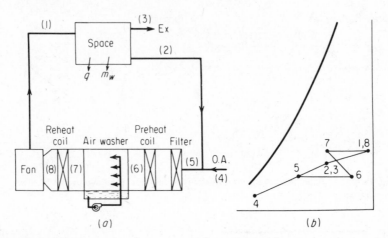

Figure 15.14 (a) Schematic diagram of a constant-volume HVAC system in a winter operating mode with a preheater coil to prevent accidental freezing, an unheated air washer for humidity control, and a reheat coil. q is heat loss and m_w is moisture loss. (b) Psychrometric process achieved with this system. (J. L. Threlkeld, *Thermal Environmental Engineering* 2nd ed., (1970) Reprinted by permission of Prentice Hall, Inc. Englewood Cliffs, N.J.)

system. Since the recirculating water has achieved the air temperature and the spray humidifier is not heated, the process is approximately adiabatic and the process line is 6–7. After the air passes through the mist impingement plates, the stream is reheated (7 to 1, 8). This process determines the dry bulb temperature of the air returned to the occupied space. Note that both the preheat and reheat coils provide sensible heating only; that is, no water vapor is added and the humidity ratio is constant.

15.2.2 Summer (Cooling) Season

The summer heat load requires both sensible cooling, that is, reduction of air temperature, and removal of moisture (Fig. 15.15a). To conserve energy, minimal outside air is mixed with the return air. The path through the central HVAC system includes the filter and the cooling coil, which produces a shift in state point from 5 to 6 (Fig. 15.15b). The actual path is difficult to determine. The state point shift occurs in two steps. Initially, sensible cooling takes place and the state point moves to the left, approaching the saturation line. Then water vapor condenses on the coil, causing the state point to move downward along the saturation line. Since the coil performance is not ideal, the state point of the departing air approaches the saturation line but is not on it. To remove sufficient moisture the temperature at point 6 frequently is too low. The reheat coil positioned downstream of the cooling coil provides control of the return air state point (6–7).

Central systems are available in a variety of configurations. The system described above is inherently inefficient since it first cools and then reheats the

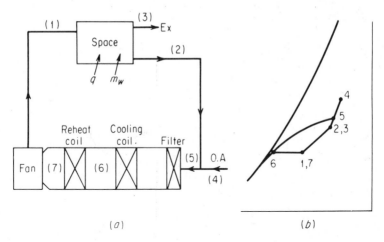

Figure 15.15 (a) Schematic diagram of a constant-volume HVAC system in a summer operational mode. q is heat gain and m_w is moisture gain. (b) Sensible and insensible cooling are achieved (5—6) and return to the desired state point is obtained by a reheat coil (6—1). (J. L. Threlkeld, *Thermal Environmental Engineering* 2nd ed., (1970) Reprinted by permission of Prentice Hall, Inc. Englewood Cliffs, N.J.)

air. Cooling systems frequently use a bypass or runaround configuration (Fig. 15.16). In these systems a reduced volume of air is processed through the cooling coil (and the reheat is accomplished with a portion of the return air), thereby improving the economy of the process.

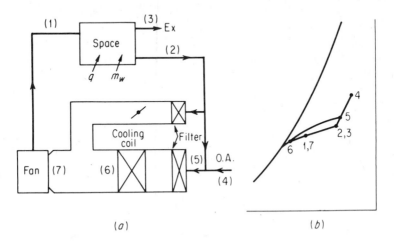

Figure 15.16 (a) Schematic diagram of a summer air conditioning system with a bypass duct that permits eliminating the reheat coil and its accompanying inefficiency. q is heat gain and m_w is moisture gain. (b) Reheating is obtained by mixing stream 2, which is at the return-air state point, and stream 6, the state point downstream of the cooling coil. (J. L. Threlkeld, *Thermal Environmental Engineering* 2nd ed., (1970) Reprinted by permission of Prentice Hall, Inc. Englewood Cliffs, N.J.)

15.3 AIR DELIVERY TO OCCUPIED SPACES

In the previous discussion several psychrometric processes were accomplished in a central air conditioning system. Air at a given individual state point was delivered to the occupied spaces at a rate calculated to handle the load. This is the simplest configuration and facilitates the understanding of the process. However, it is limited in its ability to achieve a habitable environment while conserving energy. As a result a number of variations have developed; their use depends on the type of space conditioned and the age of the HVAC system. The following five air distribution systems are frequently encountered in the United States and the occupational health specialist should be acquainted with their design and performance.

15.3.1 Zone Reheat Systems

Most air conditioning systems serve not a single large space but a number of individual occupied spaces each with different cooling and heating loads. The common solution to this problem is a multiple-duct system with zone reheat (Fig. 15.17). The reheat coil is controlled locally by a thermostat placed in the individual space. Sensible heat control (dry bulb temperature) is thus possible but moisture control is not. This limitation is acceptable in most offices since precise control of humidity normally is not absolutely necessary.

15.3.2 Induction Units

Conventional central HVAC systems handle large air volumes at low velocity. Such systems require large ducts which occupy valuable interior space. In the 1960s a common alternative was to condition small volumes of primary air in a central system and distribute it to individual office induction units (Fig. 15.18). The primary airflow induces the flow of room air over finned heat exchangers served by hot water in the winter and cold water in the summer. Primary air, and therefore the resulting induced airflow, is controlled by a local thermostat.

The trade-offs in this system are numerous. The large duct runs are eliminated, but hot and cold water must be supplied to each occupied space. Additionally,

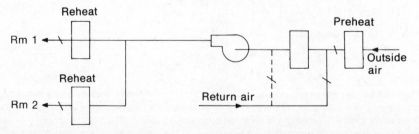

Figure 15.17 Schematic of a system with individual reheat coils positioned in the delivery duct at the individual conditioned space.

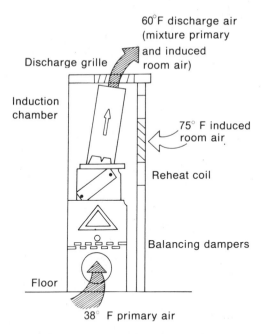

Figure 15.18 Induction units, widely used in the 1960s, are frequently the cause of complaints of comfort conditions and malodors due to poor maintenance.

the conditioned primary air, albeit in reduced volume, cannot be returned to the central system for energy conservation. This approach also requires servicing and maintenance of a large number of induction units. Frequently, the filters and finned heat exchangers are not cleaned and performance of the units degrades, prompting occupant complaints. These systems have also been associated with respiratory sensitization resulting from bacterial growth in condensate collection trays (Bernstein et al., 1983).

15.3.3 Dual-Duct Systems

In the dual-duct system, two conditioned airstreams are delivered to the occupied space. One duct delivers air at a temperature above that required for space conditioning; a second provides a cold airstream. These hot and cold ducts or decks terminate at a mixing box (Fig. 15.19a), where flow metering defines the state point of the mixed air. The simplest configuration has a single supply blower and no return-air fan (Fig. 15.19b). A more complex system with two supply fans, a return-air fan, multiple bypass ducts, and optional outside air ducts (Fig. 15.20) is available.

The dual-duct system permits the plant engineer to handle a wide range of loads resulting from diversified occupancy. While some of the five systems described in this section have difficulty handling variations in insensible load, the dual-duct system can be designed to handle such a problem. When outside conditions are moderate (e.g., spring or fall) the optional outside air duct can be used to control the modest cooling loads that may be encountered. This technique, called free cooling or enthalpy control, has been used widely in the United States since 1980.

The dual-duct system is normally designed for constant-volume operation with resulting high energy consumption. To improve the economics of these systems, a proposed three-duct multizone system includes the hot and cold ducts described above in addition to a bypass duct. The latter duct bypasses all conditioning equipment and merely recirculates room air; the hot duct is wide open at maximum heating conditions, while the other ducts are closed. With a moderate heating load the hot duct is partially closed, the bypass duct modulates toward the open position, and the cold duct remains closed. The bypass is open and the hot and cold ducts are closed under no-load conditions. As cooling is called for by the controller, the cold duct damper modulates open, the bypass modulates toward the closed position, and the hot deck damper is closed.

15.3.4   Variable-Air-Volume Systems

The systems described above are constant volume and vary the state point of the incoming air to control the environmental load. The variable-air-volume (VAV)

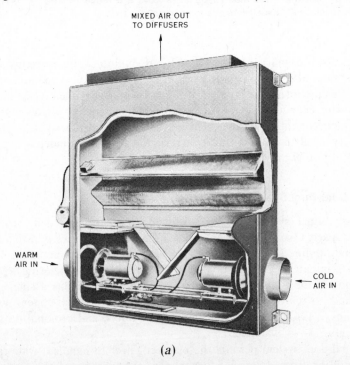

MIXED AIR OUT
TO DIFFUSERS

WARM
AIR IN

COLD
AIR IN

(a)

Figure 15.19   (*a*) The flexibility of double-duct systems has resulted in their increased popularity in the 1970s and 1980s. The key to the system is the mixing box, which included a pneumatically driven valve which meters flow from the hot and cold ducts.

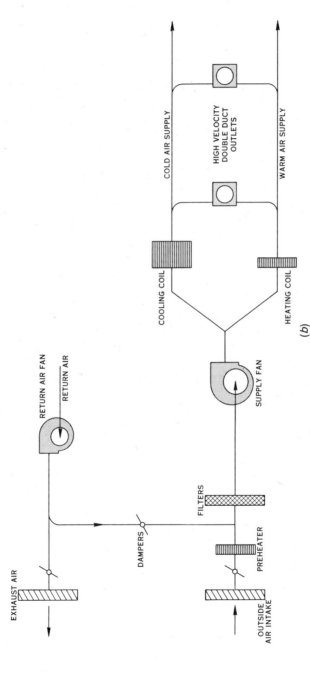

Figure 15.19 (b) Double-duct, one-fan system (Courtesy of Trane Company.)

(b)

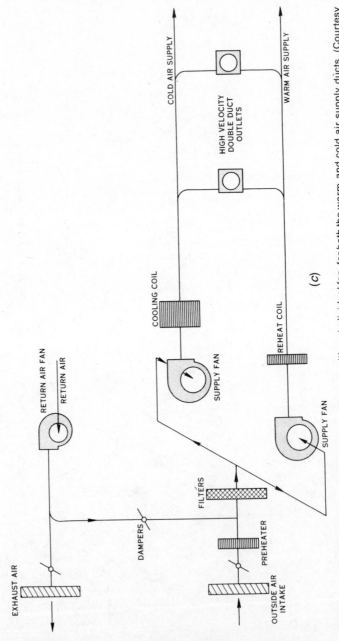

Figure 15.20 Schematic diagram of a two-duct system with an individual fan, for both the warm and cold air supply ducts. (Courtesy of Trane Company.)

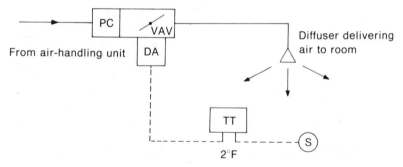

Figure 15.21 Schematic diagram of a variable-air-volume (VAV) system, probably the most commonly utilized system in the 1980s. TT, pneumatic thermostat; PC, pressure compensator; VAV variable air-volume terminal; DA, damper actuator.

delivery system employs a controller, usually a pneumatic thermostat which drives dampers in the supply duct, to vary the quantity of air delivered to the space and therefore to determine the psychrometric state point (Fig. 15.21).

Although widely used, the variable-air-volume system has a number of weaknesses. Due to the operating characteristics of the system, it may permit wide temperature excursions, which prompt occupant complaints. The minimum and maximum flows, defined by mechanical settings at the mixing box, require laborious balancing procedures. It is possible for a VAV system to reduce supply at a critical time during occupancy, with the result that the system does not provide the minimum required outside-air quantity (see Section 15.4). It has been proposed that a carbon dioxide sensor be used to sense the need for outside air supply, thereby resolving this problem.

15.3.5 Packaged Air Conditioning Units

A range of self-contained or unitary air conditioning units (Fig. 15.22) is available. These devices are frequently installed to serve a small space or used as an interim solution when a central system cannot handle the cooling load. Since these units recirculate room air, their operation may prompt complaints of poor air quality and poor air distribution from occupants.

15.4 INDOOR AIR QUALITY AND OUTSIDE AIR REQUIREMENTS

In 1981 the American Society of Heating, Refrigerating and Air Conditioning Engineers, Inc. (ASHRAE) published ASHRAE Standard 62-1981, "Ventilation for Acceptable Indoor Air Quality." This standard, a revision of Standard 62–1973, is intended to provide guidance for the provision of healthful and comfortable occupied indoor environments. The standard outlines minimum ventilation rates and should not be used in place of more stringent regulations or recommendations.

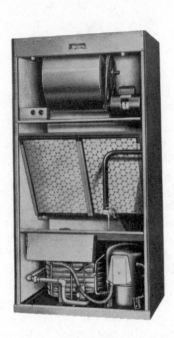

Figure 15.22 Unitary air conditioners for use with ducts or for free discharge. Self-contained air conditioning systems are now commonly used in small systems. These systems are available with both evaporator and condenser units in one envelope or as a split unit with the condenser placed outdoors. The latter system is available in a heat pump configuration operating as a cooler in summer and as a heater in the winter.

Throughout the standard, references are made to "acceptable" air quality. ASHRAE defines this as "air in which there are no known contaminants at harmful concentrations and with which a substantial majority (usually 80%) of the people exposed do not express dissatisfaction." The latter statement was apparently based on practical considerations. Studies have shown that it is difficult, if not impossible, to satisfy 100% of the people exposed to any particular environment; attempts to satisfy more than 80% of the population require large amounts of fresh air, which was felt to be economically unacceptable.

Under Standard 62-1981 the minimum ventilation rate can be defined by two approaches. The primary procedure is a specification standard calling for the provision of minimum amounts of clean, conditioned outdoor air. The alternative method is a performance standard whereby the minimum outdoor air quantities need not be met but the quality of the indoor air must conform to established guidelines. Either approach is acceptable and should provide a healthful and comfortable workplace.

15.4.1 Outdoor Air Quality Acceptable for Ventilation

A three-step procedure is outlined in Standard 62-1981 to assure the acceptability of the ambient replacement or makeup air for the air-conditioning system. The first step involves compiling data on the ambient air quality. Concentrations of carbon monoxide, lead, nitrogen dioxide, oxidants (ozone), particulates, and sulfur dioxide at the building location are determined by examining data collected by ambient monitoring stations operated by federal, state, or local authorities. If data are not available from nearby monitoring stations, air concentrations can be measured at the building site. These air concentration data are compared to the National Ambient Air Quality Standards (NAAQS) (Table 15.1).

An alternative procedure in Standard 62-1981 for determining the adequacy of the ambient air is to base the acceptability of outside air quality on data obtained from a community of similar size, meteorology, and industrial development. Finally, if the facility is located in a small community without significant air pollution sources, it may be argued that it will, ipso facto, meet the ambient air quality standards.

If the air quality for the six NAAQS pollutants is acceptable, the ambient air is then compared to a second set of standards covering 27 air contaminants for which the EPA has not set limits (Table 15.2). The selection of these contaminants and the long-term acceptable air concentrations specified by ASHRAE in Standard 62-1981 are based on standards published by municipal, state, and other jurisdictions.

If the ambient air meets both of the criteria cited above, the third step is to determine if there are any other air contaminants thought to be present at levels in excess of 10% of the OSHA permissible exposure limits (PELs).

Table 15.1 National Ambient Air Quality Standards

Contaminant	Long Term		Short Term	
	Level	Time	Level	Time
Carbon monoxide			$40 \ mg/m^3$	1 hr
			$10 \ mg/m^3$	8 hr
Lead	$1.5 \ \mu g/m^3$	3 month		
Nitrogen dioxide	$100 \ \mu g/m^3$	1 year		
Oxidants (ozone)			$235 \ \mu g/m^3$	1 hr
Particulates	$75 \ \mu g/m^3$	1 year	$260 \ \mu g/m^3$	24 hr
Sulfur dioxide	$80 \ \mu g/m^3$	1 year	$365 \ \mu g/m^3$	24 hr

Table 15.2 Additional Ambient Air Quality Guidelines

Contaminant	Long Term		Short Term	
	Level	Time	Level	Time
Acetone	7 mg/m^3	24 hr	24 mg/m^3	30 min
Acrolein			25 μg/m^3	Ceiling
Ammonia	0.5 mg/m^3	1 year	7 mg/m^3	Ceiling
Beryllium	0.01 μg/m^3	30 days		
Cadmium	2.0 μg/m^3	24 hr		
Calcium oxide (lime)			20–30 μg/m^3	Ceiling
Carbon disulfide	0.15 mg/m^3	24 hr	0.45 mg/m^3	30 min
Chlorine	0.1 mg/m^3	24 hr	0.3 mg/m^3	30 min
Chromium	1.5 μg/m^3	24 hr		
Cresol	0.1 mg/m^3	24 hr		
Dichloroethane	2.0 mg/m^3	24 hr	6.0 mg/m^3	30 min
Ethyl acetate	14 mg/m^3	24 hr	42 mg/m^3	30 min
Formaldehyde			120 μg/m^3	Ceiling
Hydrochloric acid	0.4 mg/m^3	24 hr	3 mg/m^3	30 min
Hydrogen sulfide	40–50 μg/m^3	24 hr	42 μg/m^3	1 hr
Mercaptans			20 μg/m^3	1 hr
Mercury	2 μg/m^3	24 hr		
Methyl alcohol	1.5 mg/m^3	24 hr	4.5 mg/m^3	30 min
Methylene	20 mg/m^3	1 year	150 mg/m^3	30 min
chloride	50 mg/m^3	24 hr		
Nickel	2 μg/m^3	24 hr		
Nitrogen monoxide	0.5 mg/m^3	24 hrs	1 mg/m^3	30 min
Phenol	0.1 mg/m^3	24 hr		
Sulfates	4 μg/m^3	1 year		
	12 μg/m^3	24 hr		
Sulfuric acid	50 μg/m^3	1 year	200 μg/m^3	30 min
	100 μg/m^3	24 hr		
Trichlorethylene	2 mg/m^3	1 year	16 mg/m^3	30 min
	5 mg/m^3	24 hr		
Vanadium	2 μg/m^3	24 hr		
Zinc	100 μg/m^3	24 hr		

(Copyright 1981, American Society of Heating, Refrigerating and Air Conditioning Engineers).

If the outdoor air quality is deemed unacceptable under any of these three steps, adequate air cleaning must be provided using the appropriate filters, sorption beds, or catalysts before the outdoor air is introduced into the facility.

15.4.2 Air Quantity

The required quantity of outside replacement air is determined by the type of facility, the estimated occupancy level, and whether or not smoking is allowed. For example, minimum ventilation rates for offices should be 5 cfm/person where smoking is not allowed and 20 cfm/person where it is. For conference rooms, the respective numbers are 10 and 50 cfm/person.

The minimum amount of outdoor air required for all occupied situations is never less than 5 cfm/person. This value is chosen to assure that the indoor carbon dioxide (CO_2) concentration does not exceed 0.25% (2500 ppm). Carbon dioxide is a by-product of human metabolism and is exhaled in the breath at rest at an average rate of 0.63 cubic foot per hour per person. The relationship between the concentration of CO_2 originating from human metabolism, CO_2 present in the outdoor air, and its removal by exhaust ventilation, is

$$G + CQ = C_i Q \qquad (15.1)$$

where C = outdoor CO_2 concentration = 0.0003%
$\quad$ G = metabolic CO_2 production rate, cfm
$\quad$ Q = ventilation rate of outside air, cfm
$\quad$ C_i = indoor CO_2 concentration, %

Rearranging Eq. 15.1 to solve for the indoor CO_2 concentration gives

$$C_i = C + \frac{G}{Q} \qquad (15.2)$$

This equation can be normalized by dividing Q and G by the number of people present in the room:

$$C_i = C + \frac{G'}{Q'} \qquad (15.3)$$

where Q' = ventilation rate of outside air, cfm/person
$\quad$ G' = metabolic CO_2 production rate, cfm/person
$\quad\quad$ = 0.63 cfh/person or 0.0105 cfm/person

Substituting this value of G' in Eq. 15.3 and solving for Q' gives

$$Q' = \frac{0.0105}{C_i - C} \qquad (15.4)$$

By setting a maximum level of 0.0025 (0.25%) for C_i for health reasons this equation becomes

$$Q' = \frac{0.0105}{0.0025 - 0.0003}$$

$$= 4.8 \text{ cfm/person}$$

Lower acceptable CO_2 concentrations require higher ventilation rates. For example, to maintain levels below 0.1% (1000 ppm), Q' must be greater than 15 cfm/person.

The choice of 2500 ppm as the maximum allowable CO_2 concentration can be questioned. Carbon dioxide is present in ambient air in concentrations of 300 to 350 ppm. Both the OSHA permissible exposure limit and the ACGIH threshold limit value are set at 5000 ppm (0.5%) to prevent respiratory stimulation, narcosis, and degradation of hearing acuity. These symptoms occur at CO_2 concentrations of 30,000 to 50,000 ppm, which would seem to indicate that the choice of 2500 ppm is conservative. However, several studies have cited 1000 ppm as an index of air quality which triggers complaints of poor ventilation in office spaces, and over a 10-year period the authors have found 700 ppm to be an effective trigger index point for identifying poor comfort ventilation. This value requires a minimum ventilation rate of 25 cfm/person.

The substantially higher ventilation rates required in areas where smoking is allowed (ASHRAE, 1981) reflects the annoyance factor associated with tobacco smoke. Studies at Yale University have demonstrated that ventilation rates as high as 100 cfm per smoking occupant may be required to maintain acceptable air quality in areas where smoking occurs more or less continuously (Leaderer and Cain, 1983).

The requirements listed under the Ventilation Rate Procedure in Standard 62-1981 are chosen merely to provide sufficient air to minimize problems associated with carbon dioxide and smoking. If other internally generated air contaminants are present, the minimum ventilation requirements may not be adequate. Each situation must be addressed individually to assure proper indoor air quality (Wadden and Scheff, 1983).

The ASHRAE Standard also includes criteria for the use of recirculated air and variable occupancy. If substantial recirculation of indoor air is chosen as a means of providing acceptable air quality with minimal outside air, the air-cleaning system must be adequate to provide air as clean as that provided from outdoors and to assure that the amount of fresh air being introduced is not less than 5 cfm/person.

Care must be taken to ensure that the proper air-cleaning equipment is chosen, especially if smoking is permitted in the occupied space. Filters or electrostatic precipitators are commonly used to remove smoke particles. Such devices do not collect gases and vapors produced by smoking. In addition, cigarette smoke particles collected on a filter surface will elute complex volatile organic chemicals which are carried back to the occupied space as offensive odors. A sorbent such as activated charcoal is frequently required to control this problem.

15.4.3 Indoor Air Quality Procedure

The second procedure described in Standard 62-1981 to ensure healthful indoor air requires the direct control of air contaminants generated in the space to within acceptable contaminant concentrations. Air sampling is required to define the concentration of those contaminants for which limits have been established and analytical techniques are available; these include formaldehyde, ozone, radon, and carbon dioxide. Many air contaminants that might be present in indoor air do not have widely accepted nonoccupational exposure limits. In these cases modifications of OSHA PELs must be used.

It is common practice to set as conservative standards for the general population values that are 10% of the OSHA PEL. This reflects an awareness of the sensitivity of the susceptible young and old persons in the general population and the possibility of continuous exposure to contaminants 24 hours per day and not for a single work shift of 8 hours. If the contaminant is odorous or causes sensory irritation and sampling and analytical methods are not available to define its concentration, a subjective evaluation can be made. This procedure requires assembling a panel of at least 20 untrained observers to be exposed to the air in question under established protocols. The air can be considered "acceptable" if at least 80% of the panel participants feel that it is not objectionable. Response to tobacco smoke odors may also be evaluated in this way.

Using this procedure the air quality is considered acceptable if it passes the objective and subjective tests described above, regardless of the ventilation rate. Changes in building occupancy, rearrangements of work areas, and the introduction of new furnishings and/or equipment may alter the air quality, requiring a complete reevaluation of the situation.

Although this procedure may seem straightforward, it is difficult to apply since the ventilation system designer must wait until the building is occupied before the system can be evaluated fully. Unless the system has sufficient flexibility to accommodate the unexpected, an unacceptable design may result.

15.4.4 New ASHRAE Proposal

At the time this book was being completed ASHRAE had issued a draft ASHRAE Standard 62-1981R for public review. Following are five major revisions in the original ASHRAE standard 62-1981:

1. The minimum ventilation rate is increased from 5 cfm of outdoor air per person to 15 cfm/person. This reflects recent research showing that 15 cfm is required to reduce occupant odors to a concentration acceptable to 80% of the occupants. This new outdoor air quantity will also maintain the CO_2 concentration at 1000 ppm, the level now cited by WHO. In addition, this air quantity will control tobacco smoke odor to an acceptable level in areas where smoking is at a minimum.
2. Minimum ventilation rates for various spaces are no longer differentiated based on smoking and nonsmoking. One standard is proposed.
3. A modified procedure for averaging flow rate in multiroom systems is provided in the proposal.
4. The application information on air cleaning has been expanded.
5. The proposed standard still retains the alternate air quality procedure described in Section 15.4.3.

LIST OF SYMBOLS

C_i	indoor carbon dioxide concentration
C_o	outdoor carbon dioxide concentration
G	carbon dioxide production rate by metabolism
G'	carbon dioxide production rate by metabolism per person
Q	airflow

REFERENCES

Ager, B. P., and J. A. Tickner (1983), "The Control of Microbiological Hazards Associated with Air-Conditioning and Ventilation Systems," *Ann. Occup. Hyg.* **27**: 341–358.

American Society of Heating, Refrigerating and Air Conditioning Engineers (1981), "Ventilation for Acceptable Indoor Air Quality," ASHRAE Standard 62-1981, ASHRAE, Atlanta, GA.

Bernstein, R. S., W. G. Sorenson, D. Garabrant, C. Reaux, and R. D. Treitman (1983), "Exposures to Respirable, Airborne Penicillium from a Contaminated Ventilation System: Clinical, Environmental, and Epidemiological Aspects, *Am. Ind. Hyg. Ass. J.*, **44**:161.

Centers for Disease Control Morbidity and Mortality (1984), "Outbreaks of Respiratory Illness among Employees in Large Office Buildings—Tennessee, District of Columbia," *Weekly Report* (CDC MMWR), **33** (36), September 14.

Leaderer, B. P., and W. S. Cain (1983), "Air Quality in Building during Smoking and Nonsmoking Occupancy," *ASHRAE Trans.* **89**, PS. 2A and B.

National Research Council, Committee on Indoor Pollutants, J. Spengler, Chairman (1981), *Indoor Pollutants, National Academy Press*, Washington, DC.

Sun, Tseng-Yao, (1986), "Air Conditioning and Load Calculation," *Heat. Piping/Air Cond.* **59**: 103–113, (January).

ADDITIONAL READINGS

McQuiston, F. C., and J. Parker (1982), *Heating, Ventilating, and Air Conditioning*, Wiley New York.

Threlkeld, J. L. (1970) *Thermal Environmental Engineering*, Prentice Hall, Englewood Cliffs, N.J.

Trane Company (1965), *Trane Air Conditioning Manual*, Trane Company, LaCrosse, WI.

Wadden, R., and P. Scheff (1983) *Indoor Air Pollution*, Wiley, New York.

16

Reentry

Reentry is the inadvertent return of a vented contaminant back into a nearby air intake or other building opening. This condition is illustrated in Fig. 16.1, which shows a portion of the exhaust from a stack being drawn back into the building through the roof or side intakes. Figure 16.2 shows an actual exhaust-to-intake configuration that resulted in a reentry problem.

The most common reentry problems arise from offending air contaminants with low olfactory thresholds, such as food odors from a restaurant kitchen or mercaptans from a laboratory hood. Less common, but of great importance, are cases where detrimental health effects may occur due to the quantity or nature of the reentered contaminants. Infrequently, reentered contaminants are reported to cause equipment damage, such as accelerated corrosion to a ventilation system.

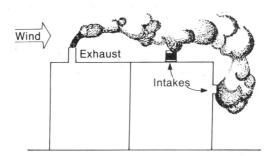

Figure 16.1 Schematic of reentry. Some portion of stack exhaust reenters building through roof and side intakes.

Figure 16.2   Actual reentry situation. Design flaws include insufficient stack height, proximity of stack to HVAC intakes, and the presence of a rain cap.

There are many documented cases of reentry, and the undocumented cases in industrial, laboratory, and medical facilities probably exceed that number manyfold. Indeed, it is the rare production or research facility that has not experienced this problem. In one case (Held, 1962) it is reported that a building was evacuated several times due to the reentry of phosgene and chlorine releases. This reference also reports an illness where the cadmium-laden exhaust from a machine shop electroplating unit was drawn directly back into the room.

At another facility (Barrett, 1963), the exhaust from a sodium dichromate tank was drawn into the plant through the building windows. The resulting indoor chromic acid concentrations were sufficient to cause respiratory irritation and perforated nasal septa. In another case at the same site, the exhaust from a trichloroethylene degreaser was drawn through an air heating system. The hydrochloric acid that resulted from pyrolysis of the vapor caused rapid corrosion of the heat exchanger.

An understanding of contaminant reentry requires knowledge of several related subjects. Initially, it is important to consider the relationship of reentry to the airflow around buildings. Within this context, exhaust system design to prevent reentry requires good engineering practices in determining stack height, location, and other exhaust and intake characteristics. This may be difficult since stack and intake designs vary greatly and often depend on considerations other than potential reentry problems.

Tracer techniques can be used to determine the potential hazard of existing exhaust-intake configurations and to investigate incidents. Wind tunnel and water channel tests have also been used to characterize systems and to offer models and data for improved design. Computational methods can be used to predict contaminant dilution for a proposed or existing system.

Each of the above subjects will be discussed. Although there is still a good deal of art in the science of stack and intake design, many of the problems that occur today can be prevented by applying available knowledge.

16.1 AIRFLOW AROUND BUILDINGS

Airflow around even simple and isolated rectangular buildings is complex (Fig. 16.3) (ASHRAE, 1985; Wilson, 1979). Wind approaching the building face creates a positive-pressure zone on the upstream side and negative zones on the roof, lee, and lateral sides of the building. As wind speed changes, the absolute pressures change but the relative magnitudes do not. At the upwind building face, flow separates on either side of a stagnation zone. A higher stagnation pressure on the upper part of the building wall may lead to downwash on the lower portion of the building. The upper wall will deflect wind upward over the edge of the roof. The area above the roof where turbulent recirculation occurs is called the cavity. The height of the cavity H_c (discussed further in Section 16.5) is important in designing stacks. Release into the cavity can lead to entrainment of contaminants into the turbulent zone above the roof and in the downwind recirculating zone. The maximum cavity height is found when the wind is normal to the upwind building face. When the wind approaches a cubical building at an approximately $45°$ angle, two counter-rotating vortices are induced at the upwind roof edges which tend to reduce the cavity height. The length of the cavity L_c will be less than the building length in many cases and the flow will reattach to the building surface (Fig. 16.4). According to ASHRAE (1985), for a building of width W and height H:

$$L_c = 1.2(HW)^{0.5} \qquad H/W \text{ or } W/H < 8$$
$$L_c = 3.4 \min(H, W) \qquad H/W \text{ or } W/H > 8 \tag{16.1}$$

Above the cavity is a zone where the streamlines are affected by the building but do not enter the downwind recirculation zone. Contaminants released into this zone are less likely to cause reentry problems but may cause downwind pollution in neighboring streets, buildings, and parking areas. Above this zone, the streamlines are unaffected by the building. When a building is located in rough terrain, is near other buildings, or has a nonideal shape, as is often the case, the flow patterns will be very complex. Figure 16.5 shows the effect of such a flow pattern on stack exhaust. Theoretical determination of stack height and location in these cases becomes more difficult; empirical measurement or visualization with scale models may be necessary.

16.2 MEASUREMENT OF REENTRY

In evaluating a potential case of reentry, it may be clear that a serious problem exists without resorting to measurements. An example is the manufacturing

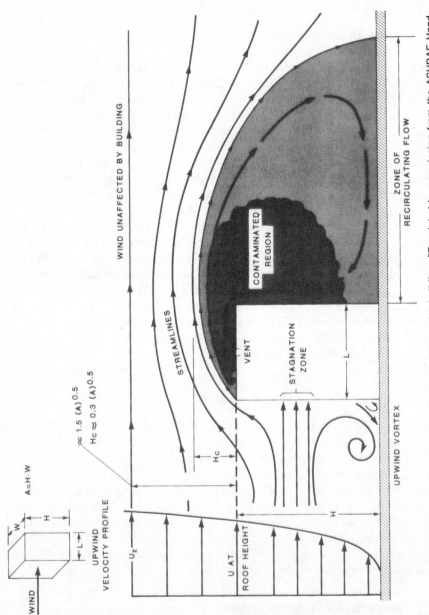

Figure 16.3 Airflow around simple and isolated rectangular building. [Reprinted by permission from the ASHRAE Handbook–1985 Fundamentals]

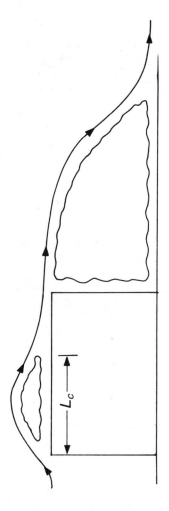

Figure 16.4 Reattachment of flow at L_c.

Figure 16.5 Effect of complex wind pattern on stack exhaust. Three photos taken within minutes of each other show widely varying flow patterns: (a) predominately downward exhaust flow; (b) lateral flow; (c) upward flow.

building shown in Fig. 16.6, where the dense nesting of exhausts and intakes on the roof inevitably led to reentry problems. In cases such as this it is possible to forgo reentry measurements and work immediately on design changes.

In most cases the extent of a reentry problem is not so clear-cut. In such cases, the extent of reentry can be measured directly with a tracer. A known amount of the tracer is released into an exhaust point, such as a laboratory hood or a local exhaust vent on a mix vessel, while measurements of the tracer concentration are concurrently made at the building's replacement air inlet or at various interior locations.

The ideal tracer (Turk et al., 1968) will be:

- Detectable at very low concentrations to facilitate measurement across several orders of magnitude
- Nontoxic
- Odorless
- Conservative (i.e., it will not thermally or chemically degrade during use)
- Not usually found in the atmosphere, so background ambient levels will not interfere with the test

Several materials have been used successfully in tracer experiments. Sulfur hexafluoride (SF_6) and other halogenated compounds have best met the criteria above (Dietz and Cote, 1973). These compounds can be detected with great sensitivity by electron-capture gas chromatography; the detection limit for SF_6 can be as low as 1 part in 10^{13}. Samples can be collected over time (e.g., in an air sampling bag) and taken to a laboratory for analysis or a variety of commercial portable gas chromatographs can be used for direct field measurements. Lamb and Cronn (1986) used an array of portable automated syringe samplers at up to 40 locations to yield isopleths on and around a laboratory building. Passive sampling devices have been used with a family of perfluorinated cyclic hydrocarbon tracers to obtain long-term average measurements (Dietz and Cote, 1982). Analysis by gas chromatography allows a six-order-of-magnitude capability with this sampling method.

Other materials, including fluorescent dyes, smoke plumes, oil fogs, and antimony oxide, have been used as tracers. Munn and Cole (1967), for example, used a fluorescent uranine dye particle to measure the dilution of a building exhaust. A time-weighted average concentration was obtained by filter collection of the tracer over a twenty-minute period.

It may be feasible to use the offending material itself as a tracer. If the release rate of a contaminant can be quantified and the downwind concentrations can be measured, this contaminant can be used in the same manner as the more common tracers. In a field study by one of the authors, a known amount of methylene chloride was released in a laboratory hood and charcoal tube vapor collection at the roof intake and inside several areas of the building was used to measure the reentry potential.

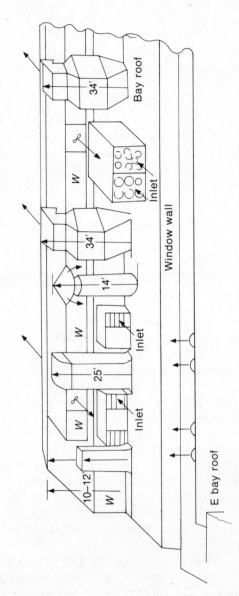

Figure 16.6 Sketch of dense nesting of exhausts and intakes on roof of manufacturing building indicate need for immediate design changes.

There are several important factors to consider in reentry measurements. Wind speed and direction are critical and should be measured on the roof with portable equipment at the time of the tracer release or, at the least, qualitatively estimated. Stack temperature and volume should be at normal operating values, and if they vary significantly with time, they should be measured at the time of the reentry measurement. If a short-term measurement is made inside a building, it is important to know the percentage of outside air in the replacement-air system. This value may vary between 10 and 100% depending on the type of HVAC system installed in the building, and on outside temperature and humidity. This measurement is especially important in variable-air-volume HVAC systems (Chapter 15).

In some cases a yes-or-no answer is adequate in defining whether reentry is occurring for a given stack-intake configuration. For this, the tracer can be released without quantification, although it must be in sufficient quantity to ensure adequate detection limits. In the more usual situation, a tracer is released at a known rate into an exhaust stream with known airflow to create a known exhaust concentration C_e:

$$C_e = Q_t/Q_e \qquad (16.2)$$

where Q_t is the released tracer flow and Q_e is the exhaust airflow.

A building dilution factor D_b can be defined as the ratio of the outside replacement airflow Q_i and the exhaust airflow containing the trace gas Q_e:

$$D_b = Q_i/Q_e \qquad (16.3)$$

Further, an observed dilution factor D_o is defined as the ratio of the exhaust tracer concentration C_e to the indoor concentration C_i resulting from reentry:

$$D_o = C_e/C_i \qquad (16.4)$$

The fraction reentry R is equal to the ratio of the amount of tracer reentering the building to the amount being exhausted:

$$R = Q_i C_i/Q_e C_e \qquad (16.5)$$

Substituting Eqs. 16.3 and 16.4 into Eq. 16.5 yields

$$R = D_b/D_o \qquad (16.6)$$

The indoor concentration C_i can be measured at the replacement-air inlet or at various locations inside the building. Using the highest measured tracer concentration as the value for C_i gives the minimum value for D_o and allows for a conservative approach to reentry evaluation.

In a series of dilution measurements, variables such as wind speed and direction are expected to produce variations in reentry, and for this reason field measurements of dilution may vary over several orders of magnitude. Drivas et al. (1972), for example, in two SF_6 tracer tests of a system with exhaust and intake in very close proximity, found a variation in D_o from 400 to 10,000, depending in large part on wind direction.

In one field study conducted by the authors, reentry was measured in a research laboratory building in response to several odor complaints. Sulfur hexafluoride (SF_6) was released in a laboratory hood at a known rate and samples were collected elsewhere inside the building by personal air sampling pumps and air sample bags. Analysis of the collected air on-site with a portable electron capture gas chromatograph showed varying SF_6 concentrations with varying meteorological conditions..

In one run of this field study, SF_6 was released at a rate of 0.04 cfm into an exhaust flow of 12,100 cfm from a building with replacement-air flow of 55,200 cfm. The computed exhaust concentration was

$$C_e = Q_t/Q_e = 0.04 \text{ cfm}/12,100 \text{ cfm} = 3.3 \times 10^3 \text{ ppb}$$

The building dilution factor was

$$D_b = Q_i/Q_e = 55,200 \text{ cfm}/12,100 \text{ cfm} = 4.6$$

The maximum concentration measured inside the building was 2.4 ppb, so the minimum observed dilution was

$$D_o = C_e/C_i = 3.3 \times 10^3 \text{ ppb}/2.4 \text{ ppb} = 1.4 \times 10^3$$

The reentry, expressed as a fraction of the released material brought back into the building, was

$$R = D_b/D_o = 4.6/1.4 \times 10^3 = 0.003$$

The worst case in a series of these measurements showed 5% reentry. With the reentry potential thus defined, a more detailed modeling study was initiated to evaluate the necessary stack height and configuration to prevent reentry (see Section 16.4).

For a given situation with reentry potential, the measured dilution factor can be used to set safe conditions of use. For example, assume that a laboratory hood exhausts phosgene during a synthesis reaction. The value of D_o (based on the maximum SF_6 value found at the inlet under a variety of meteorological conditions) is found to be 250. If the hood exhaust air flow is 500 cfm, the maximum generation rate of phosgene in the hood should be about 355 ml/min to ensure values at the intake less than the TLV of 0.1 ppm. This is an illustration of the application of this method. Of course, depending on the toxicity of the material, more or less stringent restraints may be imposed.

Reentry measurements require equipment, are time consuming, and depend on variables, such as wind direction, beyond control of the investigator. Their great advantage is that they do offer a direct measure of a given system.

16.3 CALCULATION OF EXHAUST DILUTION

Given a known contaminant concentration in an exhaust airflow as well as the exhaust-to-intake configuration, an empirically derived equation can be used to predict the contaminant concentration at an intake. The calculation gives the expected dilution D_o. The conservative approach, using the smallest dilution D_{min}, provides the most useful result. The concentration at the replacement-air intake is then calculated using Eq. 16.4. Several empirical methods are summarized below. For the complete methods, the reader is referred to the original work.

The intake concentration C_i can be calculated directly by the method of Halitsky (1982). In this model the concentration C at any point near a building exhaust will be proportional to the contaminant volume release Q_c and inversely proportional to the wind speed U and a reference building face area A:

$$C = K_c Q_c / A U \qquad (16.7)$$

where K_c is a dimensionless concentration coefficient at the coordinate location. A limited number of wind tunnel experiments (Halitsky, 1963; Wilson, 1976) provide values of K_c so that Eq. 16.7 can be solved for the concentration.

If a suitable experimental model for K_c is lacking, an equation based on a simple jet plume analogy (Halitsky, 1962) can be used to calculate the dilution:

$$D_o = m(\alpha + \beta S / \sqrt{A_e})^2 \qquad (16.8)$$

where S = shortest (stretched string) distance from exhaust to intake
     A_e = area of the exhaust opening,
m α, β = empirical constants

Experimental data (Halitsky, 1962, 1982) suggested the following form of Eq. 16.8 for D_{min}:

$$D_{min} = (1 + 0.132 S / \sqrt{A_e})^2 \qquad (16.9)$$

Wilson (1976) suggests an alternate equation for calculating D_{min}:

$$D_{min} = 0.11 U_h S^2 / Q_e \qquad (16.10)$$

where U_h = wind speed at building roof height

     S = the shortest distance from vent to receptor point

     Q_e = total exhaust airflow

Equation 16.10 should be applied only to flush-mounted exhaust vents with negligible vent exhaust penetration due to jet or buoyancy effects. As such, it can represent a worst-case approach.

Arranged in terms of Eq. 16.8, Eq. 16.10 becomes

$$D_{min} = 0.11(US^2)/(V_e A_e) \qquad (16.11)$$

where V_e is the exhaust velocity. Equations 16.10 and 16.11 may be more accurate for larger exhaust-to-intake distances because wind speed is realistically factored in. Differences in physical modeling techniques make Eq. 16.9 more applicable to a large building surrounded by small surface roughness and Eq. 16.10 more applicable to a building where surrounding topography creates significant turbulence.

Wind speed is a variable in several of the equations above. Due to two competing effects, there is a critical windspeed U_c that will result in the minimum dilution. The first effect is that the plume rise due to momentum or buoyancy will decrease with increasing wind speed, with a resulting increase in rooftop concentrations. In contrast, turbulent diffusion increases in the plume as ambient air is entrained with a resulting decrease in rooftop concentrations. Halitsky (1982) concluded that a wind speed of 13.6 mph is the critical value for minimum dilution of a typical laboratory hood. In discussing the maximum intake concentration for flush vents on a flat-roofed building, Wilson (1982) related the critical windspeed to the ratio of a plume rise factor to the downwind distance to the intake. He found that in the range $0.5U_c$ to $3.0U_c$, the roof-level concentration varied by a factor of 2.

Wilson (1984) considered the total dilution between the exhaust and intake (or other receptor point at roof level) as the product of three components: internal system dilution, wind dilution, and dilution due to stack height. If the critical windspeed is used, this approach predicts the minimum dilution. The internal system dilution is a result of in-stack dilution and is obtained by combining exhausts. The wind dilution results from the entrainment of ambient air between the exhaust and the receptor point, assuming a vent flush with roof level and no plume rise. Finally, dilution is included due to stack height and plume rise from vertical exhaust velocity. Modification of the equation for plume rise in the presence of a rain cap or a downward-facing gooseneck is also discussed.

The calculations discussed above are very useful for a quick determination of the approximate dilution for an existing or planned system. Use of the equations for the worst case (i.e., the lowest dilution) is especially valuable. Since real-world situations have many complex conditions that cannot be fully accounted for by the theoretical and empirical bases of these equations, these techniques should be used with caution.

16.4 SCALE MODEL MEASUREMENT

Scale models can be used to quantify data on flow around buildings, to validate mathematical models for reentry, and to predict reentry for a given design configuration. The great advantage of scale models is that they can include such empirical factors as rough terrain, surrounding buildings, complex building shape, and other variables not fully accounted for by theoretical analysis. Scale models can also be used to test preconstruction designs to evaluate, for example, the effect of a change in stack height on roof concentration contours when the wind is from the least favorable direction. Models are costly in both time and money and thus usually require strong justification before use.

The complex airflow around an existing or planned structure can be investigated with a geometrically scaled physical model where the flow field is simulated by air or water. Such models have been used extensively in civil engineering since being used by the French engineer, Eiffel.

General modeling criteria (Snyder, 1972; ASHRAE, 1985) are used to ensure that a group of parameters are chosen such that the model and the building are analogous. One criterion is that there must be similarity of the natural wind. This requires modeling the flow characteristics of the atmospheric boundary layer where the flow is affected by the degree of surface roughness. The vertical thermal distribution is usually uniform in wind and water tunnels so that neutral stability conditions are modeled. Thermal stratification can be modeled in a specially designed facility. There must also be geometric similarity between the building and the surrounding topography and kinematic and dynamic similarity of the exhaust effluents.

For modeling in the near field of a point source such as a stack, it is necessary to duplicate closely the local geometry in the area of the source (Plate, 1982). The interior of the model stack might require roughening, for example, to duplicate the exhaust flow properties. The bouyancy and momentum of the exhaust gas must also be scaled. Bouyancy has been modeled by using differing proportions of air–helium mixtures. The exit and intake velocities must be equal in the model and prototype to ensure dynamic similarity.

The near field of a stack is determined not only by its configuration and exhaust rate but also by the turbulence of the atmospheric flow field. To best reproduce this turbulence, it is more important to model the buildings and topographic features in the stack vicinity than to have an exactly scaled profile of the approach wind.

Several tracer techniques have been used in wind tunnel studies (Plate, 1982). Early experiments depended on smoke patterns for mapping contours. Smoke can also be used for quantitative measurements with the use of a photometer. Another early method used ammonia as a tracer with sample analysis by titration. A variety of other tracers and detection systems have been used. These include radioactive tracers such as krypton-85 with Geiger–Mueller counter detection, helium with a mass spectrometer leak detector or with a thermal conductivity meter, and hydrocarbons with a flame ionization detector.

Open-surface water channel systems have also been used to model the flow field around a stack. Colored dye can be used to visualize flow patterns around buildings of various shapes and sizes in the water channel. Electrical conductivity probes have been used to measure tracer concentrations in a water channel. Thymolphthalein, blue in basic solution, has also been used with different pH values to visualize the iso-concentration lines around an exhaust stack model.

A model was built of a research building (see Section 16.2) in an urban area (Durgin and Eberhardt, 1981). With a scale of 1:300, the 90-in.-diameter model scaled to a 2250-ft-diameter section of the urban area (Fig. 16.7). Smoke was

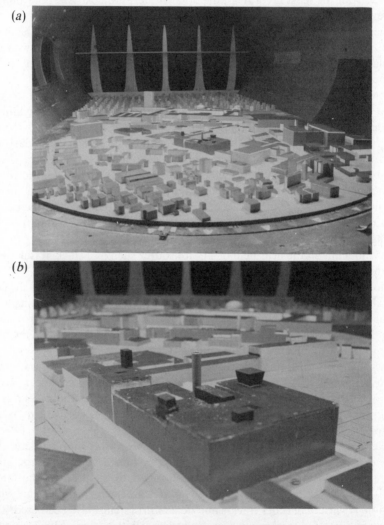

Figure 16.7 (*a*) Wind tunnel scale model, looking upstream. Note the small blocks in the background that reproduce atmospheric turbulence. Test building shown by arrow. (*b*) Close-up of test building. [From Durgin and Eberhardt (1981).]

used to visualize flow from the stacks and to define reentry qualitatively by proximity of the smoke to roof inlets. All visual data were recorded for:

1. The existing roof design at eight wind directions and four wind speeds
2. Stacks of different incremental heights at the worst-case wind direction and speed
3. The "solution" stack at eight wind directions and four speeds to verify the choice

This experimental sequence led to a compromise recommendation for stack height and design that would minimize reentry at most wind speeds and directions.

16.5 DESIGN TO PREVENT REENTRY

The removal of contaminants from an exhaust stream before they leave the stack, as described in Chapter 11, is the surest way to prevent reentry problems. In many facilities contaminant removal is not feasible; in such cases an understanding of airflow around buildings and how reentry is affected by each exhaust stack parameter makes it possible to design or modify a building to minimize reentry problems. Past experience, translated into good engineering practices and coupled with effective use of empirical and theoretical calculations, permits effective design.

16.5.1 Stack Height Determination

The aim of stack height determination is to find the minimum stack height that will exhaust contaminants into the atmosphere above the building cavity height. The prediction of this height is difficult in many situations where irregular building dimensions, the presence of roof structures such as penthouses, irregular surrounding topography, and the presence of neighboring buildings make theoretical approaches applicable only in approximation. Additional factors, often the determining ones, are cost and architectural aesthetics.

Stack height design criteria initially were based on the assumption that cavity height equaled 1.3 to 2.0 times the building height (Clarke, 1963). The *Ventilation Manual* (ACGIH, 1988) also recommends a stack height of 1.3 to 2.0 times the height of the building. A rule of thumb for power plants with tall stacks called for a stack height at least 2 1/2 times the height of the largest nearby building (Thomas et al., 1963), although this probably had little theoretical basis. For power plants and other facilities with very large discharge requirements, air pollution dispersion and not reentry is often the deciding criterion in choosing very tall stacks.

More recent estimates of cavity height are based on the area of the building facing the wind rather than just the height. Halitsky (1965) calculated the necessary stack height for short stacks such as fume hood exhausts using the maximum frontal projected area of the building and a desired dilution before sidewall intake.

Wilson (1979) used water channel laboratory tests to define the turbulence zones above the roofs of a variety of building shapes. Using both the width and height dimensions, it was found that the best experimental correlation was found for a scaling factor F to be

$$F = \min(H, W)^{0.67} \times \max(H, W)^{0.33} \tag{16.12}$$

This formulation indicates a greater effect of the smaller upwind dimension. The cavity height was found to be best approximated by

$$H_c = 0.22F \tag{16.13}$$

These water channel tests also included model buildings with discontinuities such as penthouses or sudden drops, and rules are suggested for incorporating these into the scaling factor F.

ASHRAE (1981) specifies cavity height as a function of the area of the upwind building face A:

$$H_c = 0.3(A)^{0.5} \tag{16.14}$$

This applies where the height and width of the windward side of the building are within a factor of 8 of each other. Should the ratio exceed 8, the relationship is

$$H_c = 0.85 \min(H, W) \tag{16.15}$$

An alternative to a stack height design predicated on cavity height is to design for the needed dilution. Wilson (1984) discusses this approach and points out that design using the H_c calculation format will give the same conservative treatment to minor nuisance odors that it does to highly toxic contaminants. He presents a design based on "available dilution" that has components of internal system dilution, wind dilution, and dilution from stack height (as discussed in Section 16.3). With this approach a given desired dilution can first be chosen and used to solve for the required stack height.

These stack height design estimates are most applicable in relatively simple situations. They give the designer guidance in choosing parameters and give insight in cases of existing reentry problems.

16.5.2   Good Engineering Practices for Stack Design

There are numerous references in the literature for stack and intake design to help prevent reentry problems. Stack downwash is created by the negative-pressure zone on the downwind side of a stack. When the stack exit velocity is low relative to the wind velocity, downwash can occur, increasing contaminant concentrations up to six stack diameters into the stack eddy zone. The effective stack height is decreased by this mechanism (Fig. 16.8). To prevent downwash, stack velocity should be at least 1.5 times the wind velocity (Clarke, 1965). For most areas in the United States, the maximum wind velocity is less than or equal to 20 mph (1770 fpm) 98% of the time. Using this value for the wind velocity, the stack velocity should be $1770 \times 1.5 = 2660$ fpm. The terminal velocity for a raindrop is about 2000 fpm (Laws and Parsons, 1943). Thus the design stack velocity of 2500 to 3000 fpm will prevent downwash and keep rain out of the stack.

It is sometimes desirable to add a tapered cone to a stack to increase the exit velocity. The added pressure drop, however, must be considered in the system design and fan specification. Halitsky (1982) states that plume rise h is related to exit velocity V_e and exhaust air flow Q_e by

$$h \propto V_e^{1/3} Q_e^{1/3} \qquad (16.16)$$

For a reduction in stack diameter by a factor of 2 with constant airflow, the velocity will increase by a factor of 4 and the plume rise by a factor of $4^{1/3} = 1.6$. The increased velocity obtained from a tapered tip may also be important in preventing downwash.

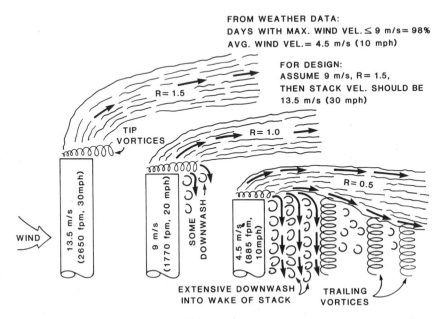

Figure 16.8 Design of stack exit velocity to prevent downwash. Stack velocities of 2500 to 3000 fpm are recommended. (Reprinted by permission from the ASHRAE Handbook–1985 Fundamentals)

ASHRAE (1981) presents recommended and poor stack designs for proper vertical discharge and rain protection (Fig. 16.9). Rain caps, very commonly used, are a poor practice in stack design since they are not very effective in keeping out the rain (Clarke, 1963) unless they are very close to the discharge point. Caps blunt the momentum benefit of a straight exhaust and deflect the exhaust horizontally and downward (Fig. 16.10). An uncapped stack will project exhaust at significant velocities at greater than 12 stack diameters upwards. The same stack with a rain cap located two-thirds of one duct diameter above the exit will have greatly decreased velocities in the vertical direction and will also project the exhaust in the horizontal and downward directions with the effect still seen in these directions at more than 5 diameters.

The roof is frequently the best location for both exhausts and intakes. On the side of the building, exhaust and intakes may be significantly affected by varying wind pressures. Vents on the side of the building may exhaust into the downward flow on the upwind face. Building contamination may arise from a side exhaust

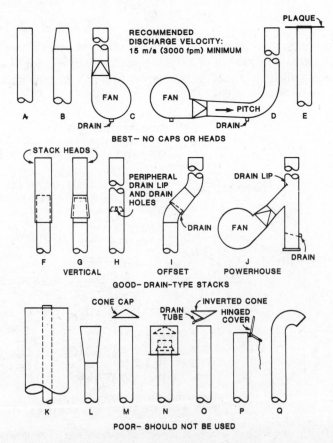

Figure 16.9 Stack design for proper vertical discharge. [Reprinted by permission from the ASHRAE Handbook–1985 Fundamentals]

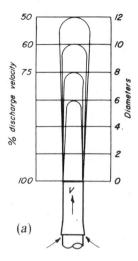

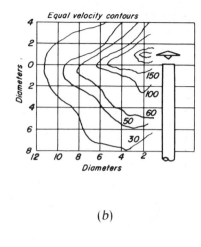

Figure 16.10 Effect of cap on exhaust upward momentum: (*a*) vertical discharge maximizes dilution; (*b*) weather cap defects discharge horizontally and downward. [From Hama (1963).]

Figure 16.11 Exhaust from side of building.

as described in Barrett (1963), with direct flow back into the building through windows or intakes (Fig. 16.11). On buildings with several roof levels, location of the stack on the highest level is usually most effective in promoting the greatest dilution. Considerable thought should be given to the spatial relation of the exhausts and intakes in the design of a new facility. Methods range from placing the intake at the base of a "well-designed" exhaust stack (Clarke, 1965) to the more usual method of maximizing the distance between the two. Prevailing wind direction should not be depended upon exclusively in planning intake location. Reentry can be a serious problem even if the potential exists only during infrequent wind directions. The effects of varying distance can be compared using the computational methods discussed in Section 16.3. Use of the geometric techniques proposed by Wilson (1979) may be helpful in this design effort.

Architectural fences, often used to improve roof appearance, should not be used to enclose exhausts and intakes. The fences create a cavity that can increase the reentry potential.

In some cases, it has proven wise to choose a large central exhaust system rather than many smaller ones. A central exhaust system has both advantages and disadvantages, as discussed in Chapter 9. With regard to reentry, having one larger stack rather than several smaller ones gives a higher release point. Also, Eq. 16.16 demonstrates that a larger exhaust volume, even with a constant velocity, results in an increase in plume rise. Another important advantage of a central system is the added dilution of unforeseen or occasional high contaminant production from a single hood in the total system. This may be a consideration, for example, in the semiconductor industry, where gas cabinets with relatively small exhaust volumes may be called upon to handle a leak of a highly toxic chemical. An example of a building exhaust system modified to release all exhaust points through one larger stack is shown in Fig. 16.12 (Bearg, 1987). As discussed in Chapter 9, however, the use of several separate fans allows individual exhaust hoods to be shut down when not needed, with a savings in energy costs. There may also be concern about exhaust flexibility and the

Figure 16.12 Separate exhaust systems from many laboratory hoods combined into one large exhaust point.

possible loss in production time at all work sites if the one central fan is down for servicing or repair.

Much has been learned from past mistakes in exhaust design that can be used, in conjunction with recent theoretical and model studies, to design and modify exhaust systems to prevent reentry problems. As knowledge and concern about chemical exposures have increased, attention to careful roof and stack design is becoming recognized as an important part of the overall design effort of a facility.

LIST OF SYMBOLS

A	Area of building face
A_e	area of exhaust opening
C	concentration
C_e	concentration in exhaust
C_i	concentration measured downstream from exhaust, at intake or indoors
D_b	building dilution factor
D_{min}	minimum observed dilution factor
D_o	observed dilution factor
F	building dimension scaling factor
H	building height
H_c	cavity height
h	plume rise
K_c	dimensionless concentration coefficient
L_c	cavity length
m	empirical constant
Q_c	contaminant flow
Q_e	exhaust airflow
Q_i	replacement airflow
Q_t	tracer flow
R	fraction reentry
S	shortest distance from exhaust to intake
U	wind speed
U_c	critical wind speed
U_h	wind speed at building height
V_e	exhaust velocity
W	building width
α, β	empirical constants

REFERENCES

American Conference of Governmental Industrial Hygienists (ACGIH), Committee on Industrial Ventilation (1988) *Industrial Ventilation Manual*, 20th ed., ACGIH, Lansing, MI.

American Society of Heating, Refrigerating and Air Conditioning Engineers (ASHRAE; 1985), "Air Flow Around Buildings," *ASHRAE, Handbook—1985; Fundamentals,* ASHRAE, Atlanta, GA. Chap. 14.

Barrett, J. C. (1963), letter in "Open for Discussion," *Heat. Piping Air Cond.* **35**: 78, August.

Bearg, D., Life Energy Associates, Concord, MA (1987), personal communication.

Clarke, J. H. (1963), "The Design of Exhaust Systems and Discharge Stacks," *Heat. Piping Air Cond.* **35**: 118–132, May.

Clarke, J. H. (1965), "The Design and Location of Building Inlets and Outlets to Minimize Wind Effect and Building Reentry of Exhaust Fumes," *Am. Ind. Hyg. Assoc. J.* **26**: 242–247.

Dietz, R. N. and E. A. Cote, (1973), "Tracing Atmospheric Pollutants by Gas Chromatographic Determination of Sulfur Hexafluoride," *Environ. Sci. Technol.* **7**: 338–342.

Dietz, R. N. and E. A. Cote, (1982), "Air Infiltration Measurements in a Home Using a Convenient Perfluorocarbon Tracer Technique," *Environ. Int.,* **8**: 419–433.

Drivas, P. J., P. G. Simmonds, and F. H. Shair, (1972), "Experimental Characterization of Ventilation Systems in Buildings," *Environ. Sci. Technol.,* **6**: 609–614.

Durgin, F. H., and D. S. Eberhardt, (1981), Wright Brothers Wind Tunnel, MIT, Report No. WBWT-TR-1141.

Halitsky, J. (1962), "Diffusion of Vented Gas around Buildings," *J. Air Pollut. Control Assoc.,* **12**: 74–80.

Halitsky, J. (1963), "Gas Diffusion Near Buildings," *ASHRAE Trans.,* **69**: 464–485.

Halitsky, J. (1965), "Estimation of Stack Height Required to Limit Contamination of Building Air Intakes," *Am. Ind. Hyg. Assoc. J.,* **26**: 106–116.

Halitsky, J. (1982), "Atmospheric Dilution of Fume Hood Exhaust Gases," *Am. Ind. Hyg. Assoc. J.,* **43**: 185–189.

Hama, G. M. (1963), letter in "Open for Discussion," *Heat. Piping Air Cond.,* **35**: 80, August.

Held, B. J. (1962), "Planning Ventilation for Nuclear Reactor Facilities," *Am. Ind. Hyg. Assoc. J.,* **23**: 83–87.

Lamb, B. K. and D. R. Cronn, (1986), "Fume Hood Exhaust Re-entry into a Chemistry Building," *Am. Ind. Hyg. Assoc. J.* **47**: 115–123.

Laws, J. O., and D. H. Parsons, (1943), "Relations of Rain Drop Size to Intensity," *Trans. Am. Geophys. Union,* Pt. II, 452.

Meroney, R. N. (1982), "Turbulent Diffusion near Buildings," in *Engineering Meteorology,* E. J. Plate, ed., Elsevier, Amsterdam.

Munn, R. E., and A. F. W. Cole, (1967), "Turbulence and Diffusion in the Wake of a Building," *Atmos. Environ.* **1**: 33–43.

Plate, E. J. (1982), "Wind Tunnel Modelling of Wind Effects in Engineering," in *Engineering Meteorology,* E. J. Plate, ed., Elsevier, Amsterdam.

Snyder, W. H. (1972), "Similarity Criteria for the Application of Fluid Models to the Study of Air Pollution Meteorology," *Boundary-Layer Meteorol.* **3**: 113–134.

Thomas, F. W., S. B. Carpenter, and F. E. Gartrell, (1963), "Stacks—How High?" *J. Air Pollut. Control Assoc.,* **13**: 198–204.

Turk, A., S. M. Edmonds, H. L. Mark, and G. F. Collins, (1968), "Sulfur Hexafluoride as a Gas-Air Tracer," *Environ. Sci. Technol.* **2**: 44–48.

Wilson, D. J., (1976), "Contamination of Air Intakes from Roof Exhaust Vents," *ASHRAE Trans.,* **82**(1): 1024–1038.

Wilson, D. J. (1979), "Flow Patterns over Flat-Roofed Buildings and Application to Exhaust Stack Design," *ASHRAE Trans.,* **85**(2): 284–295.

Wilson, D. J. (1982), "Critical Wind Speeds for Maximum Exhaust Gas Reentry from Flush Vents at Roof Level Intakes," *ASHRAE Trans.*, **88**(1): 503–513.

Wilson, D. J. (1984), "A Design Procedure for Estimating Air Intake Contamination from Nearby Exhaust Vents," *ASHRAE Trans.*, **90**: 136–152.

ADDITIONAL READING

American Society of Heating, Refrigerating and Air Conditioning Engineers (ASHRAE; 1985), *ASHRAE Handbook—1981 Fundamentals*, ASHRAE, Atlanta, GA.

Index